John Henry Pepper, Henry George Hine

The Boy's Playbook of Science

Including the various manipulations and arrangements of chemical and

philosophical apparatus required for the successful performance of scientific

experiments

John Henry Pepper, Henry George Hine

The Boy's Playbook of Science
Including the various manipulations and arrangements of chemical and philosophical apparatus required for the successful performance of scientific experiments

ISBN/EAN: 9783337068998

Printed in Europe, USA, Canada, Australia, Japan

Cover: Foto ©berggeist007 / pixelio.de

More available books at **www.hansebooks.com**

Wheatstone's telephonic concert at the Polytechnic, in which the sounds and vibrations pass *inaudible* through an intermediate hall, and are reproduced in the lecture-room unchanged in their qualities and intensities. *Frontispiece.*

THE

BOY'S PLAYBOOK OF SCIENCE:

INCLUDING THE

Various Manipulations and Arrangements

OF

CHEMICAL AND PHILOSOPHICAL APPARATUS REQUIRED
FOR THE SUCCESSFUL PERFORMANCE OF
SCIENTIFIC EXPERIMENTS.

IN ILLUSTRATION OF THE ELEMENTARY BRANCHES OF
CHEMISTRY AND NATURAL PHILOSOPHY.

BY

JOHN HENRY PEPPER,

F.C.S., A. INST. C.E.; LATE PROFESSOR OF CHEMISTRY AT THE ROYAL POLYTECHNIC,
ETC. ETC.
AUTHOR OF "THE PLAYBOOK OF METALS."

NEW EDITION.

Illustrated with 470 Engravings.

CHIEFLY EXECUTED FROM THE AUTHOR'S SKETCHES,
BY H. G. HINE.

LONDON:

GEORGE ROUTLEDGE AND SONS,
THE BROADWAY, LUDGATE.
NEW YORK: 416, BROOME STREET.

1869.

LONDON:
SAVILL, EDWARDS AND CO., PRINTERS, CHANDOS STREET,
COVENT GARDEN.

PROFESSOR LYON PLAYFAIR, C.B., F.R.S

PROFESSOR OF CHEMISTRY IN THE UNIVERSITY OF EDINBURGH.

DEAR SIR,

I DEDICATE these pages to your Children, whom I often had the pleasure of seeing at the Polytechnic during my direction of that Institution. I do so as a mark of respect and appreciation of your talent and zeal, and of your public-spirited advocacy of the Claims of Science in this great and commercial country.

Without making you responsible in any way for the shortcomings of this humble work on Elementary Science, allow me to subscribe myself,

Dear Sir,

Yours most respectfully,

JOHN HENRY PEPPER.

CONTENTS.

		PAGE
INTRODUCTION	. . . .	1

CHAPTER I.

THE PROPERTIES OF MATTER—IMPENETRABILITY 3

CHAPTER II.

CENTRIFUGAL FORCE. 17

CHAPTER III.

THE SCIENCE OF ASTRONOMY. 19

CHAPTER IV.

CENTRE OF GRAVITY 32

CHAPTER V.

SPECIFIC GRAVITY 48

CHAPTER VI.

ATTRACTION OF COHESION 59

CHAPTER VII.

ADHESIVE ATTRACTION 67

CHAPTER VIII.

CAPILLARY ATTRACTION 69

CHAPTER IX.

CRYSTALLIZATION 73

CHAPTER X.

CHEMISTRY 81

CHAPTER XI.

CHLORINE, IODINE, BROMINE, FLUORINE 129

CHAPTER XII.

CARBON, BORON, SILICON, SELENIUM, SULPHUR, PHOSPHORUS . . 151

CHAPTER XIII.

FRICTIONAL ELECTRICITY 173

CHAPTER XIV.

VOLTAIC ELECTRICITY 193

CHAPTER XV.

MAGNETISM AND ELECTRO-MAGNETISM 206

CHAPTER XVI.

ELECTRO-MAGNETIC MACHINES 211

CHAPTER XVII.

THE ELECTRIC TELEGRAPH 218

CHAPTER XVIII.

RUHMKORFF'S, HEARDER'S, AND BENTLEY'S COIL APPARATUS . . 230

CHAPTER XIX.

MAGNETO-ELECTRICITY 241

CHAPTER XX.

DIA-MAGNETISM 247

CHAPTER XXI.

LIGHT, OPTICS, AND OPTICAL INSTRUMENTS 255

CHAPTER XXII.

THE REFRACTION OF LIGHT 298

CHAPTER XXIII.

REFRACTING OPTICAL INSTRUMENTS 303

CHAPTER XXIV.

THE ABSORPTION OF LIGHT 327

CHAPTER XXV.

THE INFLECTION OR DIFFRACTION OF LIGHT 328

CHAPTER XXVI.

THE POLARIZATION OF LIGHT 335

CHAPTER XXVII.

HEAT. 352

CHÁPTER XXVIII.

THE STEAM-ENGINE 406

CHAPTER XXIX.

THE STEAM-ENGINE—*continued* 413

INTRODUCTION.

ALTHOUGH "The South Kensington Museum" now takes the lead, and surpasses all former scientific institutions by its vastly superior collection of models and works of art, there will be doubtless many thousand young people who may remember (it is hoped) with some pleasure the numerous popular lectures, illustrated with an abundance of interesting and brilliant experiments, which have been delivered within the walls of the Royal Polytechnic Institution during the last twenty years.

On many occasions the author has received from his young friends letters, containing all sorts of inquiries respecting the mode of performing experiments, and it has frequently occurred that even some years after a lecture had been discontinued, the youth, now become the young man, and anxious to impart knowledge to some "home circle" or country scientific institution, would write a special letter referring to a particular experiment, and wish to know how it was performed.

The following illustrated pages must be regarded as a series of philosophical experiments detailed in such a manner that any young person may perform them with the greatest facility. The author has endeavoured to arrange the manipulations in a methodical, simple, and popular form, and will indeed be rewarded if these experiments should arouse dormant talent in any of the rising generation, and lead them on gradually from the easy reading of the present "Boy's Book," to the study of the complete and perfect philosophical works of Leopold Gmelin, Faraday, Brande, Graham, Turner, and Fownes.

Every boy should ride "a hobby-horse" of some kind; and whilst play, and plenty of it, must be his daily right in holiday time, he ought not to forget that the cultivation of some branch of the useful Arts and Sciences will afford him a delightful and profitable recreation when

B

satiated with mere *play*, or imprisoned by bad weather, or gloomy with the unamused tediousness of a long winter's evening.

The author recollects with pleasure the half-holidays he used to devote to Chemistry, with some other King's College lads, and in spite of terrible pecuniary losses in retorts, bottles, and jars, the most delightful amusement was enjoyed by all who attended and assisted at these juvenile philosophical meetings.

It has been well remarked by a clever author, that bees are *geometricians*. The cells are so constructed as, with the least quantity of material, to have the largest sized spaces and the least possible interstices. The mole is a *meteorologist*. The bird called the nine-killer is an *arithmetician*, also the crow, the wild turkey, and some other birds. The torpedo, the ray, and the electric eel are *electricians*. The nautilus is a *navigator*. He raises and lowers his sails, casts and weighs anchor, and performs nautical feats. Whole tribes of birds are *musicians*. The beaver is an *architect, builder,* and *wood-cutter*. He cuts down trees and erects houses and dams. The marmot is a *civil engineer*. He does not only build houses, but constructs aqueducts, and drains to keep them dry. The ant maintains a regular *standing army*. Wasps are *paper manufacturers*. Caterpillars are *silk-spinners*. The squirrel is a *ferryman*. With a chip or a piece of bark for a boat, and his tail for a sail, he crosses a stream. Dogs, wolves, jackals, and many others, are *hunters*. The black bear and heron are *fishermen*. The ants are *day-labourers*. The monkey is a *rope dancer*. Shall it, then, be said that any boy possessing the Godlike attributes of Mind and Thought with Free-will can only eat, drink, sleep, and play, and is therefore lower in the scale of usefulness than these poor birds, beasts, fishes, and insects? No! no! Let " Young England" enjoy his manly sports and pastimes, but let him not forget the mental race he has to run with the educated of his own and of other nations; let him nourish the desire for the acquisition of " scientific knowledge," not as a mere school lesson, but as a treasure, a useful ally which may some day help him in a greater or lesser degree to fight " The Battle of Life."

BOY'S PLAYBOOK OF SCIENCE.

CHAPTER I.

THE PROPERTIES OF MATTER—IMPENETRABILITY.

In the present state of our knowledge it seems to be universally agreed, that we cannot properly commence even popular discussions on astronomy, mechanics, and chemistry, or on the imponderables, heat, light, electricity, and magnetism, without a definition of the general term " matter;" which is an expression applied by philosophers to every species of substance capable of occupying space, and, therefore, to everything which can be seen and felt.

The sun, the moon, the earth, and other planets, rocks, earths, metals, glass, wool, oils, water, alcohol, air, steam, and hosts of things, both great and small, all solids, liquids and gases, are included under the comprehensive term *matter*. Such a numerous and varied collection of bodies must necessarily have certain qualities, peculiarities, or properties; and hence we come in the first place to consider "The general powers or properties of matter." Thus, if we place a block of wood or stone in any position, we cannot take another substance and put it in the space filled by the wood or stone, until the latter be removed. Now this is one of the first and most simple of the properties of matter, and is called *impenetrability*, being the property possessed by all solid, liquid, and gaseous bodies, of filling a space to the exclusion of others until they be removed, and it admits of many amusing illustrations, both as regards the proof and modification of the property.

Thus, a block of wood fills a certain space: how is it (if impenetrable) that we can drive a nail into it ? A few experiments will enable us to answer this question.

Into a glass (as depicted at fig. 1) filled with spirits of wine, a quantity of cotton wool many times the bulk of the alcohol may (if the experiment is carefully performed) be pushed without causing a drop to overflow the sides of the vessel.

Here we seem to have a direct contradiction of the simple and indis-

Fig. 1.

putable truth, that "two things cannot occupy the same space at once."
But let us proceed with our experiments :—

Fig. 2.

We have now a flask full of water,
and taking some very finely-powdered
sugar, it is easy to introduce a notable quantity of that substance without increasing the bulk of the water ;
the only precaution necessary, is not
to allow the sugar to fall into the
flask in a mass, but to drop it in
grain by grain, and very slowly, allowing time for the air-bubbles (which
will cling to the particles of sugar)
to pass off, and for the sugar to dissolve.   Matter, in the experiments
adduced, appears to be penetrable,
and the property of impenetrability
seems only to be a creation of fancy :
reason, however, enables us to say
that the latter is not the case.

A nail may certainly be hammered into wood, but the particles are
thrust aside to allow it to enter.   Cotton wool may be placed in spirits
of wine because it is simply greatly extended and bulky matter, which,
if compressed, might only occupy the space of the kernel of a nut, and

if this were dropped into a half-pint measure full of alcohol, the increase of bulk would not cause the spirit to overflow. The cotton-wool experiment is therefore no contradiction of *impenetrability*. The experiment with the sugar is the most troublesome opponent to our term, and obliges us to amend and qualify the original definition, and say, that the ultimate or smallest particles or atoms of bodies only are impenetrable; and we may believe they are not in close contact with each other, because certain bulks of sugar and water occupy more space separately than when mixed.

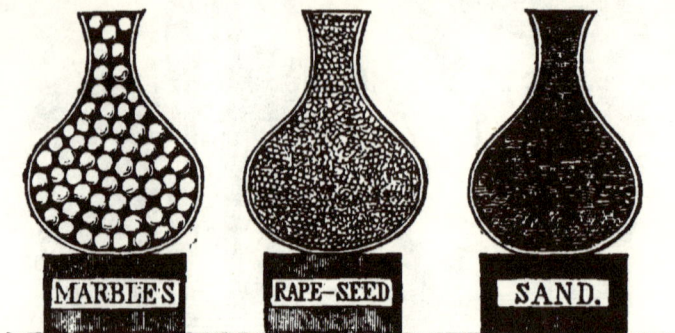

Fig. 3.

If we compare the flask of water to a flask full of marbles, and the sugar to some rape-seed, it will be evident that we may almost pour another flask full of the latter amongst the marbles, because they are not in close contact with each other, but have spaces between them; and after pouring in the rape-seed, we might still find room for some fine sand.

The particles of one body may thus enter into the spaces left between those of another without increasing its volume; and hence, as has been before stated, "The atoms only of bodies are truly impenetrable."

This spreading, as it were, of matter through matter assumes a very important function when we come to examine the constitution of the air we breathe, which is chiefly a mechanical mixture of gases: seventy-nine parts by volume or measure of nitrogen gas, twenty-one parts of oxygen gas, and four parts of carbonic acid vapour in every ten thousand parts of air having the following relations as to weight:—

	Specific gravity.
Nitrogen	972
Oxygen	1105
Carbonic acid	1524

It might be expected that these gases would arrange themselves in our atmosphere in the above order, and if that were the case, we should have the carbonic-acid *gas* (a most poisonous one) at the bottom, and touching the earth, then the oxygen, and, last of all, the nitrogen; a state of things in which *organized* life could not exist. The gases do not, however, separate: indeed, they seem to act as it were like *vacuums* to one another, and "the diffusion of gases" has become a recognised fact, governed by fixed laws. This fact is curiously illustrated, as shown in our cut, by filling a bottle with carbonic acid, and another with hydrogen; and having previously fitted corks to the bottles, perforated so as to admit a tube, place the bottle containing the carbonic acid on the table, then take the other full of hydrogen, keeping the mouth downwards, and fit in the cork and tube: place this finally into the cork of the carbonic-acid bottle, which may be a little larger than the other, in order to make the arrangement stand firmer; and after leaving them for an hour or so, the carbonic acid, which is twenty-two times heavier than the hydrogen, will ascend to the latter, whilst the hydrogen will descend to the carbonic acid. The presence of the carbonic acid in the hydrogen bottle is easily proved by pouring in a wineglassful of clear lime-water, which speedily becomes milky, owing to the production of carbonate of lime; whilst the proof of the hydrogen being present in the carbonic acid is established by absorbing the latter with a little cream of lime —*i.e.*, slacked lime mixed to the consistence of cream with some water—and setting fire to the hydrogen that remains, which burns quietly with a yellowish flame if unmixed with air; but if air be admitted to the bottle, the mixture of air and hydrogen inflames rapidly, and with some noise. One of the most elegant modes of showing the diffusion of gases is by taking a large round dry porous cell, such as would be employed in a voltaic battery, and having cemented a brass cap with a glass tube attached to its open extremity, it may then be supported by a small tripod of iron

Fig. 4.

Fig. 5.

A. The porous cell. B. The jar of hydrogen. C. The brass cap and glass tube D, the end of which dips into the tumbler containing the solution of indigo E. F F. The wire and stand supporting the porous cell and tube in tumbler.

wire, and the end of the glass tube placed in a tumbler containing a small quantity of water coloured blue with sulphate of indigo. If a tolerably large jar containing hydrogen is now placed over the porous cell, bubbles of gas make their escape at the end of the tube, because the hydrogen diffuses itself more rapidly into the porous cell than the air which it already contains passes out. When the jar is removed, the reverse occurs, hydrogen diffuses out of the porous cell, and the blue liquid rises in the tube.

This diffusive force prevents the accumulation of the various noxious gases on the earth, and spreads them rapidly through the great bulk of the atmosphere surrounding the globe.

Although air and other gases are invisible, they possess the property of impenetrability, as may be easily proved by various experiments. Having opened a pair of common bellows, stop up the nozzle securely, and it is then impossible to shut them; or, fill a bladder with air by blowing into it, and tie a string fast round the neck; you then find that you cannot, without breaking the bladder, press the sides together.

It is customary to say that a vessel is empty when we have poured out the water which it contained. Having provided two glass vessels full of water, place each of them in an empty white pan, to receive the over-

Fig. 6 represents the water overflowing, as the glass, with the orifice closed, is pressed down, proving the impenetrability of air.

Fig. 7. The orange has entered the glass vessel, and the air having passed from the orifice, no water overflows.

flow, then lay an orange upon the surface of the water of one of them, and being provided with a cylindrical glass, open at one end, with a hole in the centre of the closed end, place your finger firmly over the orifice, and endeavour, by inverting the glass over the orange, and pressing upon the surface of the water, to make it enter the interior of the glass cylinder; the resistance of the air will now cause the water to overflow into the white pan, whilst the orange will not enter. The orange may now be transferred to the other vessel of water, and on removing the finger from the orifice of the cylindrical glass, and inverting it as before over the orange, the air will rush out and the orange and water will enter, whilst there will be no overflow as in the preceding experiment. The comparison of the two is very striking, and at once teaches the fact desired.

Whilst the vessels of water are still in use, another pretty experiment may be made with the metal potassium. First throw a small piece of the metal on the surface of the water, to show that it takes fire on contact with that fluid; then, having provided a gas-jar, fitted with a cap

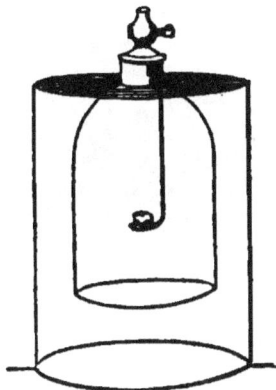

Fig. 8. Gas-jar with stop-stock closed, and potassium in ladle; air prevents the entrance of the water.

Fig. 9. Gas-jar; stop-cock open; the air passes, the water enters, and the potassium is inflamed.

and stop-cock, and a little spoon screwed into the bottom of the stopcock inside the gas-jar, place another piece of potassium in the little spoon, and, after closing the stop-cock, push the jar into one of the vessels of water: as before, the impenetrability of the air prevents the water flowing up to the potassium; but, on opening the stop-cock, the air escapes, the water rushes up, and directly it touches the potassium, combustion ensues.

Having sufficiently indicated the nature and meaning of impenetrability, we may proceed to discuss experimentally three other marked and special qualities of matter—viz., *inertia, gravity,* and *weight.*

Inertia is a power which (according to Sir Isaac Newton) is implanted in all matter of resisting any change from a state of rest. It is sometimes called *vis inertiæ*, and is that property possessed by all matter, of remaining at rest till set in motion, and *vice versá;* and it expresses, in brief terms, resistance to motion or rest.

A pendulum clock wound up and ready to go, does not commence its movements, until the inertia of the pendulum is overcome, and motion imparted to it. On the other hand, when seated in a carriage, should any obstruction cause the horse to stop suddenly, it is only perhaps by a violent effort, if at all, that we can resist the onward movement of our

Fig. 10. Tin tray, with glass bottom, full of water; candle placed underneath.

bodies. To illustrate inertia, construct a metal tray, about three feet long, two feet wide, and two inches deep, with a glass bottom, and arrange it on a framework supported by legs, like a table, and having filled it with water, let the room be darkened, and then place under the tank a lighted candle, at a sufficient distance from the glass to prevent the heat cracking it. If a piece of calico or paper, stretched on a framework, be now held over the water at an angle of about thirty degrees, all that occurs on the surface of the water will be rendered visible on such screen. Attention may now be directed to the quiescence, or the inertia of the water, while the opposite condition of movement and formation of the waves may be beautifully shown by touching the surface of the water with the finger; the miniature waves being depicted on the screen, and continuing their motion till set at rest by striking against the sides of the tin tray.

Fig. 11.   Same tray, with calico screen; showing the waves as they are produced by touching the surface of the water with the finger.

Should the above experiment be thought too troublesome or expensive to prepare, inertia may be demonstrated by filling a tea-cup or other convenient vessel with water, and after moving rapidly with it in any direction, if we stop suddenly, the rigidity of all parts of the cup we hold brings them simultaneously to a state of rest; but the mobility of the liquid particles allows of their continuing in motion in their original direction, and the liquid is spilled. Thus, carelessness in handing and spilling a cup of tea (though not to be recommended) serves to illustrate an important principle. The inertia of bodies in motion is further and lamentably illustrated by the accidents caused from the sudden stoppage of a railway train whilst in rapid motion, when heads and knees come in contact with frightful results.—It is more especially demonstrated by the earth, the moon, and the other planets continuing their motion for ever in the absence of any friction or resistance to oppose their onward progress. It is the friction arising from the roughness of the ground, the resistance of the air, and the force of the earth's attraction, which puts a stop to bodies set in motion about the surface of the earth.

GRAVITATION.

Inertia represents a passive force, *gravitation*, an active condition of matter; and this latter may truly be termed a force of attraction, because it acts between masses at sensible or insensible distances: it is illustrated by a stone, unsupported, falling to the ground; by the stone pressing with force on the earth, and requiring power to raise it from the ground: indeed, it is commonly reported that it was by an accident— "an apple falling from a tree"—that the great Newton was led to reflect on the universal law of gravitation, and to pronounce upon it in the following memorable words:—

"*Every particle of matter in the universe attracts every other particle of matter with a force or power directly proportional to the quantity of matter in each, and decreasing as the squares of the distances which separate the particles increase.*"

These words may appear very obscure to our juvenile readers; but when dissected and examined properly, they clearly define the property of gravitation. For instance, "every particle attracts every other with a force proportional to the quantity of matter in each." This statement was verified some years back by Maskelyne, who, having sought out and discovered a steep, precipitous rock in the Schichallion mountains, in Scotland, suspended from it a metal weight by a cord, and going to a convenient distance with a telescope, and observing the weight, he found that it did not hang perpendicularly, like an ordinary plumb-line, but was attracted, or impelled, to the sides of the rock by some kind of attraction, which, of course, could be no other than that indicated by Newton as the attraction of gravitation.

This truly wonderful power of attraction pervades all masses; and being, as before stated, proportional to the quantity of matter, if a man could be transported to the surface of the sun, he would become about thirty times heavier: he would be attracted, or impelled, to the sun with thirty times more gravitating force than on the surface of the earth, and would weigh about two tons. Of course, nursing a baby on the sun's surface would be a very serious affair with our ordinary strength; whilst on some of the smaller planets, such as Ceres and Pallas, we should probably gravitate with a force of a few pounds only, and with the same muscular power now possessed, we should quite emulate the exploits of those domestic little creatures sometimes

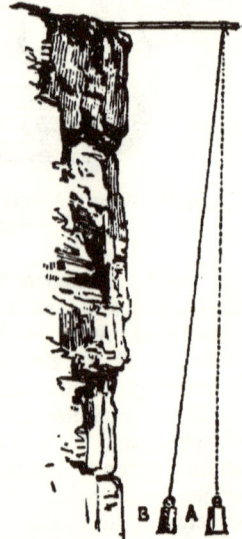

Fig. 12. The Schichallion Rocks. The dotted line and weight A represent the ordinary position of a plumb-line, whilst the line of the weight B indicates (of course, with some exaggeration) the attractive power of the mass of the rock drawing it from the perpendicular.

called "the industrious fleas," and our jumping would be something marvellous.

There is no very good lecture-table experiment that will illustrate gravitation, although attention may be directed to the fact of a piece of potassium thrown on the surface of water in a plate generally rushing to the sides, and, as if attracted, attaching itself with great force to the substance of the pottery or porcelain; or, if a model ship, or lump of wood, be allowed to float at rest in a large tank of water, and a number of light chips of wood or bits of straw be thrown in, they generally collect and remain around the larger floating mass.

A very good idea, however, may be afforded of the universal action of gravity maintaining all things in their natural position on the earth by

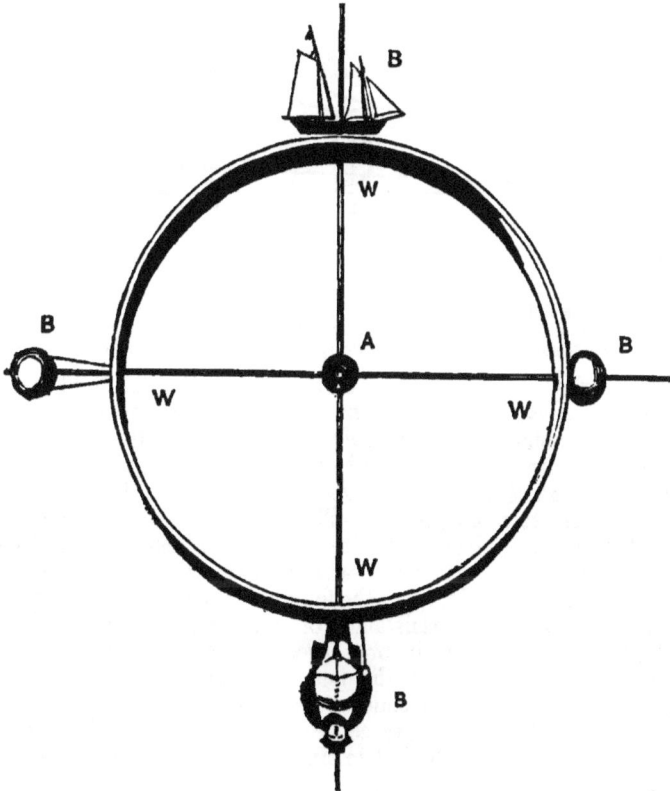

Fig. 13.

A. The centre ball, representing the earth's centre of gravity.

w w w w. Four wires fixed into centre ball, and passing through and secured in the hoop, projecting about one foot from the circumference.

B B B B. Two balls—a model ship and toy—working on the wires like beads, with vulcanized India-rubber straps attached to them and the circumference of the hoop.

taking a hoop and arranging in and upon it balls, or a model ship, or other toy, and wires, as depicted in our diagram.

With this simple apparatus we may illustrate the upward, downward, and sideway movement of bodies from the earth, and the counteraction by the force of gravitation of any tendency of matter to fall away from the globe, which is represented in the model by the india-rubber springs pulling the balls and toys back again to the circumference of the hoop.

The attraction of gravitation decreases (quoting the remainder of Newton's definition) as the squares of the distances which separate the particles increase—*i.e.*, it obeys the principle called "inverse proportion"—viz., the greater the distance, the less gravitating power; the less the distance, the greater the power of gravitation. Gravitation is like the distribution of light and other radiant forces, and may be thus illustrated.

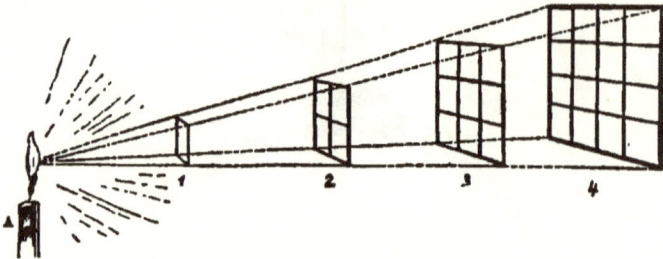

Fig. 14. Place a lighted candle, marked A, at a certain distance from No. 1, a board one foot square; at double the distance the latter will shadow another board, No. 2, four feet square; at three times, No. 3, nine feet square; at four, No. 4, sixteen feet; and so on.

To make the comparison between the propagation of light and the attraction of gravitation, we have only to imagine the candle, *a*, to represent the point where the force of gravity exists in the highest degree of intensity; suppose it to be the sun—the great centre of this power in our planetary system. A body, as at No. 1, at any given distance will be attracted (like iron-filings to a magnet) with a certain force; at twice the distance, the square of two being four, and by inverse proportion, the attraction will be four times less; at thrice the distance, nine times less; at the fourth distance, sixteen times less; and so on. With the assistance of this law, we may calculate, roughly, the depth of a well, or a precipice, or a column, by ascertaining the time occupied in the fall of a stone or other heavy substance. A falling body descends about 16 feet in one second, 64 feet in two seconds, 144 feet in three seconds, 256 feet in four seconds, 400 feet in five seconds, 576 feet in six seconds; the spaces passed over being as the squares of the times.

Suppose a stone takes three seconds in falling to the surface of the water in a well, then $3 \times 3 = 9 \times 16 = 144$ feet would be a rough estimate of the depth. The calculation will exceed the truth in consequence of the stone being retarded in its passage by the resistance of the air.

All bodies gravitate equally to the earth: for instance, if an open box, say one foot in length, two inches broad, and two inches deep, be provided with a nicely-fitted bottom, attached by a hinge, a number of substances, such as wood, cork, marble, iron, lead, copper, may be arranged in a row; and directly the hand is withdrawn, the moveable flap flies open, and if the manipulation with the disengagement of the trap-door is good, the whole of the substances are seen to proceed to the earth in a straight line, as shown in our drawing.

Fig. 15.

Fig. 16.

If a heavy substance, like gold, be greatly extended by hammering and beating into thin leaves, and then dropped from the hand, the resistance of the air becomes very apparent; and a gold coin and a piece of gold-leaf would not reach the earth at the same time if allowed to fall from any given height. This fact is easily displayed by the assistance of a long glass cylindrical vessel placed on the air-pump, with suitable apparatus arranged with little stages to carry the different substances; upon two of them may be placed a feather and a gold coin, and on the third, another gold coin and a piece of gold-leaf.

In arranging the experiment, great care ought to be taken that the little stages are all nicely cleaned, and free from any oil, grease, or other matter which might cause the feathers or the gold-leaf to cling to the stages when they are disengaged, by moving the brass stop round that works in the collar of leathers. Sometimes these leathers are oiled, and in that case, when the vacuum is made, the oil, by the pressure, is squeezed out, and, passing down, may reach the stages and spoil the experiment, by causing the feathers and gold-leaf to stick to the brass, producing great disappointment, as the illustration, usually called the *"guinea and feather glass experiment"* takes some time to prepare. The air-pump being in good order, the long glass is first greased on the lower welt or edge, and then placed firmly on the air-pump plate. The top edge, or welt, may now be greased, and the gold coins, feathers, and gold-leaf arranged in the drop-apparatus; this is carefully placed on the top of the glass, and firmly squeezed down. The author has always found a tallow candle, rolled in a sheet of paper (so as to leave about half the candle exposed), the best grease to smear the glass with for air-pump experiments; if the weather is cold, the candle may

Fig. 17.

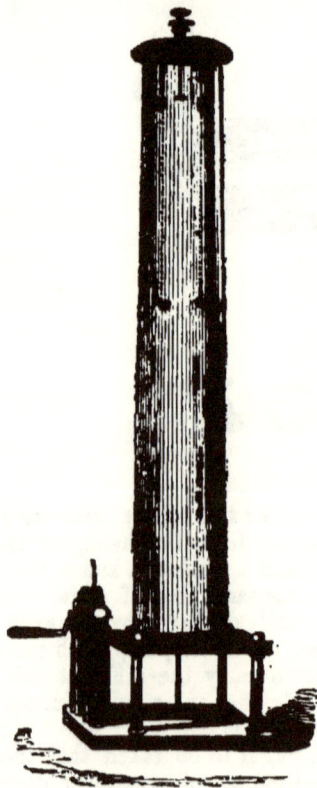

Fig. 18.

be placed for a few minutes before an ordinary fire to soften the tallow. Pomatum answers perfectly well when the surfaces of glass and brass are all nicely ground; but as air-pumps and glasses by use get scratched and rubbed, the tallow seems to fill up better all ordinary channels by which air may enter to spoil a vacuum.

The apparatus being now arranged, the air is pumped out; and here, again, care must be taken not to shake the gold off the stages. When a proper vacuum has been obtained, which will be shown by the pump-gauge, the stop is withdrawn from one of the stages, and the gold and feather are seen to fall simultaneously to the air-pump plate. Another stage, with the gold-leaf and coin, may now be detached; both showing distinctly, that when the resistance of the air is withdrawn, all bodies, whether called *light* or *heavy*, gravitate equally to the earth. Then, the screw at the bottom of the pump-barrels being opened, attention may be directed to the whizzing noise the air makes on entering the vacuum, and when the air is once more restored to the long glass vessel, the last stage may be allowed to fall; and now, the gold coin reaches the pump-plate first, and the feather, lingering behind, loses (as it were) the race, and touches the plate after the gold coin; thus demonstrating clearly the resistance of the air to falling bodies.

Another, and perhaps less troublesome, mode of showing the same fact, is to use a long glass tube closed at each end with brass caps cemented on. One cap should have the largest possible aperture closed by a brass screw, and the other may fit a small hand-pump.

If a piece of gold and a small feather are placed in the tube, it may be shown that the former reaches the bottom of the tube first, whilst it is full of air, and when the air is withdrawn by means of the pump, and the tube again inverted, both the gold and the feather fall in the same time.

Fig. 19. A B. Glass tube containing a piece of gold and a feather, which are placed in at the large aperture A. C. Small hand-pump.

For this reason, all attempts to measure heights or depths by observing the time occupied by a falling body in reaching the earth must be incorrect, and can only be rough approximations. An experiment tried at St. Paul's Cathedral, with a stone, which was allowed to fall from the cupola, indicated the time occupied in the descent to be four and a half seconds: now, if we square this time, and multiply by 16, a height of 324 feet is denoted; whereas the actual height is only 272 feet, and the difference of 52 feet shows how the stone was retarded in its passage through the air; for, had there been no obstacle, it would have reached the ground in $4\frac{3}{16}$ths seconds.

The force of gravitation is further demonstrated by the action of the sun and moon raising the waters of the ocean, and producing the tides; and also by the earth and moon, and other planets and satellites, being prevented from flying from their natural paths or orbits around the sun. It is also very clearly proved that there must be some kind of attractive force resident in the earth, or else all moveable things, the water, the air, the living and dead matters, would fly away from the surface of the earth in obedience to what is called "centrifugal force." Our earth is twenty-four hours in performing one rotation on its axis, which is an imaginary line drawn from pole to pole, and represented by the *wire* round which we cause a sphere to rotate. All objects, therefore, on the earth are moving with the planet at an enormous velocity; and this movement is called the earth's diurnal, or daily rotation. Now,

Fig. 20.

it will be remembered, that mud or other fluid matter flies off, and is not retained by the circumference of a wheel in motion: when a mop is trundled, or a dog or sheep, after exposure to rain, shake themselves, the water is thrown off by what is called centrifugal force (*centrum*, a centre, *fugio*, to fly from).

CHAPTER II.

CENTRIFUGAL FORCE.

THAT power which drives a revolving body from a centre, and it may be illustrated by turning a closed parasol, or umbrella, rapidly round on its centre, the stick being the axis—the ribs fly out, and if there is much friction in the parts, the illustration is more certain by attaching a bullet to the end of each rib, as shown in our drawing.

Fig. 21.

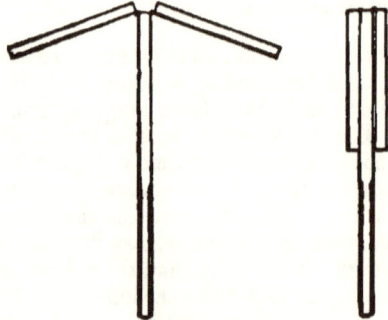

Fig. 22.

The same fact may be illustrated by a square mahogany rod, say one inch square and three feet long, with two flaps eighteen inches in length, hanging by hinges, and parallel to the sides of the centre rod, which immediately fly out on the rotation of the long centre piece.

The toy called the centrifugal railway is also a very pretty illustration of the same fact. A glass of water, or a coin, may be placed in the little carriage, and although it must be twice hanging perpendicular in a line with the earth, the carriage does not tumble away from its appointed track, and the centrifugal force binds it firmly to the interior of the circle round which it revolves.

C

Fig. 23.

Another striking and very simple illustration is to suspend a hemispherical cup by three cords, and having twisted them, by turning round the cup, it may be filled with water, and directly the hand is withdrawn, the torsion of the cord causes the cup to rotate, and the water describes a circle on the floor, flying off at a tangent from the cup, as may be noticed in the accompanying cut.

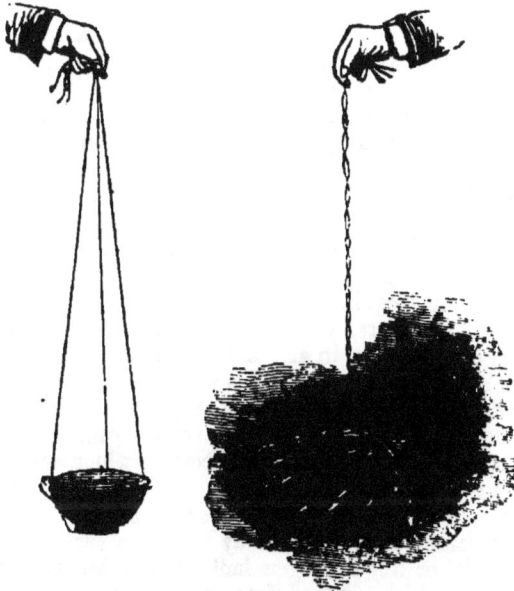

Fig. 24.

A hoop when trundled would tumble on its side if the force of gravitation was not overcome by the centrifugal force which imparts to it a motion in the direction of a tangent (*tango*, to touch) to a circle. The same principle applies to the spinning-top—this toy cannot be made to stand upon its point until set in rapid motion.

Returning again to the subject of gravitation, we may now consider it in relation to other and more magnificent examples which we discover by studying the science of astronomy.

CHAPTER III.

THE SCIENCE OF ASTRONOMY.

IN a work of this kind, professedly devoted to a very brief and popular view of the different scientific subjects, much cannot be said on any special branch of science; it will be better, therefore, to take up one subject in astronomy, and by discussing it in a simple manner, our young friends may be stimulated to learn more of those glorious truths which are to be found in the published works of many eminent astronomers, and especially in that of Mr. Hind, called "The Illustrated London Astronomy." One of the most interesting subjects is the phenomenon of the eclipse of the sun; and as 1858 is likely to be long remembered for its "annular eclipse," we shall devote some pages and illustrations to this subject.

Eclipses of the sun are of three kinds—partial, annular, and total. Many persons have probably seen large partial eclipses of the sun, and may possibly suppose that a total eclipse is merely an intensified form of a partial one; but astronomers assert that no degree of partial eclipse, even when the very smallest portion of the sun remains visible, gives the slightest idea of a total one, either in the solemnity and overpowering influence of the spectacle, or the curious appearances which accompany it.

The late Mr. Baily said of an eclipse (usually called that of Thales), which caused the suspension of a battle between the Lydians and Medes, that only a total eclipse could have produced the effect ascribed to it. Even educated astronomers, when viewing with the naked eye the sun nearly obscured by the moon in an annular eclipse, could not tell that *any part* of the sun was hidden, and this was remarkably verified in the annular eclipse of the 15th March of this year.

During the continuance of a total eclipse of the sun, we are permitted a hasty glance at some of those secrets of Nature which are not revealed at any other time—glories that hold in tremulous amazement even veteran explorers of the heavens and its starry worlds.

The general meaning of an eclipse may be shown very nicely by lighting a common oil, or oxy-hydrogen lantern in a darkened room, and throwing the rays which proceed from it on a three-feet globe. The lantern may be called the sun, and, of course, it is understood that correct comparative sizes are not attempted in this arrangement; if it were so, the globe representing the earth would have to be a mere speck, for if we make the model of the sun in proportion to a three-feet globe, no ordinary lecture hall would contain it. This being premised, attention is directed to the lantern, which, like the sun, is self-luminous, and is giving out its own rays; these fall upon the globe we have designated the earth, and illuminate one-half, whilst the other is shrouded in darkness, reminding us of the opacity of the earth, and teaching, in a familiar

c 2

manner, the causes of day and night. Another globe, say six inches in diameter, and supported by a string, may be compared to the moon, and, like the earth, is now luminous, and shines only by borrowed light : the moon is simply a reflector of light ; like a sheet of white cardboard, or a metallic mirror. When, therefore, the small globe is passed between the lantern and the large globe, a shadow is cast on the lantern : it is also seen that only the half of the small globe turned towards the lantern is illuminated, while the other half, opposite the large globe, is in shadow or darkness. And here we understand why the moon appears to be black while passing before the sun ; so also by moving the small globe about in various curves, it is shown why eclipses are only visible at certain parts of the earth's surface ; and as it would take (roughly speaking) fifty globes as large as the moon to make one equal in size to our earth, the shadow it casts must necessarily be small, and cannot obscure the whole hemisphere of the earth turned towards it. An eclipse of the sun is, therefore, caused by the opaque mass of moon passing between the sun and the earth. Whilst an eclipse of the moon is caused by the earth moving directly between the sun and the moon : the large shadow cast by the earth renders a total eclipse of the moon visible to a greater number of spectators on that half of the earth turned towards the moon. All these facts can be clearly demonstrated with the arrangement already described, of which we give the following pictorial illustration :—

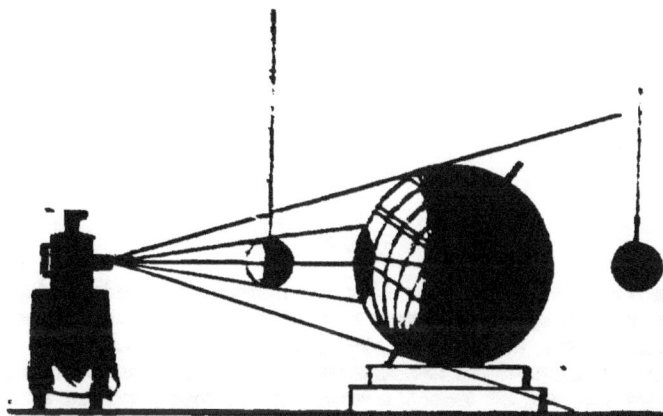

Fig. 35.

In using this apparatus, it should be explained that if the moon were as large as the sun, the shadow would be cylindrical like the figure 1, and of an unlimited length. If she were of greater magnitude, it would precisely resemble the shadow cast in the experiment already adduced with the lantern and shown at No. 2. But being so very much smaller than the sun, the moon projects a shadow which converges to a point as shown in the third diagram.

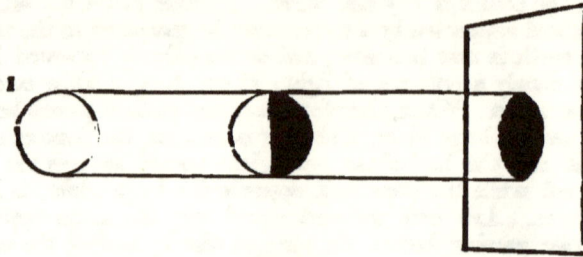

Fig. 1.

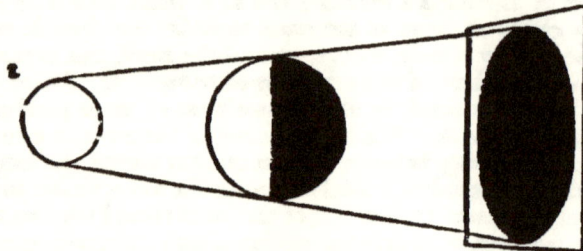

Fig. 2.

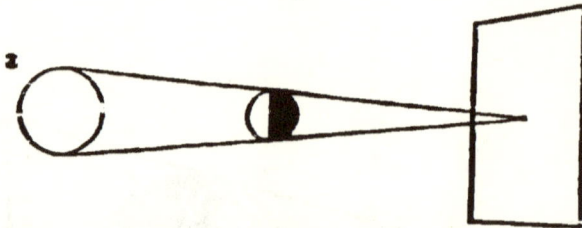

Fig. 3.

In order to comprehend the difference between an annular and a total eclipse of the sun, it is necessary to mention the apparent sizes of the sun and moon: thus, the former is a very large body—viz., eight hundred and eighty-seven thousand miles in diameter; but then, the sun is a very long way off from the earth, and is ninety millions of miles distant from us; therefore, he does not appear to be very large: indeed, the sun seems to be about the same size as the moon; for, although the sun's diameter is (roughly speaking) four hundred times greater than that of the moon, he is four hundred times farther away from us, and, consequently, the sun and moon appear to be the same size, and when they come in a straight line with the eye, the nearer and smaller body, the moon, covers the larger and more distant mass, the sun; and hence, we have either an annular, or a total eclipse, showing how a small body may come between the eye and a larger body, and either partially or completely obscure it.

With respect to an annular eclipse, it must be remembered, that the paths of all bodies revolving round others are elliptical; *i.e.*, they take place in the form of an ellipse, which is a figure easily demonstrated; and is, in fact, one of the conic sections.

If a slice be taken off a cone, parallel with the base, we have a circle thus—

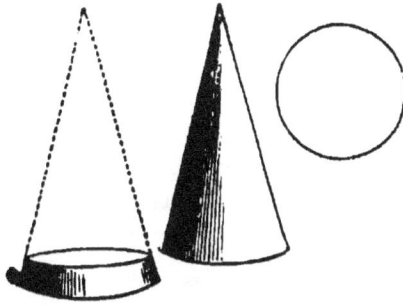

Fig. 29.

If it be cut obliquely, or slanting, we see at once the figure spoken of, and have the ellipse as shown in this picture.

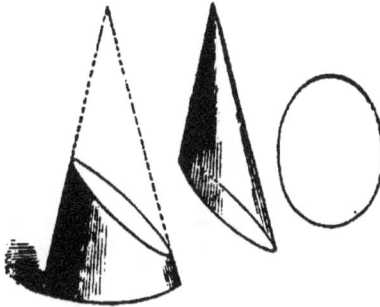

Fig. 30.

Now, the ellipse has two points within it, called "the foci," and these are easily indicated by drawing an ellipse on a diagram-board, in which two nails have been placed in a straight line, and about twelve inches apart. Having tied a string so as to make a loop, or endless cord, a circle may first be drawn by putting the cord round one of the nails, and holding a piece of chalk in the loop of the string, it may be extended to its full distance, and a circle described; here a figure is produced round one point, and to show the difference between a circle and an ellipse, the endless cord is now placed on the two nails, and the chalk being carried round inside the string, no longer produces the circle, but that familiar form called the oval. As a gardener would say, an oval has been struck; and the two points round which it has been described,

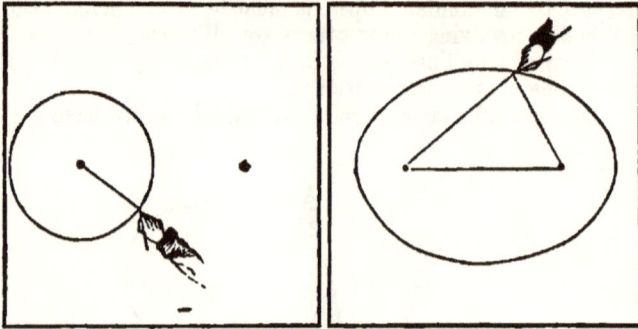

Fig. 31.

are called the *foci*. This explanation enables us to understand the next diagram, showing the motion of the earth round the sun; the latter being placed in one of the foci of a very moderate ellipse, and the various points of the earth's orbit designated by the little round globes marked A, B, C, D, where it is evident that the earth is nearer to the sun at B than at D. In this diagram the ellipse is exaggerated, as it ought, in fact, to be very nearly a circle.

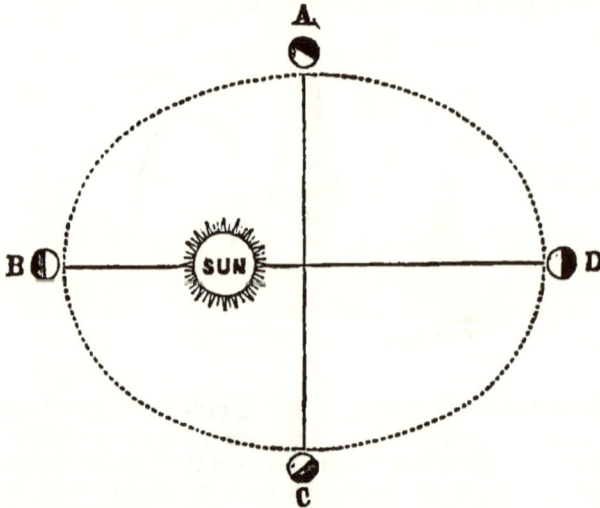

Fig. 32.

We are about three millions of miles nearer to the sun in the winter than we are in the summer; but from the more oblique or slanting direction of the rays of the sun during the winter season, we do not derive any increased heat from the greater proximity. The sun, there-fore, apparently varies in size; but this seeming difference is so trifling that it is of no importance in the discussion: and here we may ask, why does the earth move round the sun? Because it is impelled by *two forces,* one of which has already been fully explained, and is called the *centrifugal* power, and the other, although termed the *centripetal* force, is only another name for the " attraction of gravitation."

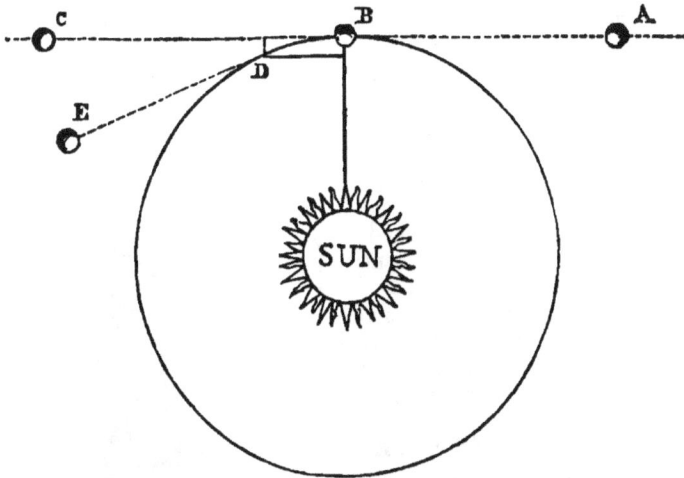

Fig. 33.

To show their mutual relations, let us suppose that, at the creation of the universe, the earth, marked A, was hurled from the hand of its Maker; according to the law of inertia, it would continue in a straight line, A c, for ever through space, provided it met with no resistance or obstruction. Let us now suppose the earth to have arrived at the point B, and to come within the sphere of the attraction of the sun s;

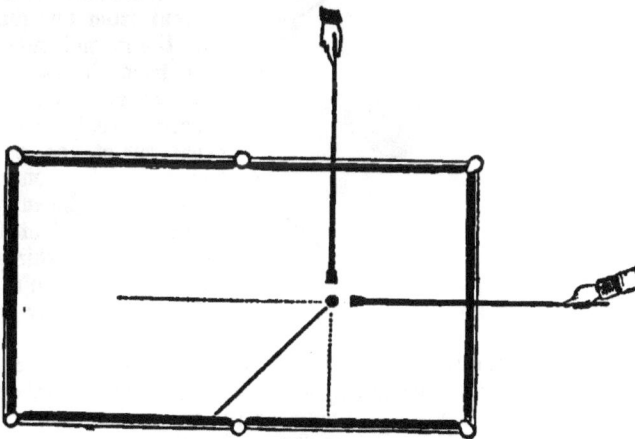

Fig. 34.

THE SCIENCE OF ASTRONOMY. 25

here we have at once contending forces acting at right angles to each other; either the earth must continue in its original direction, A C, or fall gradually to the sun. But, mark the beauty and harmony of the arrangement: like a billiard-ball, struck with equal force at two points at right angles to each other, it takes the mean between the two, or what is termed the diagonal of the parallelogram (as shown in our drawing of a billiard-table), and passes in the direction of the curved line, B D; having reached D, it is again ready to fly off at a tangent; the centrifugal force would carry it to E, but again the gravitating force controls the centripetal, and the earth pursues its elliptical path, or orbit, till the Almighty Author who bade it move shall please to reverse the command.

The mutual relations of the centripetal and centrifugal forces may be illustrated by suspending a tin cylindrical vessel by two strings, and having filled it with water, the vessel may be swung round without spilling a single drop; of course, the movement must be commenced carefully, by making it oscillate like a pendulum. The cord which binds it to the finger may be compared to the centripetal force, whilst the centrifugal power is illustrated by the water pressing against the sides and remaining in the vessel. Upon the like principles the moon

Fig. 35.

revolves about the earth, but her orbit is more elliptical than that of the earth around the sun; and it is evident from our diagram that the moon is much further from the earth at A than at B. As a natural consequence, the moon appears sometimes a little larger and sometimes smaller than the sun; the apparent mean diameter of the latter being thirty-two minutes, whilst the moon's apparent diameter varies from twenty-nine and a

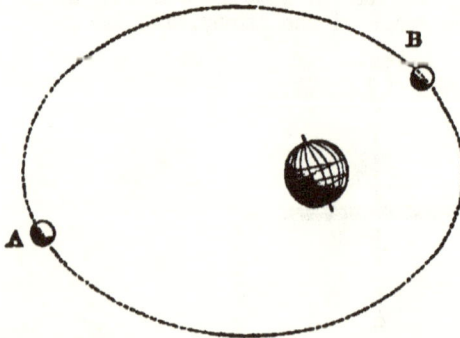

Fig. 36.

half to thirty-three and a half minutes. Now, if the moon passes exactly between us and the sun when she is apparently largest, then a total eclipse takes place; whereas, if she glides between the sun and ourselves when smallest—i.e., when furthest off from the earth—then she is not suffi-

ciently large to cover the sun entirely, but a ring of sunlight remains
visible around her, and what is called an annular eclipse of the sun
occurs. This fact may be shown in an effective manner by placing the
oxy-hydrogen lantern before a sheet, or other white surface, and throw-

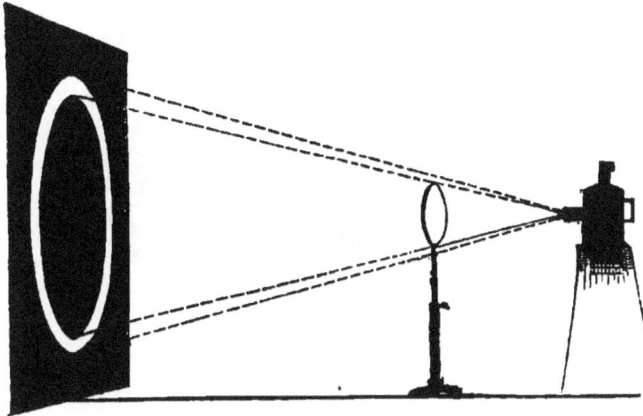

Fig. 37.

ing a bright circle of light upon it, which may be called the sun; then,
if a round disc of wood be passed between the lantern and the sheet, at
a certain distance from the nozzle of the lantern, all the light is cut off,
the circle of light is no longer apparent, and we have a resemblance to
a total eclipse.

By taking the round disc of wood further from the lantern, and repeating the experiment, it will be found that the whole circle of light
is not obscured, but a ring of light appears around the dark centre, corresponding with the phenomenon called the annular (ring-shaped)
eclipse.

If a bullet be placed very near to one eye whilst the other remains
closed, a large target may be wholly shut out from vision; but if the
bullet be adjusted at a greater distance from the eye, then the centre
only will be obscured, and the outer edge or ring of the target remains
visible.

When the advancing edge, or first *limb*, as it is termed, of the moon
approaches very near to the second limb of the sun, the two are joined
together for a time by alternations of black and white points, called
Baily's beads.

This phenomenon is supposed to be caused partly by the uneven and
mountainous edge of the moon, and partly by that inevitable fault of
telescopes, and of the nervous system of the eye, which tends to enlarge
the images of luminous objects, producing what is called irradiation. It
is exceedingly interesting to know that, although the clouds obscured
the annular eclipse of 1858, in many parts of England, we are yet

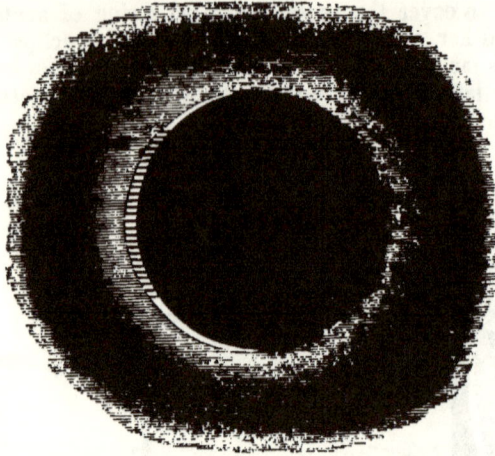

Fig. 38.

left the recorded observations of one fortunate astronomer, Mr. John Yeats, who states that—

"All the phenomena of an annular eclipse were clearly and beautifully visible on the Fotheringay-Castle-mound, which is a locality easily identified. Baily's beads were perfectly plain on the completion of the *annulus*, which occurrence took place, according to my observation, at about seventy seconds after 1 o'clock; it lasted about eighty seconds. The 'beads,' like drops of water, appeared on the upper and under sides of the moon, occupying fully three-fourths of her circumference.

"Prior to this, the upper edge of the moon seemed dark and rough, and there were no other changes of colour. At 12·43, the cusps, for a few moments, bore a very black aspect.

"There was nothing like intense darkness during the eclipse, and less gloom than during a thunderstorm. Bystanders prognosticated rain; but it was the shadow of a rapidly-declining day. At 12 o'clock, a lady living on the farm suddenly exclaimed, 'The cows are coming home to be milked!' and they came, all but one; that followed, however, within the hour. Cocks crowed, birds flew low or fluttered about uneasily, but every object far and near was well defined to the eye.

"A singular broadway of light stretched north and south for upwards of a quarter of an hour; from about 12·54 to 1·10 P.M.

If the annular eclipse of the sun be a matter for wonderment, the total eclipse of the same is much more surprising; no other expression than that of *awfully grand*, can give an idea of the effects of totality, and of the suddenness with which it obscures the light of heaven. The darkness, it is said, comes dropping down like a mantle, and as the moment of full obscuration approaches, people's countenances become livid, the horizon is indistinct and sometimes invisible, and there is a general appearance of horror on all sides. These are not simply the inventions of active human imaginations, for they produce equal, if not greater effects, upon the brute creation. M. Arago quotes an instance of a half-starved dog, who was voraciously devouring some food, but dropped it the instant the darkness came on. A swarm of ants, busily engaged, stopped when the darkness commenced, and remained motionless till the light reappeared. A herd of oxen collected themselves into a circle and stood still, with their horns outward, as if to resist a common enemy; certain plants, such as the convolvulus and silk-tree acacia, closed their leaves. The latter statement was corroborated during the annular eclipse of the 15th of March, 1858, by Mr. E. S. Lane, who states, that crocuses at the Observatory, Beeston, had their blossoms expanded before the eclipse; they commenced closing, and were quite shut at about one minute previous to the greatest darkness; and the flowers opened partially about twenty minutes afterwards. A " *total eclipse*" of the sun has always impressed the human mind with terror and wonder in every age : it was always supposed to be the forerunner of evil; and not only is the mind powerfully impressed, as darkness gradually shuts out the face of the sun, but at the moment of totality, a magnificent corona, or glory of light, is visible, and prominences, or flames, as they are often termed, make their appearance at different points round the circle of the dark mass. This glory does not flash suddenly on the eye; but commencing at the first limb of the sun, passes quickly from one limb to the other. Our illustration shows

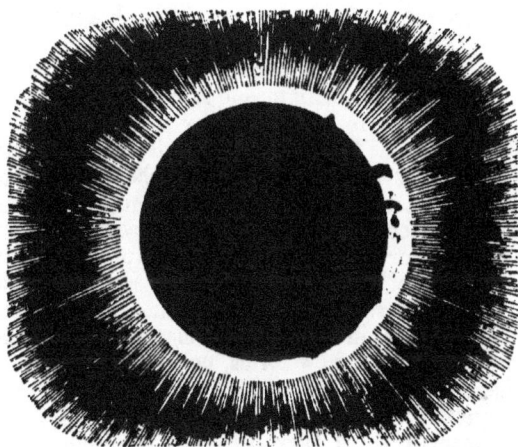

Fig. 39.

"the corona" and the "rose-coloured prominences," whose nature we shall next endeavour to explain. Professor Airy describes the change from the last narrow crescent of light to the entire dark moon, surrounded by a ring of faint light, as most curious, striking, and magical in effect. The progress of the formation of the corona was seen distinctly. It commenced on the side of the moon opposite to that at which the sun disappeared, and in the general decay and disease which seemed to oppress all nature, the moon and the corona appeared almost like a local sore in that part of the sky, and in some places were seen double. Its texture appeared as if fibrous, or composed of entangled threads; in other places brushes, or feathers of light proceeded from it, and one estimate calculated the light at about one-seventh part of a full moon light. The question, whether the corona is concentric with the sun and moon, was specially mooted by M. Arago, and Professor Baden Powell has produced such excellent imitations of the "corona" by making opaque bodies occult, or conceal, very bright points, that it cannot be considered as material or real, although it ought to be remembered that the best theory of the zodiacal light represents it to be a nebulous mass, increasing in density towards the sun, and yet no portion of this nebulous mass was seen during the totality. But by far the most remarkable of all the appearances connected with a "total eclipse" are the rose-coloured prominences, mountains, or flames, projecting from the circumference of the moon to the inner ring of the corona; and, although they had been observed by Vaserius (a Swedish astronomer) in 1733, they took the modern astronomers entirely by surprise in 1842, and they were not prepared with instruments to ascertain the nature of these strange and almost portentous forms. In 1851, however, great preparations were made to throw further light on the subject. Professor Airy went to make his observations, and he says, "That the suddenness of the darkness in 1851 appeared much more striking than in 1842, and the forms of the rose-coloured mountains were most curious. One reminded him of a boomerang (that curious weapon thrown so skilfully by the aborigines of Australia); this same figure has been spoken of by others as resembling a Turkish scimitar, strongly coloured with rose-red at the borders, but paler in the centre. Another form was a pale-white semicircle based on the moon's limbs; a third figure was a red detached cloud, or balloon, of nearly circular form, separated from the moon by nearly its own breadth; a fourth appeared like a small triangle, or conical red mountain, perhaps a little white in the interior;" and the Professor proceeds to say, "I employed myself in an attempt to draw roughly the figures, and it was impossible, after witnessing the increase in height of some, and the disappearance of another, and the arrival of new forms, not to feel convinced that the phenomena belonged to the sun, and *not* to the moon."

Still the question remains unanswered, what are these "rose-coloured prominences?" If they belong to the sun, and are mountains in that luminary, they must be some thirty or forty thousand miles in height.

M. Faye has formally propounded the theory, that they are caused by refraction, or a kind of mirage, or the distortion of objects caused by heated air. This phenomenon is not peculiar to any country, though most frequently observed near the margin of lakes and rivers, and on hot sandy plains. M. Monge, who accompanied Buonaparte in his expedition to Egypt, witnessed a remarkable example between Alexandria and Cairo, where, in all directions, green islands appeared surrounded by extensive lakes of pure, transparent water. M. Monge states that "Nothing could be conceived more lovely or picturesque than the landscape. In the tranquil surface of the lake, the trees and houses with which the islands are covered were strongly reflected with vivid and varied hues, and the party hastened forward to enjoy the refreshment apparently proffered them; but when they arrived, the lake, on whose bosom the images had floated—the trees, amongst whose foliage they arose, and the people who stood on the shore, as if inviting their approach, had all vanished, and nothing remained but the uniform and irksome desert of sand and sky, with a few naked and ragged Arabs."

If M. Monge and his party had not been undeceived, by actually going to the spot, they would, one and all, have been firmly convinced that these visionary trees, lakes, and buildings had a real existence. This kind of mirage is known in Persia and Arabia by the name of "serab" or miraculous water, and in the western districts of India by that of "scheram." This illusion is the effect of unusual refraction, and M. Faye attempts to account for the rose-coloured mountains by something of a similar nature.

It is right, however, to mention, that learned astronomers do not consider this theory of any value.

Lieutenant Patterson, one of the observers of the eclipse of 1851, says, that "It is very remarkable that the flames or prominences correspond exactly (at least as far as he could judge) with the spots on the sun's surface." Taking this statement with that of M. Faye, it may be assumed, as a new idea, and nothing more, that these prominences are, after all, mere aerial pictures of these openings in the sun's atmosphere, or what are called "sun spots." In the "Edinburgh Philosophical Journal," it is said, that although it has lately been shown in the Edinburgh Observatory that it is possible to produce, by certain optical experiments, red flames on the sun's limb of precisely the rose-coloured tint described, yet, on weighing the whole of the evidence, there does seem a great preponderance in favour of the eclipse flames being real appendages of the sun, and in that case they must be masses of such vast size as to play no unimportant part in the economy of that stupendous orb.

During the last eclipse great disappointment was felt that the darkness was so insignificant, although, when we consider the enormous light-giving power of the sun, and know that it was not wholly obscured, we could hardly have expected any other result. There can be no doubt that a decided change in the amount of light is only to be observed during a total eclipse of the sun, one of which occurred on the 7th of September, 1858; but, unfortunately, it was only visible in South America; we must therefore content ourselves with the descriptions of those astronomers who can be fully relied on. From the graphic account given by Professor Piazzi Smyth, the astronomer-royal for Scotland, of a total eclipse as seen by him on the western coast of Norway, we may form some notion of the imposing appearance of the surrounding country when obscured during the occurrence of this rare astronomical phenomenon.

The Professor remarks, "To understand the scene more fully, the reader must fancy himself on a small, rocky island on a mountainous coast, the weather calm, and the sky at the beginning of the eclipse seven-tenths covered with thin and bright cirro-strati clouds. As the eclipse approaches, the clouds gradually darken, the rays of the sun are no longer able to penetrate them through and through, and drench them with living light as before, but they become darker than the sky against which they are seen. The air becomes sensibly colder, the clouds still darker, and the whole atmosphere murkier.

"From moment to moment as the totality approaches, the cold and darkness advance apace; and there is something peculiarly and terribly convincing in the two different senses, so entirely coinciding in their indications of an unprecedented fact being in course of accomplishment. Suddenly, and apparently without any warning (so immensely greater were its effects than those of anything else which had occurred), the totality supervenes, and darkness *comes down*. Then came into view lurid lights and forms, as on the extinction of candles. This was the most striking point of the whole phenomenon, and made the Norse peasants about us flee with precipitation, and hide themselves for their lives.

"Darkness reigned everywhere in heaven and earth, except where, along the north-eastern horizon, a narrow strip of unclouded sky presented a low burning tone of colour, and where some distant snow-covered mountains, beyond the range of the moon's shadow, reflected the faint mono-chromatic light of the partially eclipsed sun, and exhibited all the detail of their structure, all the light, and shade, and markings of their precipitous sides with an apparently supernatural distinctness. After a little time, the eyes seemed to get accustomed to the darkness, and the looming forms of objects close by could be discerned, all of them exhibiting a dull-green hue; seeming to have exhaled their natural colour, and to have taken this particular one, merely by force of the red colour in the north.

"Life and animation seemed, indeed, to have now departed from everything around, and we could hardly but fear, against our reason, that if such a state of things was to last much longer, some dreadful calamity must happen to us all; while the lurid horizon, northward, appeared so like the gleams of departing light in some of the grandest paintings by Danby and Martin, that we could not but believe, in spite of the alleged extravagances of these artists, that Nature had opened up to the constant contemplation of their mind's-eye some of those magnificent revelations of power and glory which others can only get a glimpse of on occasions such as these."

It can be easily imagined, that under such peculiar and awful circumstances, the careful observation of these effects must be somewhat difficult, and the only wonder is that the astronomical observations are conducted with any certainty at all.

In the eclipse of 1842, it was not only the vivacious Frenchman who was carried away in the impulse of the moment, and had afterwards to plead that "*he was no more than a man*" as an excuse for his unfulfilled part in the observations, but the same was the case with the grave Englishman and the more stolid German. In 1851, much the same failure in the observations occurred; and on some person asking a worthy American, who had come with his instruments from the other side of the world expressly to observe the eclipse, what he had succeeded in doing? he merely answered, with much quiet impressiveness, "*That if it was to be observed over again, he hoped he would be able to do something, but that, as it was, he had done nothing: it had been too much for him.*" This is not quite so bad as the fashionable lady who had been invited to look at an eclipse of the sun through a grand telescope, but arriving too late, inquired whether "it could not be shown *over again.*"

With this brief glance at the science of astronomy, we once more return to the term "gravity," which will introduce to us some new and interesting facts, under the head of what is called "centre of gravity."

CHAPTER IV.

CENTRE OF GRAVITY.

That point about which all the parts of a body do, in any situation, exactly balance each other.

THE discovery of this fact is due to Archimedes, and it is a point in every solid body (whatever the form may be) in which the *forces* of *gravity* may be considered as *united.* In our globe, which is a sphere, or rather an oblate spheroid, the centre of gravity will be the centre. Thus, if a plummet be suspended on the surface of the earth, it points directly to the centre of gravity, and, consequently, two plummet-lines suspended side by side cannot, strictly speaking, be parallel to each other.

If it were possible to bore or dig a gallery through the whole substance of the earth from pole to pole, and then to allow a stone or the fabled Mahomet's coffin to fall through it, the momentum—*i.e.*, the force of the moving body, would carry it beyond the centre of gravity. This force, however, being exhausted, there would be a retrograde movement, and after many oscillations it would gradually come to rest, and then, unsupported by anything material, it would be suspended by the force of gravitation, and now enter into and take part in the general attracting force; and being equally attracted on every side, the stone or coffin must be totally without weight. *Momentum* is prettily illustrated by a series of inclined planes

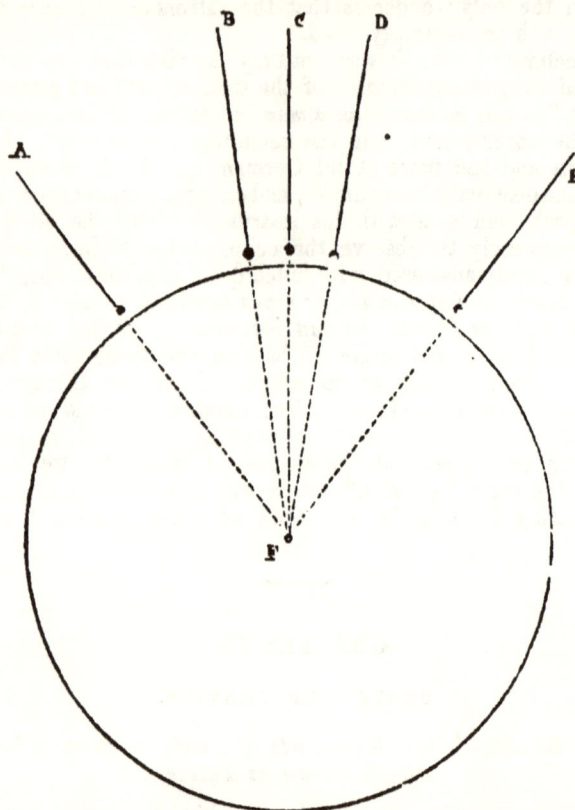

Fig. 40. F. The centre. A B C D E. Plummet-lines, all pointing to the centre, and
therefore diverging from each other.

cut in mahogany, with a grooved channel at the top, in imitation of the
famous Russian ice mountains: and if a marble is allowed to run down

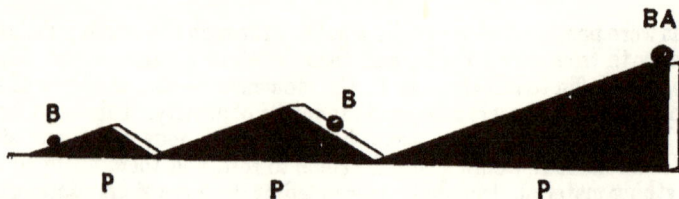

Fig. 41. P P P. Inclined planes, gradually decreasing in height, cut out of inch mahogany,
with a groove at the top to carry an ordinary marble. B B B. Different positions of the
marble, which starts from B A.

D

the first incline, the momentum will carry it up the second, from which it will again descend and pass up and down the third and last miniature mountain.

In a sphere of uniform density, the centre of gravity is easily discovered, but not so in an irregular *mass;* and here, perhaps, an explanation of terms may not be altogether unacceptable.

Mass, is a term applied to solids, such as a mass of lead or stone.

Bulk, to liquids, such as a bulk of water or oil.

Volume, to gases, such as a volume of air or oxygen.

To find the centre of gravity of any mass, as, for example, an ordinary school-slate, we must first of all suspend it from any part of the frame; then allow a plumb-line to drop from the point of suspension, and mark its direction on the slate. Again, suspend the slate at various other points, always marking the line of direction of the plummet, and at the point where the lines intersect each other, there will be the centre of gravity.

If the slate be now placed (as shown in Fig. 43) on a blunt

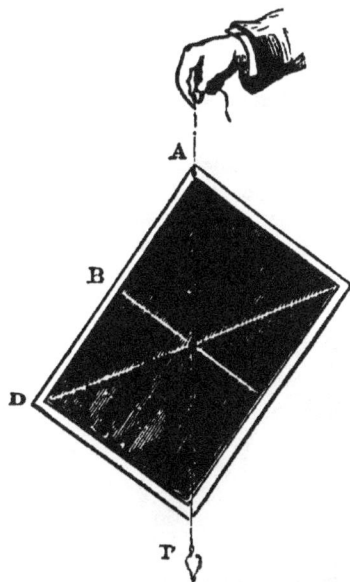

Fig. 43.

Fig. 42. A B D. The three points of suspension. o, The point of intersection, and, therefore, the centre of gravity. P, The line of plummet.

wooden point at the spot where the lines cross each other, it will be found to balance exactly, and this place is called the *centre of gravity,* being the point with which all other particles of the body would move with parallel and equable motion during its fall. The equilibrium of bodies is therefore much affected by the position of the centre of gravity. Thus, if we cut out an elliptical figure from a board one inch in thickness, and rest it on a flat surface by one of its edges (as at No. 1, fig. 44), this point of contact is called the point of support, and the centre of gravity is immediately above it.

In this case, the body is in a state of secure equilibrium, for any motion on either side will cause the centre of gravity to ascend in these directions, and an oscillation will ensue. But if we place it upon the smaller end, as shown at No. 2 (fig. 44), the position will be one of equilibrium, but not stable or secure; although the centre of gravity is directly above the point of support, the slightest touch will displace

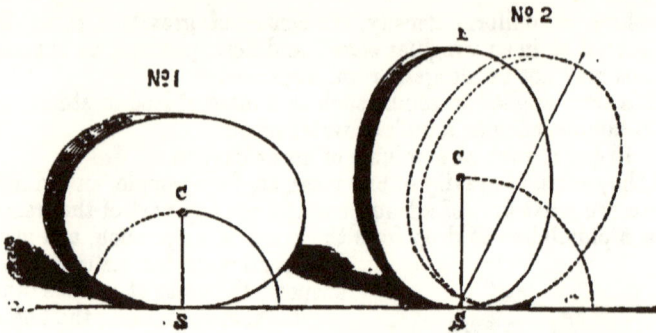

Fig. 44. The point of support. c, The centre of gravity.

the oval and cause its overthrow. The famous story of Columbus and the egg suggests a capital illustration of this fact; and there are two modes in which the egg may be poised on either of the ends.

The one usually attributed to the great discoverer, is that of scraping or slightly breaking away a little of the shell, so as to flatten one of the ends, thus ·——

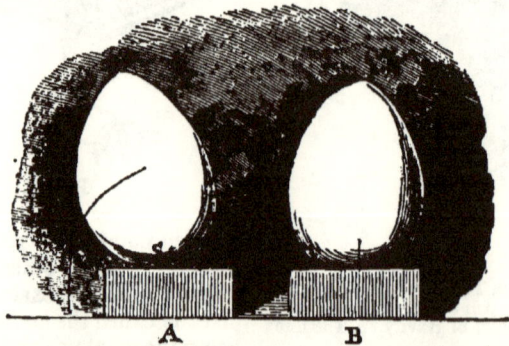

Fig. 45. A Represents the egg in its natural state, and, therefore, in unstable equilibrium; B, another egg, with the surface, s, flattened, by which the centre of gravity is lowered, and if not disturbed beyond the extent of the point of support the equilibrium is stable.

The most philosophical mode of making the egg stand on its end and without disturbing the exterior shell is to alter the position of the yolk, which has a greater density than the white, and is situated about the centre. If the egg is now shaken so as to break the membrane enclosing the yolk, and thus allow it to sink to the bottom of the smaller end, the centre of gravity is lowered; there is a greater proportion of weight concentrated in the small end, and the egg stands erect, as depicted at fig. 46.

It is this variable position of the centre of gravity in ivory balls (one part of which may be more dense than another) that so frequently annoys even the best billiard-players; and on this account a ball will deviate from the line in which it is impelled, not from any fault of the player, but in consequence of the ivory ball being of unequal density, and, therefore, not having the centre corresponding with the centre of gravity. A good billiard-player should, therefore, always try the ball before he engages to play for any large sum.

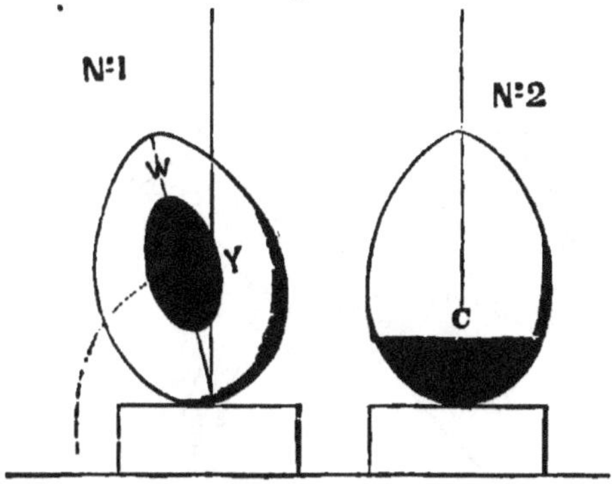

Fig. 46.—No. 1. Section of egg. c. Centre of gravity. y. The yolk. w. The white.
No. 2. o. Centre of gravity, much lowered. y. The yolk at the bottom of the egg.

The toy called the "tombola" reminds us of the egg-experiment, as there is usually a lump of lead inserted in the lower part of the hemi-

Fig. 47.—No. 1. c. Centre of gravity in the lowest place, figure upright.
No. 2. c. Centre of gravity raised as the figure is inclined on either side, but falling again into the lowest place as the figure gradually comes to rest.

sphere, and when the toy is pushed down it rapidly assumes the upright position because the centre of gravity is not in the lowest place to which it can descend; the latter position being only attained when the figure is upright.

There is a popular paradox in mechanics—viz., "a body having a tendency to fall by its own weight, may be prevented from falling by adding to it a weight on the same side on which it tends to fall," and the paradox is demonstrated by another well-known child's toy as depicted in the next cut.

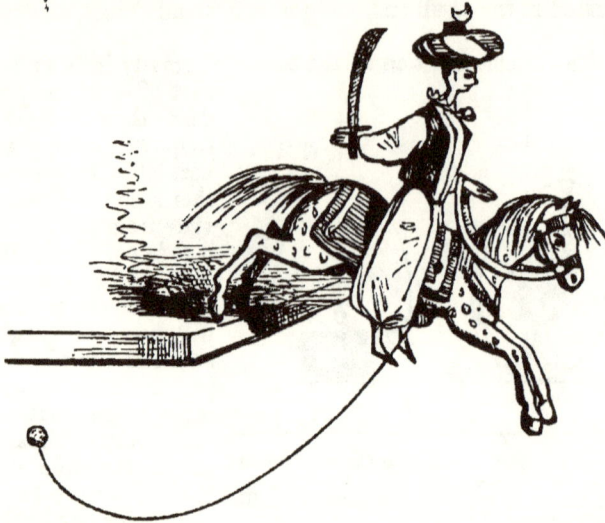

Fig. 48. The line of direction falling beyond the base; the bent wire and lead weight throwing the centre of gravity under the table and near the leaden weight; the hind legs become the point of support, and the toy is perfectly balanced.

After what has been explained regarding the improvement of the stability of the egg by lowering the situation of the centre of gravity, it may at first appear singular that a stick loaded with a weight at its upper extremity can be balanced perpendicularly with greater ease and precision than when the weight is lower down and nearer the hand; and that a sword can be balanced best when the hilt is uppermost;

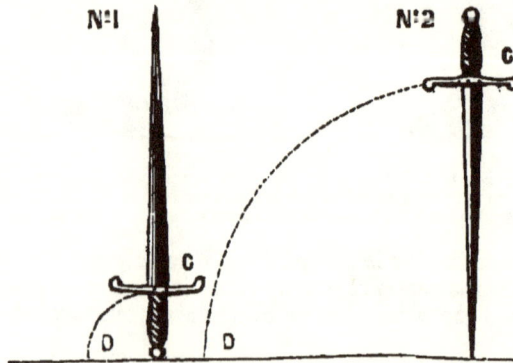

Fig. 49.—No. 1. Sword balanced on handle: the arc from c to d is very small, and if the centre, c, falls out of the line of direction it is not easily restored to the upright position. No 2. Sword balanced on the point: the arc from c to d much larger, and therefore the sword is more easily balanced.

but this is easily explained when it is understood that with the
handle downwards a much smaller arc is described as it falls than when
reversed, so that in the former case the balancer has not time to re-
adjust the centre, whilst in the latter position the arc described is so
large that before the sword falls the centre of gravity may be restored
within the line of direction of the base.

For the same reason, a child tripping against a stone will fall
quickly; whereas, a man can recover himself; this fact can be very
nicely shown by fixing two square pieces of mahogany of different

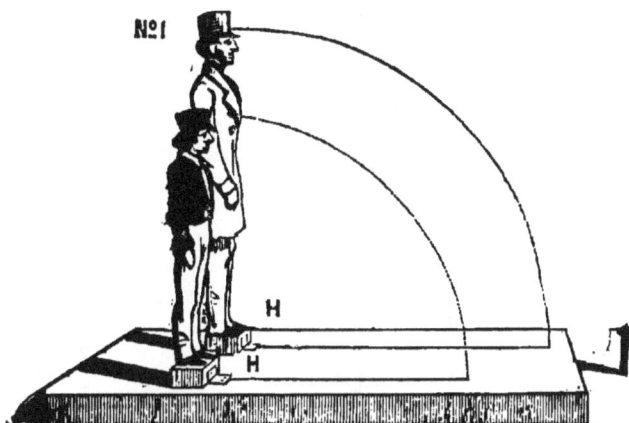

Fig. 50.—No. 1. The two pieces of mahogany, carved to represent a man and a boy, one
being 10 and the other 5 inches long, attached to board by hinges at H H.

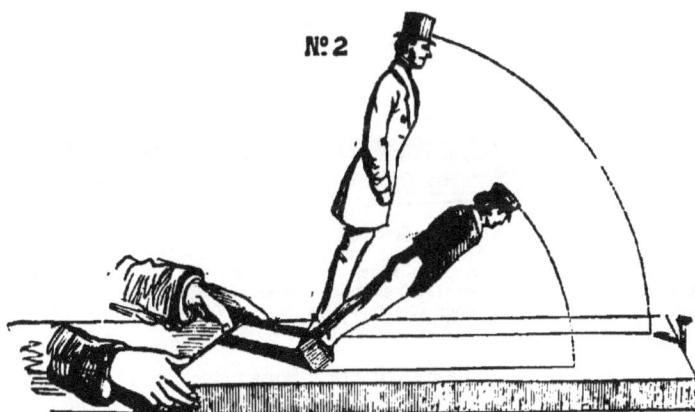

Fig. 51.—No. 2. The board pushed forward, striking against a nail, when the
short piece falls first, and the long one second.

lengths, by hinges on a flat base or board, then if the board be pushed
rapidly forward and struck against a lead weight or a nail put in the

table, the short piece is seen to fall first and the long one afterwards; the difference of time occupied in the fall of each piece of wood (which may be carved to represent the human figure) being clearly denoted by the sounds produced as they strike the board.

Boat-accidents frequently arise in consequence of ignorance on the subject of the centre of gravity, and when persons are alarmed whilst sitting in a boat, they generally rise suddenly, raise the centre of gravity, which falling, by the oscillation of the frail bark, outside the line of direction of the base, cannot be restored, and the boat is upset; if the boat were fixed by the keel, raising the centre of gravity would be of little consequence, but as the boat is perfectly free to move and roll to one side or the other, the elevation of the centre of gravity is fatal, and it operates just as the removal of the lead would do, if changed from the base to the head of the " tombola" toy.

A very striking experiment, exhibiting the danger of rising in a boat, may be shown by the following model, as depicted at Nos. 1 and 2, figs. 52 and 53.

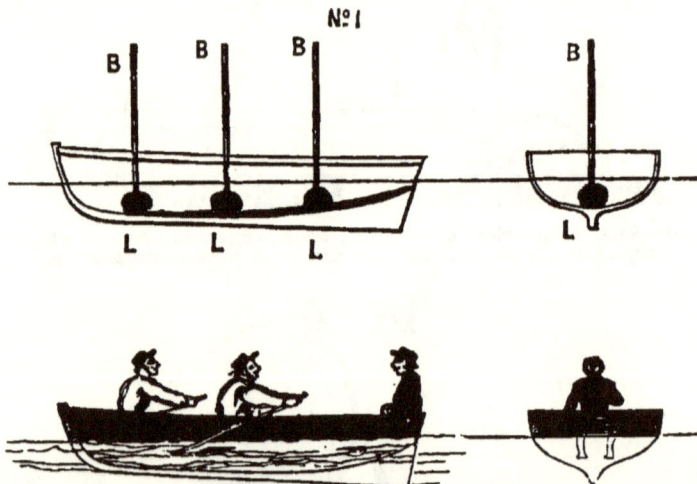

Fig. 52.—No. 1. Sections of a toy-boat floating in water. B B B. Three brass wires placed at regular distances and screwed into the bottom of the boat, with cuts or slits at the top so that when the leaden bullets, L L L, which are perforated and slide upon them like beads, are raised to the top, they are retained by the brass cuts springing out; when the bullets are at the bottom of the lines they represent persons sitting in a boat, as shown in the lower cuts, and the centre of gravity will be within the vessel.

We thus perceive that the stability of a body placed on a base depends upon the position of the line of direction and the height of the centre of gravity.

Security results when the line of direction falls within the base. Instability when just at the edge. Incapability of standing when falling without the base.

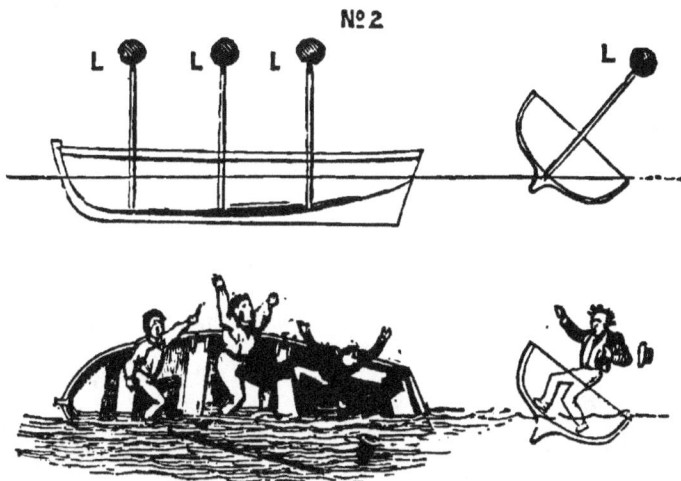

Fig. 53.—No. 2. The leaden bullets raised to the top now show the result of persons suddenly rising, when the boat immediately turns over, and either sinks or floats on the surface with the keel upwards.

The leaning-tower of Pisa is one hundred and eighty-two feet in height, and is swayed thirteen and a half feet from the perpendicular,

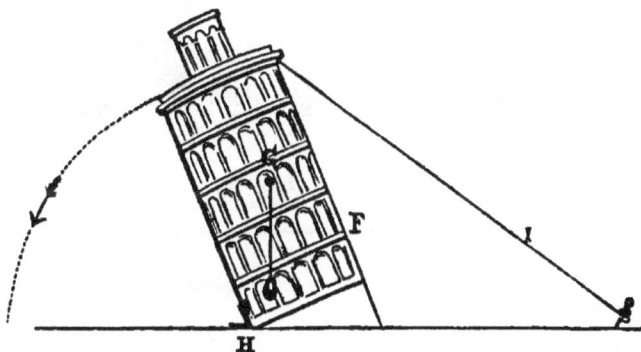

Fig. 54. F. Board cut and painted to represent the leaning-tower of Pisa. G. The centre of gravity and plummet-line suspended from it. H. The hinge which attaches it to the base board. I. The string, sufficiently long to unwind and allow the plummet to hang outside the base, so that, when cut, the model falls in the direction of the arrow.

but yet remains perfectly firm and secure, as the line of direction falls considerably within the base. If it was of a greater altitude it could no longer stand, because the centre of gravity would be so elevated that the line of direction would fall outside the base. This fact may be illustrated by taking a board several feet in length, and having cut it out to represent the architecture of the leaning-tower of Pisa, it may then be painted in distemper, and fixed at the right angle with a hinge to another board representing the ground, whilst a plumb-line may be dropped from the centre of gravity; and it may be shown that as long as the plummet falls within the base, the tower is safe; but directly the model tower is brought a little further forward by a wedge so that the plummet hangs outside, then, on removing the support, which may be a piece of string to be cut at the right moment, the model falls, and the fact is at once comprehended.

The leaning-towers of Bologna are likewise celebrated for their great inclination; so also (in England) is the hanging-tower, or, more correctly, the massive wall which has formed part of a tower at Bridge-north, Salop; it deviates from the perpendicular, but the centre of gravity and the line of direction fall within the base, and it remains secure; indeed, so little fears are entertained of its tumbling down, that a stable has been erected beneath it.

One of the most curious paradoxes is displayed in the ascent of a billiard-ball from the thin to the thick ends of two billiard-cues placed

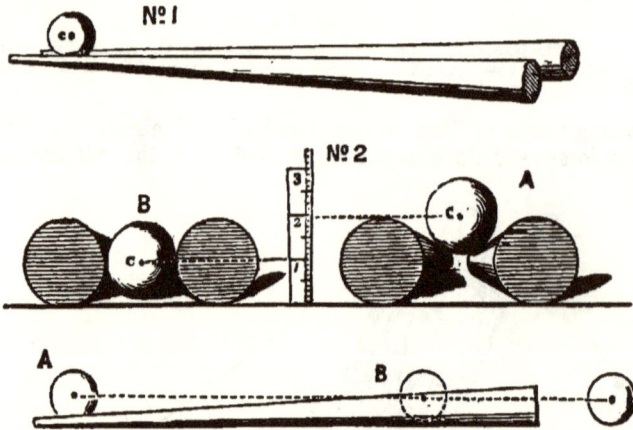

Fig. 55.—No. 1. Two billiard-cues arranged for the experiment and fixed to a board: the ball is rolling *up*.
No. 2. Sections showing that the centre of gravity, c, is higher at A than at B, which represents the thick end of the cues; it therefore, in effect, rolls down hill.

at an angle, as in our drawing above; here the centre of gravity is raised at starting, and the ball moves in consequence of its actually *falling* from the high to the low level.

Much of the stability of a body depends on the height through which the centre of gravity must be elevated before the body can be overthrown. The greater this height, the greater will be the immovability of the mass. One of the grandest examples of this fact is shown in the ancient Pyramids; and whilst gigantic palaces, with vast columns, and all the solid grandeur belonging to Egyptian architecture, have succumbed to time and lie more or less prostrate upon the earth, the

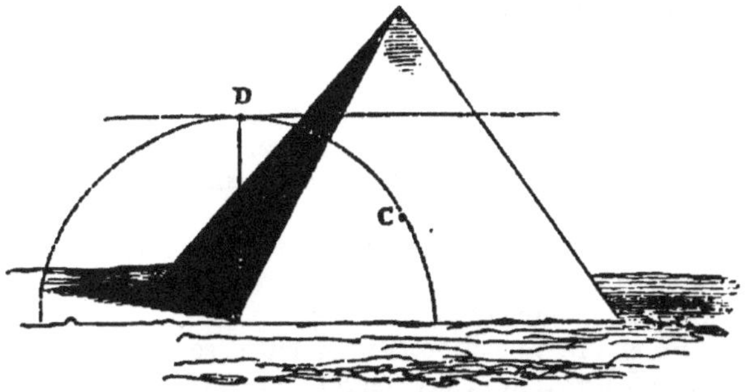

Fig. 56. c. Centre of gravity, which must be raised to D before it can be overthrown.

Pyramids, in their simple form and solidity, remain almost as they were built, and it will be noticed, in the accompanying sketch, how difficult, if not impossible, it would be to attempt to overthrow bodily one of these great monuments of ancient times.

The principles already explained are directly applicable to the construction or secure loading of vehicles; and in proportion as the centre of gravity is elevated above the point of support (that is, the wheels), so is the insecurity of the carriage increased, and the contrary takes place if the centre of gravity is lowered. Again, if a waggon be loaded

No. 1. No. 2.

Fig. 57. No. 1. The centre of gravity is near the ground, and falls within the wheels. No. 2. The centre of gravity is much elevated, and the line of direction is outside the wheels,

with a very heavy substance which does not occupy much space, such as iron, lead, or copper, or bricks, it will be in much less danger of an overthrow than if it carries an equal weight of a lighter body, such as pockets of hops, or bags of wool or bales of rags.

In the one instance, the centre of gravity is near the ground, and falls well within the base, as at No. 1, fig. 57. In the other, the centre of gravity is considerably elevated above the ground, and having met with an obstruction which has raised one side higher than the other, the line of direction has fallen outside the wheels, and the waggon is overturning as at No. 2.

The various postures of the human body may be regarded as so many experiments upon the position of the centre of gravity which we are every moment unconsciously performing.

To maintain an erect position, a man must so place his body as to cause the line of direction of his weight to fall within the base formed by his feet.

The more the toes are turned outwards, the more contracted will be the base, and the body will be more liable to fall backwards or forwards ; and the closer the feet are drawn together, the more likely is the body to fall on either side. The acrobats, and so-called "India-Rubber

Fig. 58.

Brothers," dancing dogs, &c., unconsciously acquire the habit of accurately balancing themselves in all kinds of strange positions; but as these accomplishments are not to be recommended to young people, some other marvels (such as balancing a pail of water on a stick laid upon a table) may be adduced, as illustrated in fig. 59.

Let A B represent an ordinary table, upon which place a broomstick, C D, so that one-half shall lay upon the table and the other extend from it; place over the stick the handle of an empty pail (which may possibly require to be elongated for the experiment) so that the handle touches or falls into a notch at H; and in order to bring the pail well under the

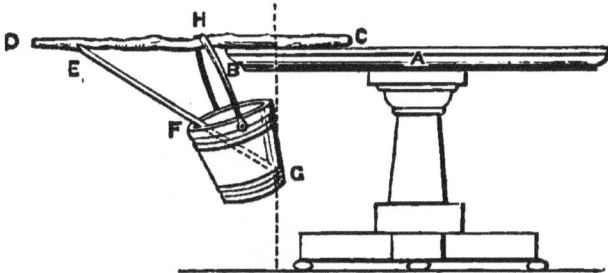

Fig. 59.

table, another stick is placed in the notch E, and is arranged in the line G F E, one end resting at G and the other at E. Having made these preparations, the pail may now be filled with water; and although it appears to be a most marvellous result, to see the pail apparently balanced on the end of a stick which may easily tilt up, the principles already explained will enable the observer to understand that the centre of gravity of the pail falls within the line of direction shown by the dotted line; and it amounts in effect to nothing more than carrying a pail on the centre of a stick, one end of which is supported at E, and the other through the medium of the table, A B.

This illustration may be modified by using a heavy weight, rope, and stick, as shown in our sketch below.

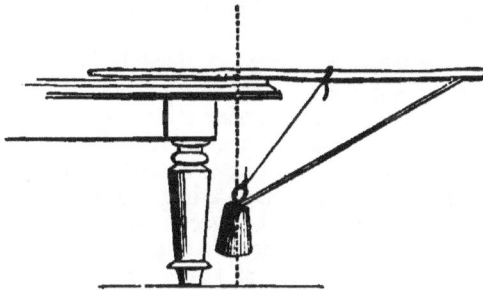

Fig. 60.

Before we dismiss this subject it is advisable to explain a term referring to a very useful truth, called the centre of percussion; a knowledge of which, gained instinctively or otherwise, enables the workman to wield his tools with increased power, and gives greater force to the cut of the swordsman, so that, with some physical strength, he may perform the feat of cutting a sheep in half, cleaving a bar of lead, or neatly dividing, *à la Saladin*, in ancient Saracen fashion, a silk hand-kerchief floating in the air. There is a feat, however, which does not require any very great strength, but is sufficiently startling to excite much surprise and some inquiry—viz., the one of cutting in half a broom-stick supported at the ends on tumblers of water without spilling the water or cracking or otherwise damaging the glass supports.

Fig. 61.

These and other feats are partly explained by reference to time: the force is so quickly applied and expended on the centre of the stick that it is not communicated to the supports; just as a bullet from a pistol may be sent through a pane of glass without shattering the whole square, but making a clean hole through it, or a candle may be sent through a plank, or a cannon-ball pass through a half opened door without causing it to move on its hinges. But the success of the several feats depends in a great measure on the attention that is paid to the delivery of the blows at the *centre of percussion* of the weapon; this is a point in a moving body where the percussion is the greatest, and about which the impetus or force of all parts is balanced on every side. It may be better understood by reference to our drawing below. Applying this principle to a model sword made of wood, cut in half in the centre of the blade, and then united with an elbow-joint, the handle being fixed to a board by a wire passed through it and the two upright pieces of wood, the fact is at once apparent, and is well shown in Nos. 1, 2, 3, fig. 62.

AB N° 1

AB N° 2

N° 3

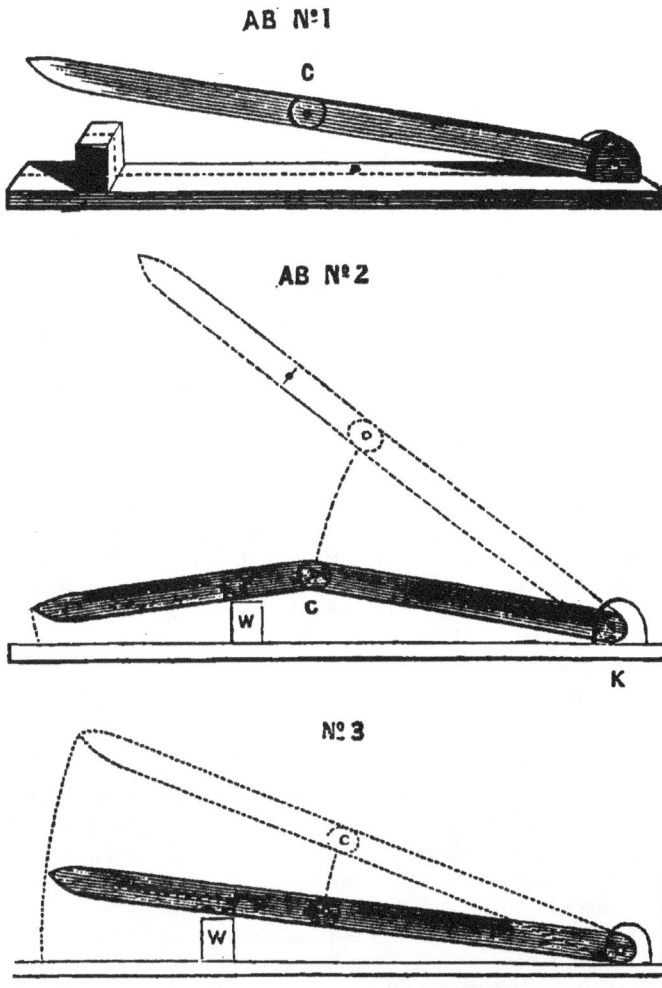

Fig. 62. No. 1, is the wooden sword, with an elbow-joint at c. No. 2. Sword attached to board at x, and being allowed to fall from any angle shown by dotted-line, it strikes the block, w, outside the centre of percussion, P, and as there is unequal motion in the parts of the sword it bends down (or, as it were, breaks) at the elbow-joint, c.

No. 3 displays the same model; but here the blow has fallen on the block, w, precisely at the centre of percussion of the sword, P, and the elbow-joint remains perfectly firm.

When a blow is not delivered with a stick or sword at the centre of percussion, a peculiar jar, or what is familiarly spoken of as a *stinging* sensation, is apparent in the hand; and the cause of this disagreeable result is further elucidated by fig. 63, in which the post, A, corresponds with the handle of the sword.

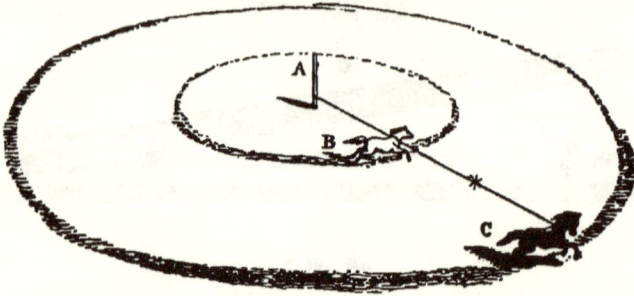

Fig. 63. A. The post to which a rope is attached. B and C are two horses running round in a circle, and it is plain that B will not move so quick as C, and that the latter will have the greatest moving force; consequently, if the rope was suddenly checked by striking against an object at the centre of gravity, the horse C would proceed faster than B, and would impart to B a backward motion, and thus make a great strain on the rope at A. But if the obstacle were placed so as to be struck at a certain point nearer C, viz., at or about the little star, the tendency of each horse to move on would balance and neutralize the other, so that there would be no strain at A. The little star indicates the *centre of percussion*.

All military men, and especially those young gentlemen who are intended for the army, should bear in mind this important truth during their sword-practice; and with one of Mr. Wilkinson's swords, made only of the very best steel, they may conquer in a chance combat which might otherwise have proved fatal to them. To Mr. Wilkinson, of Pall Mall, the eminent sword-cutler, is due the great merit of improving the quality of the steel employed in the manufacture of officers' swords; and with one of his weapons, the author has repeatedly thrust through an iron plate about one-eighth of an inch in thickness without injuring the point, and has also bent one nearly double without fracturing it, the perfect elasticity of the steel bringing the sword straight again. These, and other severe tests applied to Wilkinson's swords, show that there is no reason why an officer should not possess a weapon that will bear comparison with, nay, surpass, the far-famed *Toledo* weapon, instead of submitting to mere army-tailor swords, which are often little better than hoops of beer barrels; and, in dire combat with Hindoo or Mussulman fanatics' Tulwah, may show too late the folly of the owner.

Fig. 64.

CHAPTER V.

SPECIFIC GRAVITY.

IT is recorded of the great Dr. Wollaston, that when Sir Humphry Davy placed in his hand, what was then considered to be *the* scientific wonder of the day—viz., a small bit of the metal potassium, he exclaimed at once, "How heavy it is," and was greatly surprised, when Sir Humphry threw the metal on water, to see it not only take fire, but actually *float* upon the surface; here, then, was a philosopher possessing the deepest learning, unable, by the sense of touch and by ordinary handling, to state correctly whether the new substance (and that a metal), was heavy or light; hence it is apparent that the property of specific gravity is one of importance, and being derived from the Latin, means *species*, a particular sort or kind; and *gravis*, heavy or weight—i.e., the particular weight of every substance compared with a fixed standard of water.

We are so constantly in the habit of referring to a standard of perfection in music and the arts of painting and sculpture, that the youngest will comprehend the office of water when told that it is the philosopher's unit or starting-point for the estimation of the relative weights of solids and liquids. A good idea of the scope and meaning of the term specific gravity, is acquired by a few simple experiments, thus: if a cylindrical

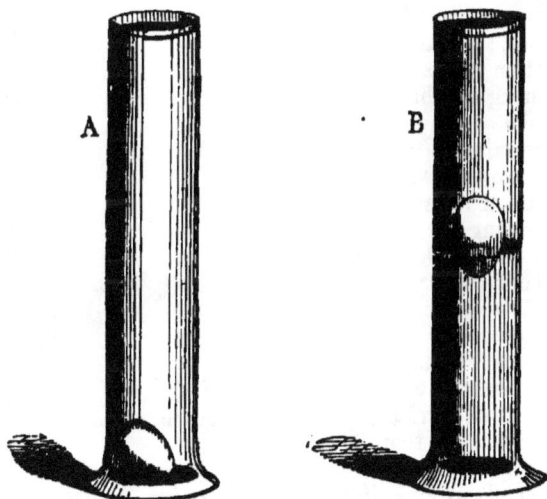

Fig. 65. A. A large cylindrical vessel containing water, in which the egg sinks till it reaches the bottom of the glass. B. A similar glass vessel containing half brine and half water, in which the egg floats in the centre—viz., just at the point where the brine and water touch.

glass, say eighteen inches long, and two and a half wide, is filled with water, and another of the same size is also filled, one half with water and the other half with a saturated solution of common salt, or what is commonly termed brine, a most amusing comparison of the relative weights of equal bulks of water and brine, can be made with the help of two eggs; when one of the eggs is placed in the glass containing water, it immediately sinks to the bottom, showing that it has a greater specific gravity than water; but when the other egg is placed in the second glass containing the brine, it sinks through the water till it reaches the strong solution of salt, where it is suspended, and presents a most curious and pretty appearance; seeming to float like a balloon in air, and apparently suspended upon nothing, it provokes the inquiry, "whether magnetism has anything to do with it?" The answer, of course, is in the negative, it merely floats in the centre, in obedience to the common principle, that all bodies float in others which are heavier than themselves; the brine has, therefore, a greater weight than an equal bulk of water, and is also heavier than the egg. A pleasing sequel to this experiment may be shown by demonstrating how the brine is placed in the vessel without mixing with the water above it; this is done by using a glass tube and funnel, and after pouring away half the water contained in the vessel (Fig. 65), the egg can be floated from the bottom to the centre of the glass, by pouring the brine down the funnel and tube. The saturated solution of salt remains in the lower part of the vessel and displaces the water, which floats upon its surface like oil on water, carrying the egg with it.

The water of the Dead Sea is said to contain about twenty-six per cent. of saline matter, which chiefly consists of common salt. It is perfectly clear and bright, and in consequence of the great density, a person may easily float on its surface, like the egg on the brine, so that if a ship

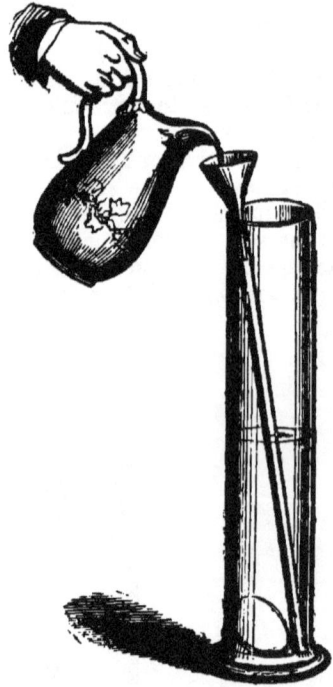

Fig. 66. A vessel half full of water, and as the brine is poured down the tube the egg gradually rises.

could be heavily laden whilst floating on the water of the Dead Sea, it would most likely sink if transported to the Thames. This illustration of specific gravity is also shown by a model ship, which being first floated on the brine, will afterwards sink if conveyed to another vessel containing water. One of the tin model ships sold as a magnetic

E

toy answers nicely for this experiment, but it must be weighted or adjusted so that it just floats in the brine, A; then it will sink, when placed, in another vessel containing only water.

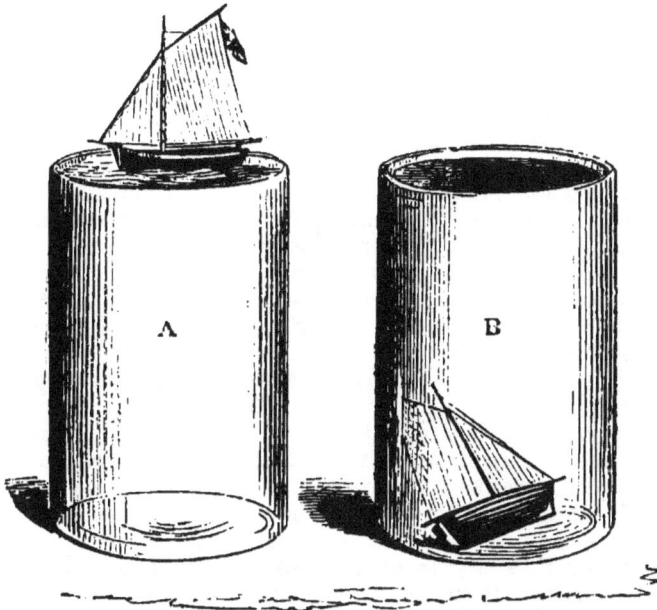

Fig. 67. A. Vessel containing brine, upon which the little model floats.
B. Vessel containing water, in which the ship sinks.

Another amusing illustration of the same kind is displayed with gold fish, which swim easily in water, floating on brine, but cannot dive to the bottom of the vessel, owing to the density of the saturated solution of salt.   If the fish are taken out immediately after the experiment, and placed in fresh water, they will not be hurt by contact with the strong salt water.

These examples of the relative weights of equal bulks, enable the youthful mind to grasp the more difficult problem of ascertaining the specific gravity of any solid or liquid substance; and here the strict meaning of terms should not be passed by.   *Specific* weight must not be confounded with *Absolute* weight; the latter means the entire amount of ponderable matter in any body: thus, twenty-four cubic feet of sand weigh about one ton, whilst specific weight means the *relation* that subsists between the *absolute weight* and the *volume* or *space* which that *weight* occupies.   Thus a cubic foot of water weighs sixty-two and a half pounds, or 1000 ounces avoirdupois, but changed to gold, the cubic foot weighs more than half a ton, and would be equal to about 19,300 ounces—hence the relation between the cubic foot of water and that of gold is nearly as 1 to 19·3; the latter is therefore called the specific gravity of gold.

Such a mode of taking the specific gravity of different substances— viz., by the weight of equal bulks, whether cubic feet or inches, could not be employed in consequence of the difficulty of procuring exact cubic inches or feet of the various substances which by their peculiar properties of brittleness or hardness would present insuperable obstacles to any attempt to fashion or shape them into exact volumes. It is therefore necessary to adopt the method first devised by Archimedes, 600 B.C., when he discovered the admixture of another metal with the gold of King Hiero's crown.

This amusing story, ending in the discovery of a philosophical truth, may be thus described:—King Hiero gave out from the royal treasury a certain quantity of gold, which he required to be fashioned into a crown; when, however, the emblem of power was produced by the goldsmith, it was not found deficient in weight, but had that appearance which indicated to the monarch that a surreptitious addition of some other metal must have been made.

It may be assumed that King Hiero consulted his friend and philosopher Archimedes, and he might have said, "Tell me, Archimedes, without pulling my crown to pieces, if it has been adulterated with any other metal?" The philosopher asked time to solve the problem, and going to take his accustomed bath, discovered then specially what he had never particularly remarked before—that, as he entered the vessel of water, the liquid rose on each side of him—that he, in fact, displaced a certain quantity of liquid. Thus, supposing the bath to have been full of water, directly Archimedes stepped in, it would overflow. Let it be assumed that the water displaced was collected, and weighed 90 pounds, whilst the philosopher had weighed, say 200 pounds. Now, the train of reasoning in his mind might be of this kind:—"My body displaces 90 pounds of water; if I had an exact cast of it in lead, the same *bulk* and *weight* of liquid would overflow; but the weight of my body was, say 200 pounds, the cast in lead 1000 pounds; these two sums divided by 90 would give very different results, and they would be the specific gravities, because the rule is thus stated:—'Divide the gross weight by the loss of weight in water, the water displaced, and the quotient gives the specific gravity.'" The rule is soon tested with the help of an ordinary pair of scales, and the experiment made more interesting by taking a model crown of some metal, which may be nicely gilt and burnished by Messrs. Elkington, the celebrated electro-platers of Birmingham. For convenience, the pan of one scale is suspended by shorter chains than the other, and should have a hook inserted in the middle; upon this is placed the crown, supported by very thin copper wire. For the sake of argument, let it be supposed that the crown weighs 17½ ounces avoirdupois, which are duly placed in the other scale-pan, and without touching these weights, the crown is now placed in a vessel of water. It might be supposed that directly the crown enters the water, it would gain weight, in consequence of being wetted, but the contrary is the case, and by thrusting the crown into the water, it may be seen to rise with great buoyancy so long as the $17\frac{1}{2}$ ounces are retained in the other scale-pan; and it will be found necessary to place at least two ounces in the scale-pan to which the crown is attached before the latter sinks in the water; and thus it is distinctly shown that the crown weighs only about $15\frac{1}{2}$ ounces in the water, and has therefore *lost* instead of *gaining* weight whilst immersed in the liquid. The rule may now be worked out:

Ounces.

Weight of crown in air . . . . $17\frac{1}{2}$
Ditto in water . . . . $15\frac{1}{2}$

Less in water 2

2) $17\frac{1}{2}$

$8{\cdot}75$

The quotient $8\frac{3}{4}$ demonstrates that the crown is manufactured of copper, because it would have been about $19\frac{1}{4}$ if made of pure gold.

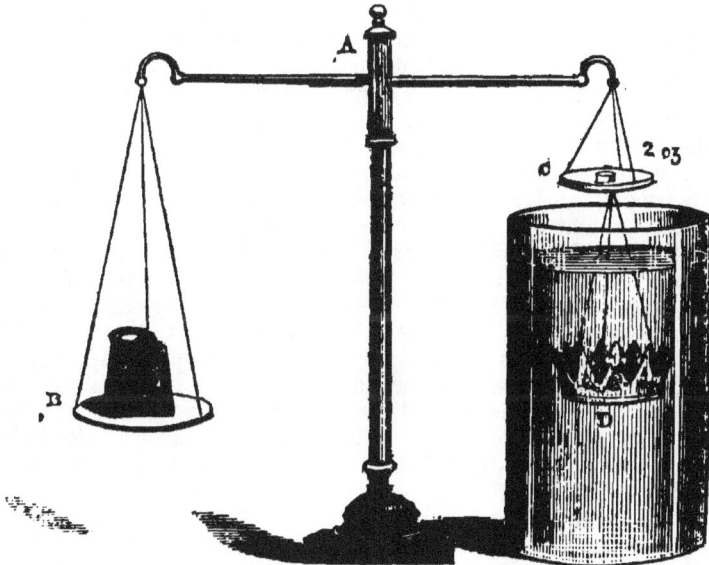

Fig. 69. A. Ordinary pair of scales. B. Scale-pan, containing $17\frac{1}{2}$ ounces, being the weight of the crown in air. c. Pan, with hook and crown attached, which is sunk in the water contained in the vessel D; this pan contains the two ounces, which must be placed there to make the crown sink and exactly balance B.

Table of the Specific Gravities of the Metals in common use.

Platinum . . . . . . 20·98
Gold . . . . 19·26 to 19·3 and 19·64
Mercury . . . . . . 13·57
Lead . . . . 11·35
Silver . . . . 10·47 to 10·5
Bismuth . . . . . . 9·82
Copper . . . . . . 8·89
Iron . . . . 7·79
Tin . . . . 7·29
Zinc . . . . 6·5 to 7·ᴉ.

The simple rule already explained may be applied to all metals of any size or weight, and when the mass is of an irregular shape, having various cavities on the surface, there may be some difficulty in taking the specific gravity, in consequence of the adhesion of *air-bubbles;* but this may be obviated either by brushing them away with a feather, or, what is frequently much better, by dipping the metal or mineral first into alcohol, and then into water, before placing it in the vessel of water, by which the actual specific gravity is to be taken.

The mode of taking the specific gravity of liquids is very simple, and is usually performed in the laboratory by means of a thin globular bottle which holds exactly 1000 grains of pure distilled water at 60° Fahrenheit. A little counterpoise of lead is made of the exact weight of the dry globular bottle, and the liquid under examination is poured into the bottle and up to the graduated mark in the neck; the bottle is then placed in one scale-pan, the counterpoise and the 1000-grain weight in the other; if the liquid (such as oil of vitriol) is heavier than water, then more weight will be required—viz., 845 grains—and these figures added to the 1000 would indicate at once that the specific gravity of oil of vitriol was 1·845 as compared with water, which is 1·000. When the liquid, such as alcohol, is lighter than water, the 1000-grain weight will be found too much, and grain weights must be added to the same scale-pan in which the bottle is standing, until the two are exactly balanced. If ordinary alcohol is being examined, it will be found necessary to place 180 grains with the bottle, and these figures deducted from the 1000 grains in the other scale-pan, leave 820, which, marked with a dot before the first figure (*sic* ·820), indicates the specific gravity of alcohol to be less than that of water.

The difference in the gravities of various liquids is displayed in a very pleasing manner by an experiment devised by Professor Griffiths, to whom chemical lecturers are especially indebted for some of the most ingenious and beautiful illustrations which have ever been devised. The experiment consists in the arrangement of five distinct liquids of various densities and colours, the one resting on the other, and distinguished not only by the optical line of demarcation, but by little balls of wax, which are adjusted by leaden shot inside, so as to sink through the upper strata of liquids, and rest only upon the one that it is intended to indicate.

The manipulation for this experiment is somewhat troublesome, and is commenced by procuring some pure bright quicksilver, upon which an iron bullet (black-leaded, or painted of any colour) is placed, or one of those pretty glass balls which are sold in such quantities at the Crystal Palace.

Secondly. Put as much white vitriol (sulphate of zinc) into a half pint of boiling water as it will dissolve, and, when cold, pour off the clear liquid, make up a ball of coloured wax (say red), and adjust it by placing little shot inside, until it sinks in a solution of sulphate of copper and floats on that of the white vitriol.

Thirdly. Make a solution of sulphate of copper in precisely the same manner, and adjust another wax ball to sink in water, and float on this solution.

Fourthly. Some clear distilled water must be provided.

Fifthly. A little cochineal is to be dissolved in some common spirits of wine (alcohol), and a ball of cork painted white provided.

Finally. A long cylindrical glass, at least eighteen inches high, and two and a half or three inches diameter, must be made to receive these five liquids, which are arranged in their proper order of specific gravity by means of a long tube and funnel.

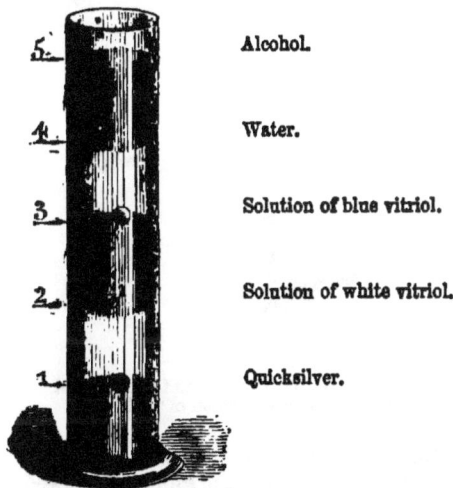

Alcohol.

Water.

Solution of blue vitriol.

Solution of white vitriol.

Quicksilver.

Fig. 69. Long cylindrical glass, 18 × 3 inches, containing the five liquids.

The four balls—viz., the iron, the two wax, and the cork balls, are allowed to slide down the long glass, which is inclined at an angle; and then, by means of the tube and funnel, pour in the tincture of cochineal, and all the balls will remain at the bottom of the glass. The water is poured down next, and now the cork ball floats up on the water, and marks the boundary line of the alcohol and water. Then the solution of blue vitriol, when a wax ball floats upon it. Thirdly, the solution of white vitriol, upon which the second wax ball takes its place; and lastly, the quicksilver is poured down the tube, and upon this heavy metallic fluid the iron or glass ball floats like a cork on water.

The tube may now be carefully removed, pausing at each liquid, so that no mixture take place between them; and the result is the arrangement of five liquids, giving the appearance of a cylindrical glass painted with bands of crimson, blue, and silver; and the liquids will not mingle with each other for many days.

A more permanent arrangement can be devised by using liquids which have no affinity, or will not mix with each other—such as mercury, water, and turpentine.

The specific weight or weights of an equal measure of air and other gases is determined on the same principle as liquids, although a different apparatus is required. A light capped glass globe, with stop-cock, from 50 to 100 cubic inches capacity, is weighed full of air, then exhausted by an air-pump, and weighed empty, the loss being taken as the weight of its volume of air; these figures are carefully noted, because *air* instead of *water* is the standard of comparison for all gases. When the specific gravity of any other gas is to be taken, the glass globe is again exhausted, and screwed on to a gas jar provided with a proper stop-cock, in which the gas is contained; and when perfect accuracy is required, the gas must be dried by passing it over some asbestos moistened with oil of vitriol, and contained in a glass tube, and the gas jar should stand in a mercurial trough. (Fig. 70.) The stop-

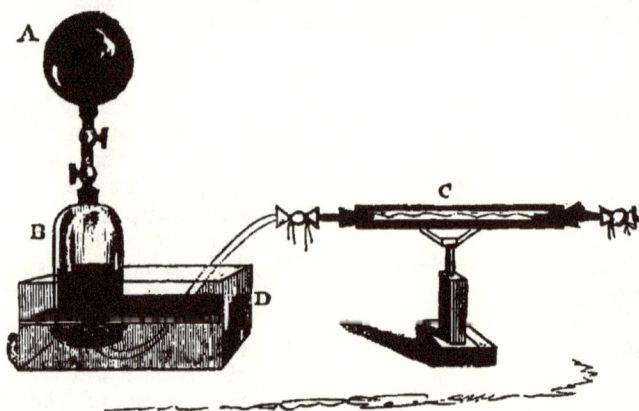

Fig. 70. A. Glass globe to contain the gas. B. Gas jar standing in the mercurial trough, D. C. Tube containing asbestos moistened with oil of vitriol.

cocks are gradually turned, and the gas admitted to the exhausted globe from the gas jar; when full, the cocks are turned off, the globe unscrewed, and again weighed, and by the common rule of proportion, as the weight of the air first found is to the weight of the gas, so is unity (1·000, the density of air) to a number which expresses the density of the gas required. If oxygen had been the gas tried, the number would be 1·111, being the specific gravity of that gaseous element. If chlorine, 2·470. Carbonic acid, 1·500. Hydrogen being much less than air, the number would only be 69, or decimally 0·069.

A very good approximation to the correct specific gravity (particularly where a number of trials have to be made with the same gas, such as

ordinary coal gas) is obtained by suspending a light paper box, with holes
at one end, on one arm of a balance, and a counterpoise on the other.
The box can be made carefully, and should have a capacity equal to a

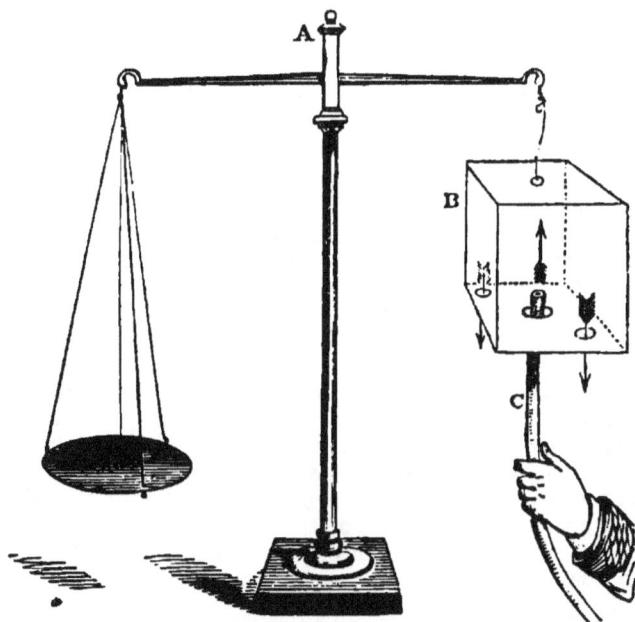

Fig. 71. A. The balance. B. The paper box, of a known capacity. C. Gas-pipe blowing
in coal-gas, the arrows showing entrance of gas and exit of the air.

half or quarter cubic foot; it is suspended with the holes downward, and is
filled by blowing in the coal gas until it issues from the apertures, and can
be recognised by the smell. The rule in this case would be equally simple:
as the known weight of the half or quarter cubic foot of common air is
to the weight of the coal gas, so is 1·000 to the number required.
(Fig. 71.)
    As an illustration of the different specific weights of the gases, a
small balloon, containing a mixture of hydrogen and air, may be so
adjusted that it will just sink in a tall glass shade inverted and supported on a pad made of a piece of oilcloth shaped round and bound
with list. On passing in quickly a large quantity of carbonic acid, the
little balloon will float on its surface; and if another balloon, containing
only hydrogen, is held in the top part of the open shade, and a sheet of
glass carefully slid over the open end, the density of the gases (although
they are perfectly invisible) is perfectly indicated; and, as a climax to
the experiment, a third balloon can be filled with laughing gas, and
.may be placed in the glass shade, taking care that the one full of pure
hydrogen does not escape; the last balloon will sink to the bottom of the

jar, because laughing gas is almost as heavy as carbonic acid, and the weight of the balloon will determine its descent. (Fig. 72.)

A soap-bubble will rest most perfectly on a surface of carbonic acid gas, and the aerial and elastic cushion supports the bubble till it bursts. The experiment is best performed by taking a glass shade twelve inches broad and deep in proportion, and resting it on a pad; half a pound of sesquicarbonate of soda is then placed in the vessel,

Balloon.
Pure hydrogen.

(Air.)

Balloon.
Hydrogen & air.

(Carbonic acid.)

Balloon.
Laughing gas.

Fig. 72. Inverted large glass shade, containing half carbonic acid and half common air.

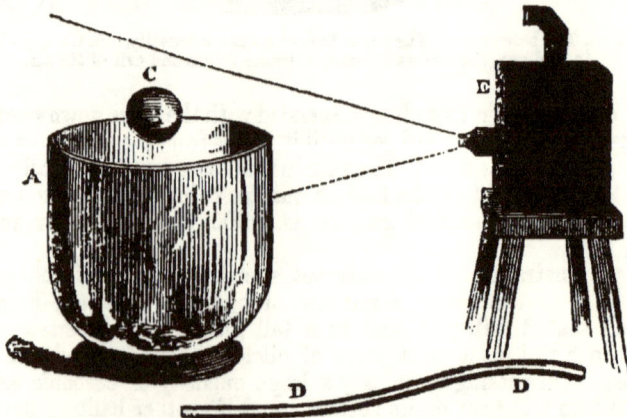

and upon this is poured a mixture of half a pint of oil of vitriol and half a pint of water, the latter being previously mixed and allowed to cool before use. An enormous quantity of carbonic acid gas is suddenly generated, and rising to the edge, overflows at the top of the glass shade. A well-formed soap-bubble, detached neatly from the end of a glass-tube, oscillates gently on the surface of the heavy gas, and presents a most curious and pleasing appearance. The soapy water is prepared by cutting a few pieces of yellow soap, and placing them in a two-ounce

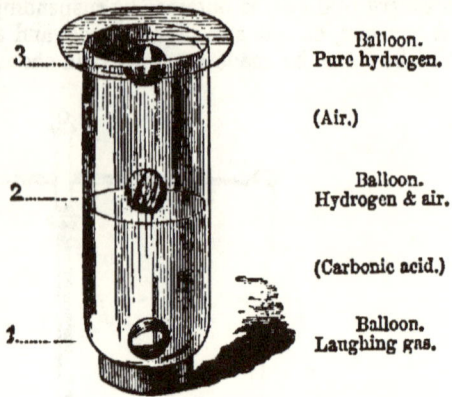

Fig. 73. A. Inverted glass shade, containing the material, B, for generating carbonic acid gas. C. The soap-bubble. D D. The glass tube for blowing the bubbles. E. Small lantern, to throw a bright beam of light from the oxy-hydrogen jet upon the thin soap-bubble, which then displays the most beautiful iridescent colours.

bottle containing distilled water. (Fig. 73.) The specific gravity of the gases, may therefore be either greater, or less than atmospheric air, which has been already mentioned as the standard of comparison, and examined by this test the vapours of some of the compounds of carbon and hydrogen are found to possess a remarkably high gravity; in proof of which, the vapour of ether may be adduced as an example, although it does not consist only of the two elements mentioned, but contains a certain quantity of oxygen. In a cylindrical tin vessel, two feet high and one foot in diameter, place an ordinary hot-water plate, of course full of boiling water; upon this warm surface pour about half an ounce of the best ether; and, after waiting a few minutes until the whole is converted into vapour, take a syphon made of half-inch pewter tube, and warm it by pouring through it a little hot water, taking care to allow the water to drain away from it before use. After placing the syphon in the tin vessel, a light may be applied to the extremity of the long leg outside the tin vessel, to show that no ether is passing over until the air is sucked out as with the water-syphon; and after this has been done, several warm glass vessels may be filled with this heavy vapour of ether, which burns on the application of flame. Finally, the remainder of the vapour may be burnt at the end of the syphon tube, demonstrating in the most satisfactory manner that the vapour is flowing through the syphon just as spirit is removed by the distillers from the casks into cellars of the public-houses. (Fig. 74.)

Fig. 74. A. Tin vessel containing the hot-water plate, B, upon which the ether is poured. C. The syphon. D. Glass to receive the vapour. E. Combustion of the ether vapour in another vessel.

Before dismissing the important subject of specific gravity (or, as it is termed by the French *savants*, "density"), it may be as well to state that astronomers have been enabled, by taking the density of the earth and by astronomical observations, to calculate the gravity of the planets belonging to our solar system ; and it is interesting to observe that the density of the planet Venus is the only one approaching the gravity of the earth :—

The Earth	1·000
The Sun	·254
The Moon	·742
Mercury	2·583
Venus	1·037
Mars	·650
Jupiter	·258
Saturn	·104
Herschel	·220

CHAPTER VI.

ATTRACTION OF COHESION.

In previous chapters one kind of attraction—viz., that of gravitation, has been discussed and illustrated in a popular manner, and pursuing the examination of the invisible, active, and real forces of nature, the attraction of cohesion will next engage our attention. There is a peculiar satisfaction in pursuing such investigations, because every step is attended by a reasonable proof; there is no ghostly mystery in philosophic studies ; the mind is not suddenly startled at one moment with that which seems more than natural ; it is not carried away in an ecstasy of wonder and awe, as in the so-called *spirit-rapping* experiments, to be again rudely brought back to the material by the disclosure of trickeries of the most ludicrous kind, such as those lately exposed by M. Jobert de Lamballe, at the Academy of Sciences at Paris. This gentleman has unmasked the effrontery of the spirit-rappers by merely stripping the stocking from the heel of a young girl of fourteen. M. Velpeau declares that the rapping is produced by the muscles of the heel and knee acting in concert, and quotes the case of a lady once celebrated as a medium, who has the power of producing the most curious and interesting music with the tendons of the thigh. This music is said to be loud enough to be heard from one end of a long room to the other, and has often played a conspicuous part in the revelations made by the medium. M. Jules Clocquet also explained the method by which the famous girl pendulum had so long abused the credulity of the Paris public. This girl, whose self-styled faculty is that of striking the hour at any time of the day or night, was attended at the Hospital St. Louis by M. Clocquet, who states that the vibrations in this case were produced by a rotatory motion in the lumbar regions of the vertebral column. The sound of these (à la rattlesnake) was so powerful, that they might be distinctly heard at a distance of twenty-five feet.

In studying the powers of nature, which the most sceptical mind allows must exist, there is an abundant field for experiment without attempting the exploits of Macbeth's witches, or the fanciful powers of Manfred; and, returning to the theme of our present chapter, it may be asked, how is cohesion defined? and the answer may be given, by directing attention to the three physical conditions of water, which assumes the form of ice, water, or steam.

In the Polar regions, and also in the Alpine and other mountains where glaciers exist, there the traveller speaks of ice twenty, thirty, forty, nay, three hundred feet in thickness. Here the withdrawal of a certain quantity of heat from the water evidently allows a new force to come into full play. We may call it what we like; but cohesion, from the Latin *cum*, together, and *hæreo*, I stick or cleave, appears to be the best and most rational term for this power which tends to make the atoms or particles of the same kind of matter move towards each other, and to prevent them being separated or moved asunder. That it is not merely hypothetical is shown by the following experiments.

If two pieces of lead are cast, and the ends nicely scraped, taking care not to touch the surfaces with the fingers, they may by simple pressure be made to cohere, and in that state of attraction may be lifted from the table by the ring which is usually inserted for convenience in the upper piece of lead; they may be hung for some time from a proper support, and the lower bit of lead will not break away from the upper one; they may even be suspended, as demonstrated by Morveau, in the vacuum of an air-pump, to show that the cohesion is not mistaken for the pressure of the atmosphere, and no separation occurs. And when the union is broken by physical force, it is surprising to notice the limited number of points, like pin points, where the cohesion has occurred; whilst the weight of the lump of lead upheld against the force of gravitation reminds one forcibly of the attraction of a mass of soft iron by a powerful magnet, and leads the philosophic inquirer to speculate on the principle of cohesion being only some masked form of magnetic or electrical attraction. (Fig. 75.)

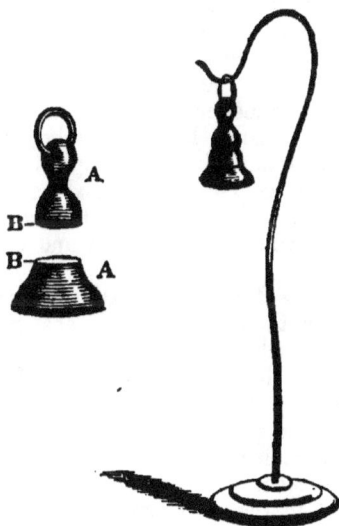

Fig. 75. A A. Two pieces of lead, scraped clean at the surfaces B B. C. Stand, supporting the two pieces of lead attached to each other by cohesion.

A fine example of the same force is shown in the use of a pair of flat iron surfaces, planed by the celebrated Whitworth, of Manchester.

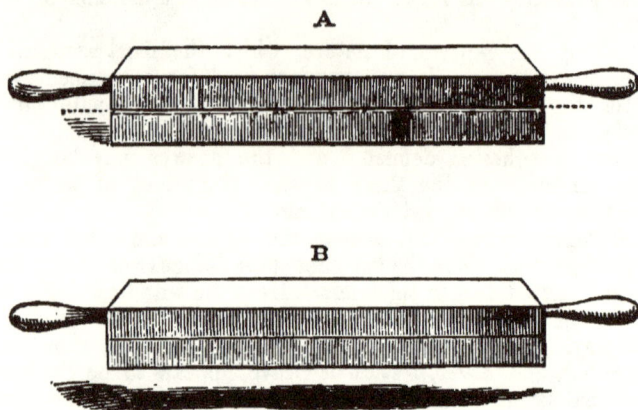

Fig. 76. A. Whitworth's planes, with film of air between them.
B. Film of air excluded when cohesion occurs.

These surfaces are so true, that when placed upon each other, the upper one will freely rotate when pushed round, in consequence of the thin film of air remaining between the surfaces, which acts like a cushion, and prevents the metallic cohesion. When, however, the upper plate is slid over the lower one gradually, so as to exclude the air, then the two may be lifted together, because cohesion has taken place. (Fig. 76.)

A glass vessel is a good example of cohesion. The materials of which it is composed have been soft and liquid when melted in the fire, and on the removal of the excess of heat it has become hard and solid, in consequence of the attractive force of cohesion binding the particles together; in the absence of such a power, of course, the material would fall into the condition of dust, and a mere shapeless heap of silicates of potash and lead would indicate the place where the moulded and co-herent glass would otherwise stand.

A lump of lead, six inches long by four broad, and half an inch thick, may be supported by dexterously taking off a thick shaving with a proper plane, and after pressing an inch or more of the strip on the planed surface of the large lump of lead, the cohesion is so powerful that the latter may be lifted from the table by the strip of metal.

The bullets projected from Perkins' steam-gun, at the rate of three hundred per minute, are thrown with such violence, that, when received on a thick plate of lead backed up with sheet iron, a cold welding takes place between the two surfaces of metal in the most perfect manner, just as two soft pieces of the metal potassium may be squeezed and welded together. The surfaces of an apple torn asunder will not readily cohere, but if cut with a sharp knife, cohesion easily occurs; so with a wound produced by a jagged surface, it is difficult to make the parts heal, whereas some of the most desperate sabre-cuts have been healed, the cohesion of the surfaces of cut flesh being very rapid; hence, if the top of a finger is cut off, it may be replaced, and will grow, in consequence of the natural cohesion of the parts.

The art of plating copper with silver, which is afterwards gilt, and then drawn out into flattened wire for the manufacture of gold lace and epaulets, usually termed bullion, is another example of the wonderful cohesion of the particles of gold, of which a single grain may be extended over the finest plate wire measuring 345 feet in length.

The process of making wax candles is a good illustration of the attraction of cohesion; they are not generally cast in moulds, as most persons suppose, but are made by the successive applications of melted wax around the central plaited wick. Other examples of cohesion are shown by icicles, and also stalagmites; which latter are produced by the gradual dropping of water containing chalk (carbonate of lime) held in solution by the excess of carbonic acid gas; the solvent gradually evaporates, and leaves a series of calcareous films, and these cohere in succession, producing the most fantastic forms, as shown in various remarkable caverns, and especially in the cave of Arta, in the island of Majorca.

In metallic substances the cohesion of the particles assumes an important bearing in the question of relative toughness and power of resisting a strain; hence the term cohesion is modified into that of the property of "tenacity."

The tenacity of the different metals is determined by ascertaining the weight required to break wires of the same length and guage. Iron

Fig. 77. B. Pan supported by leaden wire broken by a weight which the iron wire at A easily supports.

appears to possess the property of tenacity in the greatest, and lead in the least degree. (Fig. 77.)

The tenacity of iron is taken advantage of in the most scientific manner by the great engineers who have constructed the Britannia Tube, and that eighth wonder of the world, the *Leviathan*, or *Great Eastern* steam-ship. In both of these sublime embodiments of the genius and industrial skill of Great Britain the advantage of the cellular principle is fully recognised. The magnitude of this colossal ship is better realized when it is remembered that the *Great Eastern* is six times the size of the *Duke of Wellington* line-of-battle ship, that her length is more than three times that of the height of the Monument, while in breadth it is equal to the width of Pall Mall, and that a promenade round the deck will afford a walk of more than a quarter of a mile. Up to the water-mark the hull is constructed with an inner and outer shell, two feet ten inches apart, each of three-quarter-inch plate; and between them, at intervals of six feet, run horizontal webs of iron plates, which convert the whole into a series of continuous cells or iron boxes. (Fig.78.)

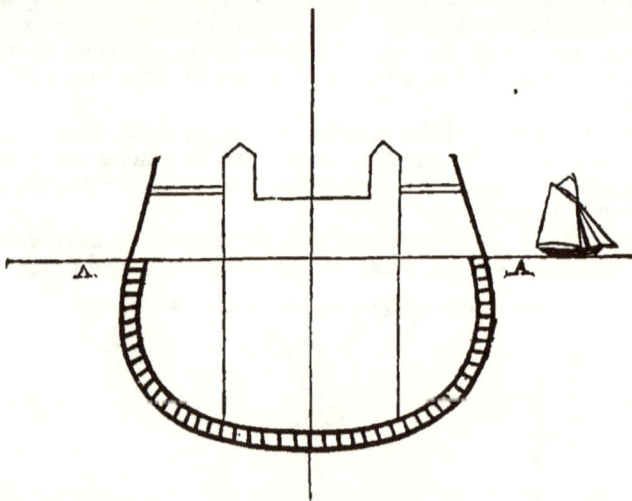

Fig. 78. Transverse section of *Great Eastern*, showing the cellular construction from keel to water-line, A A.

This double ship is useful in various ways; in the first place, the danger arising from collision is diminished, as it is supposed that the outer web only would be broken through or damaged; so that the water would not then rush into the steam-ship, but merely fill the space between the shells. In the second place, if there should be any difficulty in procuring ballast, the space can be filled with 2500 tons of water, or again pumped out, according to the requirements of the vessel. The strength of a continued cellular construction can be easily imagined, and may be well illustrated by a thin sheet of common tin plate. If the ends be rested on blocks of wood, so as to lap over the wood about one inch, they are easily displaced, and the mimic bridge broken down from its supports by the addition to the centre of a few ounce weights; whilst the same tin plate rolled up in the figure of a tube, and again rested on the same blocks, will now support many pounds weight without bending or breaking down. (Fig. 79.)

Fig. 79. A. Flat tin plate, breaking down with a few ounce weights.
B. Same tin plate rolled up supports a very heavy weight.

The deck of the ship is double or cellular, after the plan of Stephenson in the Britannia Tubular Bridge, and is 692 feet in length. The tonnage register is 18,200 tons, and 22,500 tons builder's measure; the hull of the *Great Eastern* is considered to be of such enormous tenacity, that, if it were supported by massive blocks of stone six feet square, placed at each end, at stem and stern, it would not deflect, curve, or bend down in the middle more than *six inches* even with all her machinery, coals, cargo, and living freight.

In adducing remarkable instances of the adhesive power and tenacity of inorganic matter, it may not be altogether out of place to allude to the strength and force of living matter, or muscular power. It is stated that Dr. George B. Winship, of Roxbury in America, a young physician, twenty-five years old, and weighing 143 pounds, is the strongest man alive; in fact, quite the Samson of the nineteenth century. He can raise a barrel of flour from the floor to his shoulders; can raise himself with either *little* finger till his chin is half a foot above it; can raise 200 pounds with either little finger; can put up a church bell of 141 pounds; can lift with his hands 926 pounds dead weight without the aid of straps or belts of any kind. As compared with Topham, the Cornish strong man, who could raise 800 pounds, or the Belgic one, his power is greater; and as the use of straps and belts increases the power of lifting by about four times, it is stated that Winship could lift at least 2500 pounds weight.

With these illustrations of cohesion we may return again to the abstract consideration of this power with reference to water, in which we have noticed that the antagonist to this kind of attraction is the force or power termed caloric or heat. The latter influence removes the frozen bands of winter and converts the ice to the next condition, water. In this state cohesion is almost concealed, although there is just a slight excess to hold even the particles of water in a state of unity, and this fact is beautifully illustrated by the formation of the brilliant diamond drops of dew on the surfaces of various leaves, as also in the force and power exercised by great volumes of water, which exert their mighty strength in the shape of breaker-waves, dashing against rocks and lighthouses, and making them tremble to their very base by the violence of the shock; here there must be some unity of particles, or the collective strength could not be exerted, it would be like throwing a handful of sand against a window—a certain amount of noise is produced, but the glass is not fractured; whilst the same sand united by any glutinous material, would break its way through, and soon fracture the brittle glass. It is so usual to see the particles of water easily separated, that it becomes difficult to recognise the presence of cohesion; but this bond of union is well illustrated in the experiment of the water hammer. The little instrument is generally made of a glass tube with a bulb at one end; in this bulb the water which it contains is boiled, and as the steam issues from the other extremity, drawn out to a capillary tube, the opening is closed by fusion with the heat of a blowpipe flame. As the water cools the steam condenses, and a vacuum, so far as air is concerned, is produced; if now the tube is suddenly inverted, the whole of the water falls *en masse*, collectively, and striking against the bottom of the tube, produces a metallic ring, just as if a piece of wood or metal were contained within the tube. If the end to which the water falls is not well cushioned by the palm of the hand, the water hammers itself through and breaks away that part of the glass tube. Hence it is better to construct the water hammer of copper tube, about three-quarters of an inch in diameter and three feet long; at one end a female screw-piece is inserted, into which a stop-cock is fitted; when the tube is filled to the height of about six inches with water, and shaken, the air divides the descending volume of water, and the ordinary splashing sound is heard; there is no unity or cohesion of the parts; if, however, the end of the copper tube is thrust into a fire and the water boiled so that steam issues from the cock, which is then closed, and the tube removed and cooled, a smart blow is given, and distinctly heard when the copper tube is rapidly inverted or shaken so as to cause the water to rise

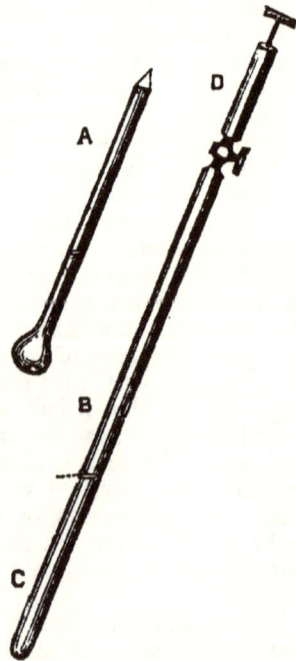

Fig. 80. A. Ordinary glass water hammer. B. Copper tube ditto, showing exhausting syringe at D, the height of the water at B, and the end to be placed in the fire at C

F

and fall. The experiment may be rendered still more instructive by turning the cock and admitting the air, which rushes in with a whizzing sound, and on shaking the tube the metallic ring is no longer heard, but it may be again restored by attaching a small air syringe or hand pump, and removing the air by exhaustion. (Fig. 80.)

In the fluid condition water still possesses a surplus of cohesion over the antagonistic force of heat; when, however, the latter is applied in excess, then the quasi-struggle terminates; the heat overpowers the cohesive attraction, and converts the water into the most willing slave which has ever lent itself to the caprices of man—viz., into steam—glorious, useful steam: and now the other end of the chain is reached, where heat triumphs; whilst in solids, such as ice, cohesion is the conqueror, and the intermediate link is displayed in the fluid state of water. If any fact could give an idea of the gigantic size of the *Great Eastern*, it is the force of the steam which will be employed to move it at the rate of about eighteen miles per hour with a power estimated at the nominal rate of 2600 horses, but absolutely of at least 12,000 horses. This steam power, coupled with the fact that she has been enormously strengthened in her sharp, powerful bows, by laying down three complete iron decks forward, extending from the bows backward for 120 feet, will demonstrate that in case of war the *Great Eastern* may prove to be a powerful auxiliary to the Government. These decks will be occupied by the crew of 300 or 400 men, and with this large increase of strength forward, the *Great Eastern*, steaming full power, could overtake and cut in two the largest wooden line-of-battle ship that ever floated. Should war unhappily spread to peaceful England, and the enormous power of this vessel be realized, her name would not inappropriately be changed from the *Great Eastern* to the *Great Terror* of the ocean. The *Times* very properly inquires, "What fleet could stand in the way of such a mass, weighing some 30,000 tons, and driven through the water by 12,000 horse-power, at the rate of twenty-two or twenty-three miles per hour. To produce the steam, 250 tons of coal per diem will be required, and great will be the honourable pride of the projectors when they see her fairly afloat, and gliding through the ocean to the Far West."

A good and striking experiment, displaying the change from the liquid to the vapour state, is shown by tying a piece of sheet caoutchouc over a tin vessel containing an ounce or two of water. When this boils, the india-rubber is distended, and breaks with a loud noise; or in another illustration, by pouring some ether through a funnel carefully into a flask placed in a ring stand. If flame is applied to the orifice, no vapour issues that will ignite, provided the neck of the flask has not been wetted with the ether. When, however, the heat of a spirit-lamp is applied, the ether soon boils, and now on the application of a lighted taper, a flame some feet in length is produced, which is regulated by the spirit-lamp below, and when this is removed, the length of the flame diminishes immediately, and is totally extinguished if the bottom of the flask is plunged into cold water; the withdrawal of the heat restores the power of cohesion. Another illustration of the vast power of steam will be shortly displayed in the Steam Ram ; and, "Supposing," says the *Times*, "the new steam ram to prove a successful design, the finest specimens of modern men-of-war will be reduced by comparison to the helplessness of cock boats. Conceive a monstrous fabric floating in mid-channel, fire proof and ball proof, capable of hurling broadsides of 100 shot to a distance of six miles ; or of clapping on steam at pleasure and running down everything on the surface of the sea with a momentum utterly irresistible.

"This terrible engine of destruction is expected to be itself indestructible. We are told that she may be riddled with shot (supposing any shot could pierce her sides), that she may have her stem and her stern cut to pieces, and be reduced apparently to a shapeless wreck, without losing her buoyancy or power. Supposing that she relies upon the shock of her impact instead of fighting her guns, it is calculated that she would sink a line-of-battle ship in three minutes, so that a squadron as large as our whole fleet now in commission would be destroyed in about one hour and a quarter."

CHAPTER VII.

ADHESIVE ATTRACTION.

The term cohesion must not be confounded with that of adhesion, which refers to the clinging to or attraction of bodies of a dissimilar kind. The late Professor Daniell defines cohesion to be an attraction of homogeneous (ὁμὸς, like, and γένος, kind) or similar particles ; adhesion to be an attraction subsisting between particles of a heterogeneous, ἕτερος, different, and γένος, kind.

There are numerous illustrations of adhesion, such as mending china, and the use of glue, or paste, in uniting different surfaces, or mortar, in building with bricks ; it is also well shown at the lecture table by means of a pair of scales, one scale-pan of which being well cleaned with alkali at the bottom, may then be rested on the surface of water contained in a plate ; the adhesion between the water and the metal is so perfect, that many grain weights may be placed in the other pan before the adhesion is broken ; and after breakage, if the pan be again placed on the water, and a few grains removed from the other, so as to adjust the two pans, and make them nearly equal, a drop of oil of turpentine being added, instantly spreads itself over the water, and breaking the adhesion between the latter and the metal, the scale-pan is immediately and again broken away, as the adhesion between the turpentine and the metal is not so great as that of water and metal. The adhesion of air and water is well displayed in an apparatus recommended for ventilating mines, in which a constant descending stream of water carries with it a quantity of air, which being disengaged, is then forced out of a proper orifice. The same kind of adhesion between air and water is displayed in the ancient

F 2

Spanish Catalan forge, where the blast is supplied to the iron furnace on a similar principle, only, a natural cascade is taken advantage of instead of an artificial fall of water through a pipe.

The adhesion of air and water becomes of some value when a river flows through a large and crowded city, because the water in its passage to and fro, must necessarily drag with it, a continuous column of air, and assist in maintaining that constant agitation of the air which is desirable as a preventive to any accumulation of noxious air charged with fœtid odours, arising from mud banks or from other causes. The fact of adhesion, existing between water and air, is readily shown, by resting one end of a long glass tube, of at least one inch diameter, on a block of wood one foot high. If water is allowed to flow down the tube, so as to leave a sufficient space of air above it, the adhesion between the two ancient elements becomes apparent, directly a little smoke is produced, near the top end of the glass tube resting on the block of wood. The smoke, which has a greater tendency to rise than to fall, is dragged down the glass tube, and accompanies the water as it flows from the higher to the lower level. The same truth is also illustrated in horizontal troughs or tubes through which water is caused to flow.

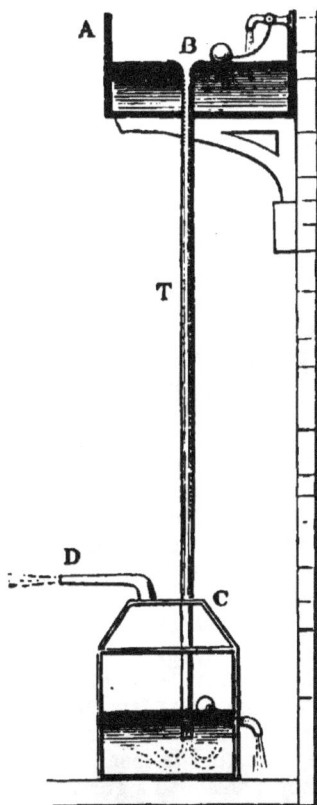

Fig. 81. Model of the apparatus for drawing down air. A, cistern of water, supplied by ball-cock, and kept at one level, so that the water just runs down the sides of the tube, and draws down the air in the centre, B C. The vessel to which the air and water are conveyed by a gutta-percha tube, T. There is another ball-cock to permit the waste water to run away when it reaches a certain level; the end of the pipe always dips some inches into this water, whilst the air escapes from the jet, D.

The adhesion between air and glass is so great, that it is absolutely necessary to boil the mercury in the tubes of the best barometers; and if this is not carefully attended to, the adhering air between the glass and mercury gradually ascends to, and destroys, the Torricellian vacuum at the top of the barometer tube. Even after the mercury is boiled, the air will creep up in course of years; and in order to prevent its passage between the glass and quicksilver, it has been recommended, that a platinum ring should be welded on to the end of the glass tube, because mercury has the power of wetting or enfilming the metal platinum, and the two being in close contact, would, as it were, shut the only door by which the air could enter the barometer tube.

CHAPTER VIII.

This kind of attraction is termed capillary, in consequence of tubes, of a calibre, or bore, as fine as hair, attracting and retaining fluids.

If water is poured into a glass, the surface is not level, but cupped at the edges, where the solid glass exerts its adhesive attraction for the liquid, and draws it from the level. If the glass be reduced to a very narrow tube, having a hair-like bore, the attraction is so great that the water is retained in the tube, contrary to the force of gravitation. Two pieces of flat glass placed close together, and then opened like a book, draw up water between them, on the same principle. A mass of salt put on a plate containing a little water coloured with indigo displays this kind of attraction most perfectly, and the water is quickly drawn up, as shown by the blue colour on the salt. A little solution of the ammonio-sulphate of copper imparts a finer and more distinct blue colour to the salt. A piece of dry Honduras mahogany one inch square, placed in a saucer containing a little turpentine, is soon found to be wet with the oil at the top, which may then be set on fire.

Almost every kind of wood possesses capillary tubes, and will float, on account of these minute vessels being filled with air; if, however, the air is withdrawn, then the wood sinks, and by boiling a ball made of beech wood in water, and then placing it under the vacuum of an air pump in other cold water, it becomes so saturated with water that it will no longer float. A remarkable instance of the same kind is mentioned by Scoresby, in which a boat was pulled down by a whale to a great depth in the ocean, and after coming to the surface it was found that the wood would neither swim nor burn, the capillary pores being entirely filled with salt water.

A piece of ebony sinks in water on account of its density, closeness, and freedom from air. A gauge made of a piece of oak, with a hole bored in it of one inch diameter, accurately receives a dry plug of willow wood which will not enter the orifice after it is wetted. Millstones are split by inserting wedges of dry hard wood, which are afterwards wetted and swelled, and burst the stone asunder. One of the most curious instances of capillary attraction is shown in the currying of leather, a process which is intended to impart a softness and suppleness to the skin, in order that it may be rendered fit for the manufacture of boots, harness, machine bands, &c. The object of the currier is to fill the pores of the leather with oil, and as this cannot be done by merely smearing the surface, he prepares the way for the oil by wetting the leather thoroughly with water, and whilst the skin is damp, oil is rubbed on, and it is then exposed to the air; the water evaporates at ordinary temperatures, but oil does not; the consequence is that the pores of the leather give up the water, which disappears in evaporation, and the oil by capillary attraction is then drawn into the body of the leather, the oil in fact takes the place vacated by the water, and renders the material very supple, and to a considerable extent waterproof. In paper making, the pores of this material, unless filled up or sized, cause the ink to blot or spread by capillary attraction. The porosity of soils is one of the great desideratums of the skilful agriculturist, and drainage is intended to remove the excess of water which would fill the pores of the earth, to the exclusion of the more valuable dews and' rains conveying nutritious matter derived from manures and the atmosphere.

A cane is an assemblage of small tubes, and if a piece of about six inches in length (cut off, of course, from the joints) be placed in a bottle of turpentine, the oil is drawn up and may be burnt at the top; it is on this principle that indestructible wicks of asbestos, and wire gauze rolled round a centre core, are used in spirit lamps. Oil, wax, and tallow, all rise by capillary attraction in the wicks to the flame, where they are boiled, converted into gas, and burnt.

The capillary attraction of skeins of cotton for water was known and appreciated by the old alchemists; and Geber, one of the most ancient of these pioneers of science, and who lived about the seventh century, describes a filter by which the liquid is separated from the solid. This experiment is well displayed by putting a solution of acetate of lead into a glass, which is placed on the highest block of a series of three, arranged as steps. Into this glass is placed the short end

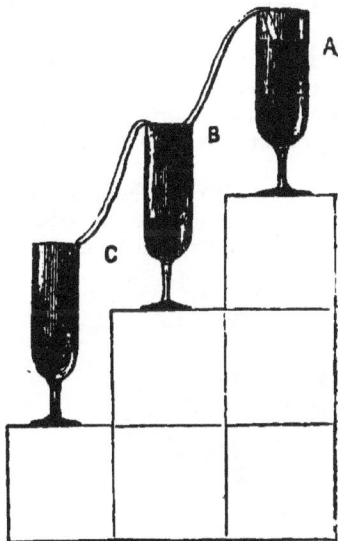

Fig. 82. Geber's filter. A. The solution of acetate of lead. B. The dilute sulphuric acid. c. The clear liquid, separated from the sulphate of lead in B.

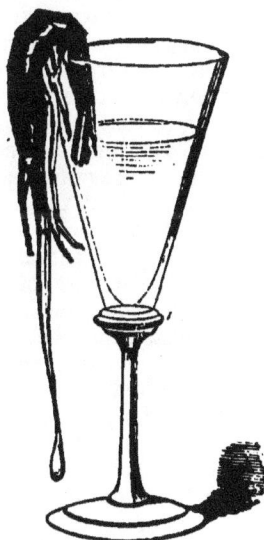

Fig. 83. Prawn syphon.

of a skein of lamp cotton, previously wetted with distilled water; the long
end dips into another glass below, containing dilute sulphuric acid, and
as the solution of lead passes into it, a solid white precipitate of sulphate
of lead is formed; then another skein of wetted cotton is placed in
this glass, the long end of which passes into the last glass, so that the
clear liquid is separated and the solid left behind. (Fig. 82.)

In this filter the lamp cotton acts as a syphon through the capillary
pores which it forms. On the same principle, a prawn may be washed
in the most elegant manner (as first shown by the late Duke of Sussex),
by placing the tail, after pulling off the fan part, in a tumbler of water,
and allowing the head to hang over, when the water is drawn up by
capillary attraction, and continues to run through the head. (Fig. 83.)
The threads of which linen, cotton, and woollen cloths are made are small
cords, and the shrinkage of such textile fabrics, is well known and
usually inquired about, when a purchase is made; here again capillary
attraction is exerted, and the fabric contracts in the two directions of
the warp and woof threads; thus, twenty-seven yards of common Irish
linen will permanently shrink to about twenty-six yards in cold water.
In these cases the water is attracted into the fibres of the textile
material, and causing them to swell, must necessarily shorten their
length, just as a dry rope strained between two walls for the purpose of
supporting clothes, has been known to draw the hooks after being suddenly wetted and shortened by a shower of rain.

In order to tighten a bandage, it is only necessary to wind the dry
linen round the limbs as close as possible, and then wet it with water,
when the necessary shrinkage takes place.

If a piece of dry cotton cloth is tied over one end of a lamp glass, the
other may be thrust into, or removed from the basin of water very easily,
but when the cotton is wetted, the fibres contract and prevent air from
entering, so that the glass retains water just as if it were an ordinary
gas jar closed with a glass stopper.

A Spanish proverb, expressing contempt, says, "go to the well with
a sieve," but even this seeming impossibility is surmounted by using a
cylinder of wire gauze, which may be filled with water, and by means of
the capillary attraction
between the meshes of
the copper-wire gauze
and the water, the whole
is retained, and may be
carefully lifted from a
basin of water; the ex-
periment only succeeds
when the air is com-
pletely driven out of the
interstices of the gauze,
and the little cylinder
completely filled with
water; this may be done

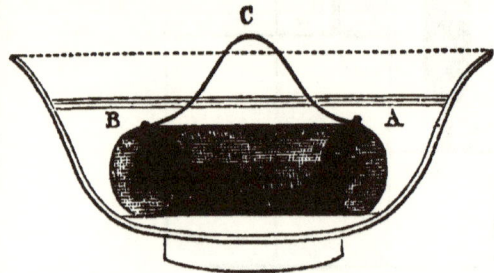

Fig. 84. A. Basin of water. B. Cylinder of wire gauze
closed at both ends with gauze. When full of water it may
be lifted from the basin by the handle, C.

by repeatedly sinking and drawing out the cylinder, or still more effectually, by first wetting it with alcohol and then dipping the cylinder in water.

A balloon, made of cotton cloth, cannot be inflated by means of a pair of bellows, but if the balloon is wetted with water, then it may be swelled out with air just as if it had been made of some air-tight material; hence the principle of varnishing silk or filling the pores with boiled oil, when it is required in the manufacture of balloons.

Biscuit ware, porous tubes for voltaic batteries, alcarrazas, or water coolers, are all examples of the same principle.

Whilst speaking most favourably of the benevolent labours of many gentlemen (beginning with Mr. Gurney) who have erected " Drinking Fountains" in London's dusty atmosphere and crowded streets, it must not be forgotten that pious Mohammedans have, in bygone times, already set us the example in this respect; and in the palmy days of many of the Moorish cities, the thirsty citizen could always be refreshed by a draught of cool water from the porous bottles provided and endowed by charitable Mussulmans, and placed in the public streets.

Fig 85. Moorish niche and porous earthenware
bottle, containing water.

CHAPTER IX.

CRYSTALLIZATION.

IT has been already stated that the force of cohesion binds the similar particles of substances together, whether they be *amorphous* or shapeless, *crystalline* or of a regular symmetrical and mathematical figure.

Fig. 86.   Crystals of snow.

The term crystal was originally applied by the ancients to silica in the form of what is usually termed rock crystal, or Brazilian pebble; and they supposed it to be water which had been solidified by a remarkable intensity of cold, and could not be thawed by any ordinary or summer heat.   Indeed, this idea of the ancients has been embodied (to a certain extent) in the shape of artificial ice made by crystallizing large quantities of sulphate of soda, which was made as flat as possible, and upon which skaters were invited to describe the figure of eight, at the usual admittance fee, representing twelve pence. A crystal is now defined to be an inorganic body, which, by the operation of affinity, has assumed the form of a regular solid terminated by a certain number of planes or smooth surfaces.

Thousands of minerals are discovered in the crystallized state—such as cubes of iron pyrites (sulphuret of iron) and of fluor spar (fluoride of calcium), whilst numerous saline bodies called salts are sold only in the form of crystals. Of these salts we have excellent examples in Epsom salts (sulphate of magnesia), nitre (nitrate of potash), alum (sulphate of alumina), and potash; the term salt being applied specially to all substances composed of an acid and a base, as also to other combinations of elements which may or may not take a crystalline form. Thus, nitre is composed of nitric acid and potash; the first, even when much diluted, rapidly changes paper, dipped in tincture of litmus and stained blue, to a red colour, whilst potash shows its alkaline nature, by changing paper, stained yellow with tincture of turmeric, to a reddish-brown. The latter paper is restored to its original yellow by dipping it into the dilute nitric acid, whilst the litmus paper regains its delicate blue colour by being passed into the alkaline solution. An acid and an alkali combine and form a neutral salt, such as nitre, which has no action whatever on litmus or turmeric; whilst the element iodine, which is not an acid, unites with the metallic element potassium, and therefore not an alkali, and forms a salt that crystallizes in cubes called iodide of potassium. Again, cane sugar, which is composed of charcoal, oxygen, and hydrogen, crystallizes in hard transparent four-sided and irregular six-sided prisms, but is not called a salt. Silica or sand is found crystallized most perfectly in nature in six-sided pyramids, but is not a salt; it is an acid termed silicic-acid. Sand has no acid taste, because it is insoluble in water, but when melted in a crucible with an alkali, such as potash, it forms a salt called silicate of potash. Magnesia, from being insoluble, or nearly so, in water, is all but tasteless, and has barely any alkaline reaction, yet it is a very strong alkaline base; 20·7 parts of it neutralize as much sulphuric acid as 47 of potash. A salt is not always a crystallizable substance, and *vice versa*. The progress of our chemical knowledge has therefore demanded a wider extension and application of the term *salt*, and it is not now confined merely to a combination of an acid and an alkali, but is conferred even on compounds consisting only of sulphur and a metal, which are termed *sulphur salts*.

So also in combinations of chlorine, iodine, bromine, and fluorine, with metallic bodies, neither of which are acid or alkaline, the term *haloid salts* has been applied by Berzelius, from the Greek (ἁλς, sea salt, and εἶδος form), because they are analogous in constitution to sea salt; and the mention of sea salt again reminds us of the wide signification of the term salt, originally confined to this substance, but now extended into four great orders, as defined by Turner:—

ORDER I. *The oxy-salts.*—This order includes no salt the acid or base of which is not an oxidised body (ex., nitrate of potash).

ORDER II. *The hydro-salts.*—This order includes no salt the acid or base of which does not contain hydrogen (ex., chloride of ammonium).

ORDER III. *The sulphur salts.*—This order includes no salt the electro-positive or negative ingredient of which is not a sulphuret (ex., hydrosulphuret of potassium).

ORDER IV. *The haloid salts.*—This order includes no salt the electro-positive or negative ingredient of which is not haloidal. (Exs., iodide of potassium and sea salt). To fix the idea of salt still better in the youthful mind, it should be remembered that alabaster, of which works of art are constructed, or marble, or lime-stone, or chalk, are all salts, because they consist of an acid and a base.

In order to cause a substance to crystallize it is first necessary to endow the particles with freedom of motion. There are many methods of doing this chemically or by the application of heat, but we cannot by any mechanical process of concentration, compression, or division, persuade a substance to crystallize, unless perhaps we except that remarkable change in wrought or fibrous iron into crystalline or brittle iron, by constant vibration, as in the axles of a carriage, or by attaching a piece of fibrous iron to a tilt hammer.

If we powder some alum crystals they will not again assume their crystalline form; if brought in contact there is no freedom of motion. It is like placing some globules of mercury on a plate. They have no power to create motion; their inertia keeps them separated by certain distances, and they do not coalesce; but incline the plate, give them motion, and bring them in contact, they soon unite and form one globule. The particles of alum are not in close contact, and they have no freedom of motion unless they are dissolved in water, when they become invisible; the water by its chemical power destroys the mechanical aggregation of the solid alum far beyond any operation of levigation. The solid alum has become liquid, like water; the particles are now free to move without let or hindrance from friction. A solution, (from the Latin *solvo*, to loosen) is obtained. The alum must indeed be reduced to minute particles, as they are alike invisible to the eye whether assisted by the microscope or not. No repose will cause the alum to separate; the solvent power of the water opposes gravitation; every part of the solution is equally impregnated with alum, and the particles are diffused at equal distances through the water; the heavy alum is actually drawn up against gravity by the water.

How, then, is the alum to be brought back again to the solid state? The answer is simple enough. By evaporating away the excess of water, either by the application of heat or by long exposure to the atmosphere in a very shallow vessel, the minute atoms of the alum are brought closer together, and crystallization takes place. The assumption of the solid state is indicated by the formation of a thin film (called a *pellicle*) of crystals, and is further and still more satisfactorily proved by taking out a drop of the solution and placing it on a bit of glass, which rapidly becomes filled with crystals if the evaporation has been carried sufficiently far (Fig. 87).

After evaporating away sufficient water, the dish is placed on one side and allowed to cool, when crystals of the utmost regularity of form

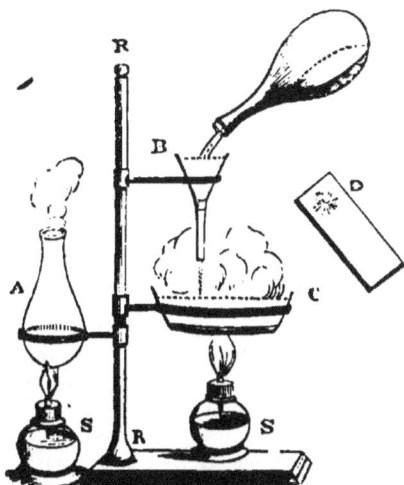

Fig. 87. ʀ ʀ. Ring-stand. ꜱ ꜱ. Spirit-lamps.
ᴀ. Flask containing boiling solution of alum.—
Solution. ʙ. Funnel, with a bit of lamp-cotton
stuffed in the bottom.—Filtration. ᴄ. Evaporating dish.—Evaporation. ᴅ. Drop on glass.—
Crystallization.

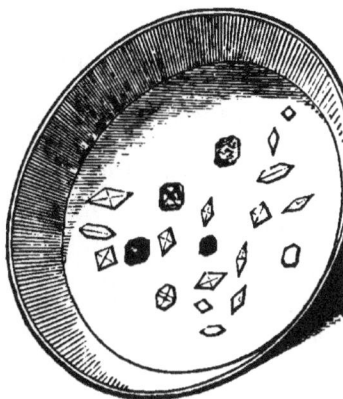

Fig. 88.

are produced, and, denoted by a geometrical term, are called octohedral
or eight-sided crystals, when in the utmost state of perfection (Fig. 88).
The science of crystallography is too elaborate to be discussed at
length in a work of this kind; the various terms connected with crystals
will therefore only be explained, and experiments given in illustration of
the formation of various crystals.

When the apices—*i.e.*, the tips or points of crystals—are cut off, they
are said to be truncated; and the same change occurs on the edges of
numerous crystals.

If some of the salt called chloride of calcium in the dry and amorphous state is exposed to the air, it soon absorbs water, or what is termed
deliquesces: the same thing occurs with the crystals of carbonate
of potash, and if four ounces are weighed out in an evaporating dish,
and then exposed for about half an hour to the air, a very perceptible
increase in weight is observed by the assistance of the scales and grain
weights. *Deliquescence* is a term from the Latin *deliqueo,* to melt, and
is in fact a gradual melting, caused by the absorption of water from
the atmosphere. The reverse of this is illustrated with various crystals,
such as Glauber's salt (sulphate of soda), or common washing soda
(carbonate of soda); if a fine clear crystal is taken out of the solution,
called the mother liquor, in which it has been crystallized, wiped dry,
and placed under a glass shade, this salt may remain for a long period

without change, but if it receive one scratch from a pin, the door is opened apparently for the escape of the water which it contains, chemically united with the salt, and called water of crystallization; the white crystal gradually swells out, the little *quasi* sore from the pin-scratch spreads over the whole, which becomes opaque, and crumbling down falls into a shapeless mass of white dust; this change is called *efflorescence*, from *effloresco*, to blow as a flower — caused by the abstraction from them of chemically-combined water by the atmosphere. With reference to the preservation of crystals, Professor Griffiths recommends them to be oiled and wiped, and placed under a glass shade, if of a deliquescent nature; or if efflorescent, they are perfectly preserved by placing them under a glass shade with a little water in a cup to keep the air charged with moisture and prevent any drying up of the crystal.

Deliquescent crystals may be preserved by placing them, when dry, in naphtha, or any liquor in which they are perfectly insoluble. Some salts, like Glauber's salts, contain so much water of crystallization that when subjected to heat they melt and dissolve in it, and this liquefaction of the solid crystal is called "watery fusion." Other salts, such as bay salt, chlorate of potash, &c., when heated, fly to pieces, with a sharp crackling noise, which is due sometimes, to the unequal expansion of the crystalline surface, or the sudden conversion of the water (retained in the crystal by capillary attraction) into steam; thus nitre behaves in this manner, and frequently retains water in capillary fissures, although it is an anhydrous salt, or salt perfectly free from combined water. The crackling sound is called *decrepitation*, and is well illustrated by throwing a handful of bay salt on a clear fire; but this property is destroyed by powdering the crystals.

Many substances when melted and slowly cooled concrete into the most perfect crystals; in these cases heat alone, the antagonist to cohesion, is the solvent power. Thus, if bismuth be melted in a crucible, and when cooling, and just as the pellicle (from *pellis*, a skin or crust) is forming on the surface, if two small holes are instantly made by a rod of iron and the liquid metal poured out from the inside (one of the holes being the entrance for the air, the other the exit for the metal); on carefully breaking the crucible, the bismuth is found to be crystallized in the most lovely cubes. Sulphur, again, may be crystallized in prismatic crystals by pursuing a similar plan; and the great blocks of spermaceti exhibited by wax chandlers in their windows, are crystallized in the interior and prepared on the same principle.

There are other modes of conferring the crystalline state upon substances—viz., by elevating them into a state of vapour by the process called sublimation (from *sublimis*, high or exalted), the lifting up and condensation of the vapour in the upper part of a vessel; a process perfectly distinct from that of *distillation*, which means to separate drop by drop. Both of these processes are very ancient, and were invented by the Arabian alchemists long antecedent to the seventh century. Examples of sublimation are shown by heating iodine, and especially

benzoic acid; with the latter, a very elegant imitation of snow is produced, by receiving the vapour, on some sprigs of holly or other ever-
green, or imitation paper snow-
drops and crocuses, placed in a
tasteful manner under a glass
vessel. The benzoic acid should
first be sublimed over the sprigs
or artificial flowers in a gas jar,
which may be removed when the
whole is cold, and a clear glass
shade substituted for it. (Fig. 89.)

All electro deposits on metals
are more or less crystalline; and
copper or silver may be deposited
in a crystalline form by placing
a scraped stick of phosphorus in
a solution of sulphate of copper
or of nitrate of silver. The phosphorus takes away the oxygen
from the metal, or deoxidizes the
solution, and the copper or silver
reappears in the metallic form.
The surface of the phosphorus
must not be scraped in the air,
but under water, when the operation is perfectly safe.

Fig. 89.   A. Gas-jar, with stopper open at
first, to be shut when the lamp is withdrawn.
B. Wooden stand, with hole to carry the cup c,
containing the benzoic acid, heated below by
the spirit-lamp, s.   F. Flowers or sprigs arranged on pieces of rock or mineral.

A singular and almost instantaneous crystallization can be
produced by saturating boiling
water with Glauber's salt, of
which one ounce and a half of
water will usually dissolve about two ounces; having done this, pour
the solution, whilst boiling hot, into clean oil flasks, or vials, or any
kind, previously warmed in the oven, and immediately cork them, or tie
strips of wetted bladder, over the orifices of the flasks or vials, or pour
into the neck a small quantity of olive oil, or close the neck with a
cork through which a thermometer tube has been passed.   When cold,
no crystallization occurs until atmospheric air is admitted; and it was
formerly believed that the pressure of the air effected this object, until
some one thought of the oil, and now the theory is modified, and crystallization is supposed to occur in consequence of the water dissolving
some air which causes the deposit of a minute crystal, and this being
the turning point, the whole becomes solid.   However the fact may be
explained, it is certain that when the liquid refuses to crystallize on the
admission of air, the solidification occurs directly a minute crystal of
sulphate of soda, or Glauber's salt, is dropped into the vessel.

When the crystallization is accomplished, the whole mass is usually
so completely solidified, that on inverting the vessel, not a drop of liquid
falls out.

It may be observed that the same mass of salt will answer any number of times the same purpose. All that is necessary to be done, is to place the vial or flask, in a saucepan of warm water, and gradually raise it to the boiling point till the salt is completely liquefied, when the vessel must be corked and secured from the air as before. When the solidification is produced much heat is generated, which is rendered apparent by means of a thermometer, or by the insertion of a copper wire into the pasty mass of crystal in the flask, and then touching an extremely thin shaving or cutting of phosphorus, dried and placed on cotton wool. Solidification in all cases produces heat. Liquefaction produces cold.

In Masters's freezing apparatus certain measured quantities of crystallized sal-ammoniac, nitre, and nitrate of ammonia, are placed in a metallic cylinder, surrounded with a small quantity of spring water contained in an outer vessel. Directly the crystals are liquefied by the addition of water, intense cold is produced, which freezes the water and forms an exact cast of the inner cylinder in ice, and this may afterwards be removed, by pouring away the liquefied salts, and filling the inner cylinder, with water of the same temperature as the air, which rapidly thaws the surrounding ice, and allows it to slip off into any convenient vessel ready to receive it. (Fig. 90.)

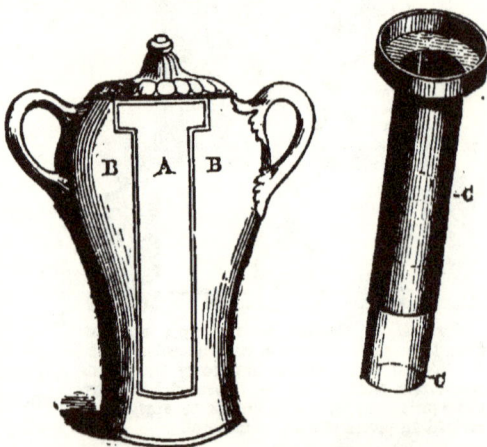

Fig. 90. A. The inner cylinder which contains the freezing mixture. B B. The outer one containing spring water. C C. The ice slipping away from the inner cylinder.

For an ingenious method of obtaining large and perfect crystals of almost any size, experimentalists are indebted to Le Blanc. His method consists in first procuring small and perfect crystals—say, octohedra of alum—and then placing them in a broad flat-bottomed pan, he pours over the crystals a quantity of saturated solution of alum, obtained by evaporating a solution of alum until a drop taken out crystallizes on cooling. The positions of the crystals are altered at least once a day with a glass rod, so that all the faces may be alternately exposed to the action of the solution, for the side on which the crystal rests, or is in contact with the vessel, never receives any increment. The crystals will thus gradually grow or increase in size, and when they have done so for some time, the best and most symmetrical, may be removed and placed separately, in vessels containing some of the same saturated solution of alum, and being constantly turned they may be obtained of almost any size desired.

Unless the crystals are removed to fresh solutions, a reaction takes place, in consequence of the exhaustion of the alum from the water, and the crystal is attacked and dissolved. This action is first perceptible on the edges and angles of the crystal; they become blunted and gradually lose their shape altogether. By this method crystals may be made to grow in length or breadth—the former when they are placed upon their sides, the latter if they be made to stand upon their bases.

On Le Blanc's principle, beautiful crystal baskets are made with alum, sulphate of copper, and bichromate of potash. The baskets are usually made of covered copper wire, and when the salts crystallize on them as a nucleus or centre, they are constantly removed to fresh solutions, so that the whole is completely covered, and red, white, and blue sparkling crystal baskets formed. They will retain their brilliancy for any time, by placing them under a glass shade, with a cup containing a little water.

The sketch below affords an excellent illustration of some of Nature's remarkable concretions in the peculiar columnar structure of basalt.

Fig. 91. The Giant's Causeway.

Fig. 92. Alchemists at work.

CHAPTER X.

CHEMISTRY.

THERE is hardly any kind of knowledge which has been so slowly acquired as that of chemistry, and perhaps no other science has offered such fascinating rewards to the labour of its votaries as the *philosopher's stone*, which was to produce an unfailing supply of gold; or *the elixir of life*, that was to give the discoverer of the gold-making art the time, the prolonged life, in which he might spend and enjoy it.

Hundreds of years ago Egypt was the great depository of all learning, art, and science, and it was to this ancient country that the most celebrated sages of antiquity travelled.

Hermes, or Mercurius Trismegistus, the favourite minister of the Egyptian king Osiris, has been celebrated as the inventor of the art of alchemy, and the first treatise upon it has been attributed to Zosymus, of Chemnis or Panopolis. The Moors who conquered Spain were remarkable for their learning, and the taste and elegance with which they designed and carried out a new style of architecture, with its love · Arabesque ornamentation. They were likewise great followers of the art of alchemy, when they ceased to be conquerors, and became more reconciled to the arts of peace. Strange that such a people, thirsting as they did in after years for all kinds of knowledge, should have destroyed, in the persons of their ancestors, the most numerous collection of books that the world had ever seen: the magnificent library of Alexandria, collected by the Ptolemies with great diligence and at an enormous expense, was burned by the orders of Caliph Omar; whilst it is stated that the alchemical works had been previously destroyed by Diocletian in the fourth century, lest the Egyptians should acquire by such means sufficient wealth to withstand the Roman power, for gold was then, as it is now, "the sinews of war."

Eastern historians relate the trouble and expense incurred by the succeeding Caliphs, who, resigning the Saracenic barbarism of their ancestors, were glad to collect from all parts the books which were to furnish forth a princely library at Bagdad. How the learned scholar sighs when he reads of seven hundred thousand books being consigned to the ignominious office of heating forty thousand baths in the capital of Egypt, and of the magnificent Alexandrian Library, a mental fuel for the lamp of learning in all ages, consumed in bath furnaces, and affording six months' fuel for that purpose. The Arabians, however, made amends for these barbarous deeds in succeeding centuries, and when all Europe was laid waste under the iron rule of the Goths, they became the protectors of philosophy and the promoters of its pursuits ; and thus we come to the seventh century, in which Geber, an Arabian prince lived, and is stated to be the earliest of the true alchemists whose name has reached posterity.

Without attempting to fill up the alchemical history of the intervening centuries, we leap forward six hundred years, and now find ourselves in imagination in England, with the learned friar, Roger Bacon, a native of Somersetshire, who lived about the middle of the thirteenth century ; and although the continual study of alchemy had not yet produced the " stone," it bore fruit in other discoveries, and Roger Bacon is said, with great appearance of truth, to have discovered gunpowder, for he says in one of his works :—" From saltpetre and *other* ingredients we are able to form a fire which will burn to any distance ;" and again alluding to its effects, " a small portion of matter, about the size of the thumb, *properly disposed*, will make a tremendous sound and coruscation, by which cities and armies might be destroyed." The exaggerated style seems to have been a favourite one with all philosophers, from the time of Roger Bacon to that of Muschenbroek of the University of Leyden, who accidentally discovered the Leyden jar in the year 1746, and receiving the first shock, from a vial containing a little water, into which a cork and nail had been fitted, states that " he felt himself struck in his arms, shoulders, and breast, so that he lost his breath, and was *two days* before he recovered from the effects of the blow and the terror;" adding, that "he would not take a *second* shock for the kingdom of France." Disregarding the numerous alchemical events occurring from the time of Roger Bacon, we again advance four hundred years—viz., to the year 1662, when, on the 15th of July, King Charles II. granted a royal charter to the Philosophical Society of Oxford, who had removed to London, under the name of the Royal Society of London for Promoting Natural Knowledge, and in the year 1665 was published the first number of the *Philosophical Transactions;* this work contains the successive discoveries of Mayow, Hales, Black, Leslie, Cavendish, Lavoisier, Priestley, Davy, Faraday; and since the year 1762 has been regularly published at the rate of one volume per annum. With this preface proceed we now to discuss some of the varied phenomena of chemical attraction, or what is more correctly termed

CHEMICAL AFFINITY.

The above title refers to an endless series of changes brought about by chemical combinations, all of which can be reduced to certain fixed laws, and admit of a simple classification and arrangement. A mechanical aggregation, however well arranged, can be always distinguished from a chemical one. Thus, a grain of gunpowder consists of *nitre*, which can be washed away with boiling water, of *sulphur*, which can be sublimed and made to pass away as vapour, of *charcoal*, which remains behind after the previous processes are complete; this mixture has been perfected by a careful proportion of the respective ingredients, it has been wetted, and ground, and pressed, granulated, and finally dried; all these mechanical processes have been so well carried out that each grain, if analysed, would be similar to the other; and yet it is, after all, only a mechanical aggregation, because the sulphur, the charcoal, and the nitre are unchanged. A grain of gunpowder moistened, crushed, and examined by a high microscopic power, would indicate the yellow particles of sulphur, the black parts of charcoal, whilst the water filtered from the grain of powder and dried, would show the nitre by the form of the crystal. On the other hand, if some nitre is fused at a dull red heat in a little crucible, and two or three grains of sulphur are added, they are rapidly oxidized, and combine with the potash, forming sulphate of potash; and after this change a few grains of charcoal may be added in a similar manner, when they burn brightly, and are oxidized and converted into carbonic acid, which also unites in like manner with the potash, forming carbonate of potash; so that when the fused nitre is cooled and a few particles examined by the microscope, the charcoal and sulphur are no longer distinguishable, they have undergone a chemical combination with portions of the nitre, and have produced two new salts, perfectly different in taste, gravity, and appearance from the original substances employed to produce them. Hence chemical combination is defined to be "*that property which is possessed by one or more substances, of uniting together and producing a third or other body perfectly dif-*

ferent in its nature from either of the two or more generating the new compound."

To return to our first experiment with the gunpowder: take sulphur, place some in an iron ladle, heat it over a gas flame till it catches fire, then ascend a ladder, and pour it gently, from the greatest height you can reach, into a pail of warm water: if this experiment is performed in a darkened room a magnificent and continuous stream of fire is obtained, of a blue colour, without a single break in its whole length, provided the ladle is gradually inclined and emptied. The substance that drops into the warm water is no longer yellow and hard, but is red, soft, and plastic: it is still sulphur, though it has taken a new form, because that element is dimorphous (δὶς twice, and μορφη a form), and, Proteus-like, can assume two forms. Take another ladle, and melt some nitre in it at a dull red heat, then add a small quantity of sulphur, which will burn as before; and now, after waiting a few minutes, repeat the same experiment by pouring the liquid from the steps through the air into water; observe it no longer flames, and the substance received into the water is not red and soft and plastic, but is white, or nearly so, and rapidly dissolves away in the water. The sulphur has united with the oxygen of the nitre and formed sulphuric acid, which combines with the potash and forms sulphate of potash; here, then, oxygen, sulphur, and potassium, have united and formed a salt in which the separate properties of the three bodies have completely disappeared; to prove this, it is only necessary to dissolve the sulphate of potash in water, and after filtering the solution, or allowing it to settle, till it becomes quite clear and bright, some solution of baryta may now be added, when a white precipitate is thrown down, consisting of sulphate of baryta, which is insoluble in nitric or other strong acids. The behaviour of a solution of sulphate of potash with a nitrate of baryta may now be contrasted with that of the elements it contains; on the addition of sulphur to a solution of nitrate of baryta no change whatever takes place, because the sulphur is perfectly insoluble. If a stream of oxygen gas is passed from a bladder and jet through the same test, no effect is produced; the nitrate of baryta has already acquired its full proportion of oxygen, and no further addition has any power to change its nature; finally, if a bit of the metal potassium is placed in the solution of nitrate of baryta it does not sink, being lighter than water, and it takes fire; but this is not in any way connected with the presence of the test, as the same thing will happen if another bit of the metal is placed in water—it is the oxygen of the latter which unites rapidly with the potassium, and causes it to become so hot that the hydrogen, escaping around the little red-hot globules, takes fire; moreover, the fact of the combustion of the potassium under such circumstances is another striking proof of the opposite qualities of the three elements—sulphur, oxygen, and potassium—as compared with the three chemically combined and forming sulphate of potash. The same kind of experiment may be repeated with charcoal: if some powdered charcoal is made red-hot, and then puffed into the air with a blowing machine, numbers of sparks are produced, and the char coal burns away and forms carbonic acid gas, a little ash being left behind; but if some more nitre be heated in a ladle, and charcoal added, a brilliant deflagration (*deflagro*, to burn) occurs, and the charcoal, instead of passing away in the air as carbonic acid, is now retained in the same shape, but firmly and chemically united with the potash of the nitre, forming carbonate of potash, or pearl-ash, which is not black and insoluble in water and acids like charcoal, but is white, and not only soluble in water, but is most rapidly attacked by acids with effervescence, and the carbon escapes in the form of carbonic acid gas. Thus we have traced out the distinction between mechanical aggregation and chemical affinity, taking for an example the difference between gunpowder as a whole (in which the ingredients are so nicely balanced that it is almost a chemical combination), and its constituents, sulphur, charcoal, and nitre, when they are chemically combined; or, in briefer language, we have noticed the difference between the mechanical mixture, and some of the chemical combinations, of three important elements. Our very slight and partial examination of three simple bodies does not, however, afford us any deep insight into the principles of chemistry; we have, as it were, only mastered the signification of a few words in a language; we might know that *chien* was the French for dog, or *cheval* horse, or *homme* man; but that knowledge would not be the acquisition of the French language, because we must first know the alphabet, and then the combination of these letters into words; we must also acquire a knowledge of the proper arrangement of these words into sentences, or grammar, both syntax and prosody, before we can claim to be a French scholar: so it is with chemistry—any number of isolated experiments with various chemical substances would be comparatively useless, and therefore the "alphabet of chemistry," or "table of simple elements," must first be acquired. These bodies are understood to be solids, fluids, and gases, which have hitherto defied the most elaborate means employed to reduce them into more than one kind of matter. Even pure light is separable into seven parts—viz., red, orange, yellow, green, blue, indigo, and violet; but the elements we shall now enumerate are not of a compound, but, so far as we know, of an absolutely simple or single nature; they represent the boundaries, not the finality, of the knowledge that may be acquired respecting them.

The elements are sixty-four in number, of which about forty are tolerably plentiful, and therefore common; whilst the remainder, twenty-four, are rare, and for that reason of a lesser utility: whenever Nature employs an element on a grand scale it may certainly be called common, but it generally works for the common good of all, and fulfils the most important offices.

CLASSIFICATION OF THE ALPHABET OF CHEMISTRY.

13 *Non-Metallic Bodies.*

Name.	Symbol.	Combining proportion or atomic weight.	Name.	Symbol.	Combining proportion or atomic weight.
1. Oxygen . .	O	= 8	8. Carbon . .	C	= 6
2. Hydrogen .	H	= 1	9. Boron . . .	B	= 10·9
3. Nitrogen . .	N	= 14	10. Sulphur . .	S	= 16
4. Chlorine . .	Cl	= 35·5	11. Phosphorus .	P	= 32
5. Iodine . . .	I	= 127·1	12. Silicon . .	Si	= 21·3
6. Bromine . .	Br	= 80·	13. *Selenium* . .	Se	= 39·5
7. Fluorine . .	F	= 18·9			

51 *Metals.*

Name.	Symbol.	Weight.	Name.	Symbol.	Weight.
1. Aluminum .	Al	= 13·7	27. Nickel. . .	Ni	= 29·6
2. Antimony .	Sb	= 129	28. *Norium* . .		
3. Arsenic . .	As	= 75	29. *Niobium* . .	Nb	
4. Barium . .	Ba	= 68·5	30. *Osmium* . .	Os	= 99·6
5. Bismuth . .	Bi	= 213	31. Platinum . .	Pt	= 98·7
6. Cadmium . .	Cd	= 56	32. Potassium .	K	= 39·2
7. Calcium . .	Ca	= 20	33. Palladium .	Pd	= 53·3
8. *Cerium* . .	Ce	= 47	34. *Pelopium* . .	Pe	
9. Chromium .	Cr	= 26·7	35. Rhodium . .	R	= 52·2
10. Cobalt. . .	Co	= 29·5	36. *Rhuthenium* .	Ru	= 52·2
11. Copper . .	Cu	= 31·7	37. Silver . . .	Ag	= 108·1
12. *Donarium* . .			38. Sodium . .	Na	= 23
13. *Didymium*. .	D		39. Strontium .	Sr	= 43·8
14. *Erbium* . .	E		40. Tin. . . .	Sn	= 59
15. Gold . . .	Au	= 197	41. *Tantalum* . .	Ta	= 184
16. *Glucinum* . .	Gl		42. *Tellurium* . .	Te	= 64·2
17. Iron . . .	Fe	= 28	43. *Terbium* . .	Tb	
18. *Ilmenium* . .	Il		44. *Thorium* . .	Th	= 59·6
19. *Iridium* . .	Ir	= 99	45. *Titanium* . .	Ti	= 25
20. Lead . . .	Pb	= 103·7	46. Tungsten . .	W*	= 95
21. *Lanthanium* .	La		47. Uranium . .	U	= 60
22. *Lithium* . .	Li	= 6·5	48. *Vanadium* . .	V	= 68·6
23. Magnesium .	Mg	= 12·2	49. *Yttrium* . .	Y	
24. Manganese .	Mn	= 27·6	50. Zinc . . .	Zn	= 32·6
25. Mercury . .	Hg	= 100	51. *Zirconium* .	Zr	= 22·4
26. *Molybdenum* .	Mo	= 46			

(N.B. The elements printed in italics are at present unimportant.

A few words will suffice to explain the meaning of the terms which head the names, letters, and numbers of the Table of Elements. The

* From the mineral Wolfran, and now exceedingly valuable, as when alloyed with iron it is harder than, and will bore through steel.

names of the elements have very interesting derivations, which it is not the object of this work to go into; the symbols are abbreviations, ciphers of the simplest kind, to save time and trouble in the frequent repetition of long words, just as the signs $+$ plus, and $-$ minus, are used in algebraic formulæ. For instance—the constant recurrence of water in chemical combinations must be named, and would involve the most tedious repetition; water consists of oxygen and hydrogen, and by taking the first letter of each word we have an instructive symbol, which not only gives us an abbreviated term for water, but also imparts at once a knowledge of its composition by the use of the letters, HO.

Again, to take a more complex example, such as would occur in the study of organic chemistry—a sentence such as *the hydrated oxide of acetule*, is written at once by $C_4H_4O_2$, the figures referring to the number of equivalents of each element—viz., 4 equivalents of C, the symbol for carbon, 4 of H (hydrogen), and 2 of O (oxygen).

The long word paranaphthaline, a substance contained in coal tar, is disposed of at once with the symbols and figures $C_{30}H_{12}$.

The figures in the third column are, however, the most interesting to the precise and mathematically exact chemist. They represent the united labours of the most painstaking and learned chemists, and are the exact quantities in which the various elements unite. To quote one example: if 8 parts by weight of oxygen—viz., the combining proportions of that element—are united with 1 part by weight of hydrogen, also its combining number, the result will be 9 parts by weight of water; but if 8 parts of oxygen and 2 parts of hydrogen were used, one only of the latter could unite with the former, and the result would be the formation again of 9 parts of water, with an overplus of 1 equivalent of hydrogen.

It is useless to multiply examples, and it is sufficient to know that with this table of numbers the figures of analysis are obtained. Supposing a substance contained 27 parts of water, and the oxygen in this had to be determined, the rule of proportion would give it at once, $9 : 27 :: 8 : 24$. 9 parts of water are to 27 parts as 8 of oxygen (the quantity contained in 9 parts of water) are to the answer required—viz., 24 of oxygen. The names, symbols, and combining proportions being understood, we may now proceed with the performance of many interesting

CHEMICAL EXPERIMENTS.

As the permanent gases head the list, they will first engage our attention, beginning with the element oxygen—Symbol O, combining proportion 8. There is nothing can give a better idea of the enormous quantity of oxygen present in the animal, vegetable, and mineral kingdoms, than the statement that it represents *one-third* of the weight of the whole crust of the globe. Silica, or flint, contains about half its weight of oxygen; lime contains forty per cent.; alumina about thirty-three per cent. In these substances the element oxygen remains inactive and powerless, chained by the strong fetters of chemical affinity to the

silicium of the flint, the calcium of the lime, and the aluminum of the
alumina. If these substances are heated by themselves they will not
yield up the large quantity of oxygen they contain.

Nature, however, is prodigal in her creation, and hence we have but
to pursue our search diligently to find a substance or mineral containing
an abundance of oxygen, and part of which it will relinquish by what
used to be called by the "old alchemists" the *torture* of heat. Such a
mineral is the black oxide of manganese, or more correctly the binoxide
of manganese, which consists of one combining proportion of the metal
manganese—viz., 27·6, and two of oxygen—viz., $8 \times 2 = 16$. If three
proportions of the binoxide of manganese are heated to redness in an
iron retort, they yield one proportion (equal to 8) of oxygen, and all
that has just been explained by so many words is comprehended in the
symbols and figures below:—

$$3\ MnO_2 = Mn_3O_4 + O.$$

Thus the $3\ MnO_2$ represent the three proportions of the binoxide of
manganese before heat is applied, whilst the sign =, the sign of equation
(equal to), is intended to show that the elements or compounds placed
before it produce those which *follow* it; hence the sequel Mn_3O_4+O
shows that another compound of the metal and oxygen is produced,
whilst the $+ O$ indicates the liberated oxygen gas. The iron retort
employed to hold the mineral should be made of cast iron in preference
to wrought iron, as the latter is very soon worn out by contact with
oxygen at a red heat. A gun-barrel will answer the purpose for an
experiment on the small scale, to which must be adapted a cock and
piece of pewter tubing. Such a make-shift arrangement may do very
well when nothing better offers; but as a question of expense, it is
probably cheaper in the end to order of Messrs. Simpson and Maule, or
of Messrs. Griffin, or of Messrs. Bolton, a cast-iron bottle, or cast-iron
retort, as it is termed, of a size sufficient to prepare two gallons of

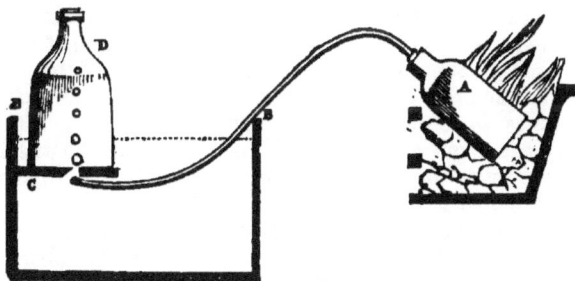

Fig. 93. A. The iron bottle, containing the black oxide of manganese, with pipe passing to
the pneumatic trough, B B, in which is fixed a shelf, C, perforated with a hole, under which
the end of the pipe is adjusted, and the gas passes into the gas-jar, D.

oxygen from the binoxide of manganese, which, with four feet of iron
conducting-pipe, and connected to the bottle with a screw, does not

cost more than six shillings—an enormous dip, perhaps, in the juvenile pocket, and therefore we shall indicate presently a still cheaper apparatus for the same purpose. (Fig. 93.)

The oxygen is conveyed to a square tin box provided with a shelf at one end, perforated with several holes at least one inch in diameter, called the pneumatic trough; any wooden trough, butter or wash-tub, foot-pan or bath, provided with a shelf, may be raised by the same title to the dignity of a piece of chemical apparatus. The gas jar must be filled with water by withdrawing the stopper and pressing it down into the trough, and when the neck is below the level of the water, the stopper is again inserted, and the jar with the water therein contained

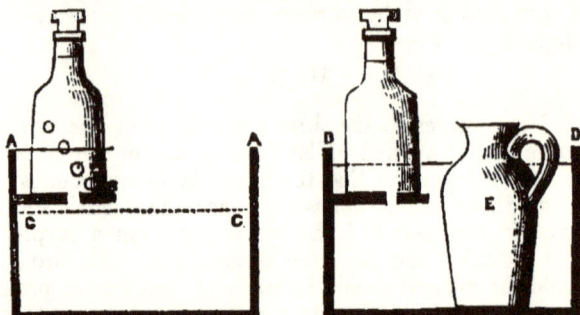

Fig. 94. A A. Pneumatic trough, with gas jar raised to shelf; bubbles of air are rushing in at B, as the level of the water is below the shelf—viz., at C C. D D. Same trough and gas jar with water kept over the shelf by the introduction of the stone pitcher E, full of water.

lifted steadily on to the shelf, the entry of atmospheric air being prevented by keeping the lower part of the gas jar, called the welt, under the water. Sometimes the pneumatic trough contains so small a quantity of water that on raising the gas jar to the shelf the liquid does not cover the bottom, and the air rushes up in large bubbles. Under these circumstances it is better to provide a gallon stone jug full of water, so that when the jar is being raised to the shelf it may be thrust into the trough (on the same principle as the crow and the pitcher in the fable), and thus by its bulk (as the stones in the pitcher) raise the water to the proper level. When the gas jar is about half filled with gas the jug may be withdrawn. This arrangement saves the trouble of constantly adding and baling out water from the pneumatic trough. (Fig. 94.)

There are other solid oxygenized bodies in which the affinities are less powerful, and hence a lower degree of heat suffices to liberate the oxygen gas, and one of the most useful in this respect is the salt termed chlorate of potash. If the substance is heated by itself, the temperature required to expel the oxygen is almost as high as that demanded for the black oxide of manganese; but, strange to say, if the two substances are reduced to powder, and mixed in equal quantities by weight, then a very moderate increase of heat is sufficient to cause the chlorate of potash to give up its oxygen, whilst the oxide of manganese undergoes no change whatever. It seems to fulfil only a mechanical office—possibly that of separating each particle of chlorate of potash from the other, so

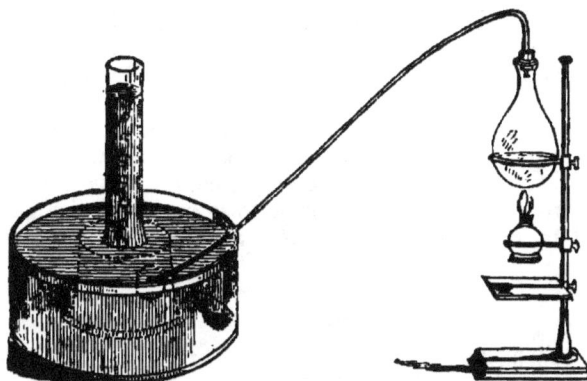

Fig. 95. Preparation of oxygen from { chlorate of potash, oxide of manganese.

$$KO. ClO_5 = \left\{ \begin{matrix} O_6 \\ KCl. \end{matrix} \right.$$

that the heat attacks the substance in detail, just as a solid square of infantry might repel almost any attack, whilst the same body dispersed over a large space might be of little use; so with the chlorate of potash, which undergoes rapid decomposition when mixed with and divided amongst the particles of the oxide of manganese; less so with the red oxide of iron, and still less with sand or brick-dust. (Fig. 95.)

This curious fact is explained usually by reference to what is called catalytic action, or *decomposition by contact* (κατα, downwards, and λυω, I unloosen), *being a power possessed by a body of resolving another into a new compound without undergoing any change itself.* To make this term still clearer, we may notice another example in linen rags, which may be exposed for any length of time to the action of water without fear of conversion into sugar; if, however, oil of vitriol is first added to the linen rags, and they are subsequently digested at a proper temperature with water, then the rags are converted into sugar (the author has seen a specimen made of an "old shirt"); but, curious to relate, the oil of vitriol is unchanged in the process, and if the process be commenced with a pound of acid, the same quantity is discoverable at the end of the chemical decomposition of the linen rags, and their conversion into sugar.

If a mixture of equal parts of oxide of manganese and chlorate of potash is placed in a clean Florence flask, with a cork, and pewter, or glass tube attached, great quantities of oxygen are quickly liberated, on the application of the heat of a spirit lamp. Such a retort would cost about fourpence, and if the flask is broken in the operation it can be easily replaced by another, value one penny, as the same cork and tube will generally suit a number of these cheap glass vessels. Corks may always be softened by using either a proper cork squeezer, or by placing them under a piece of board or a flat surface, and rolling and pressing the cork till quite elastic.

Whilst fitting the latter into the neck of a flask, it is perhaps safer to hold the thin and fragile vessel in a cloth, so that if the flask breaks the chemical experiment may not be arrested for many days by the severe cutting and wounding of the fingers. After the cork is fitted, it is to be removed from the flask and bored with a cork borer. This useful tool is sold in complete sets to suit all sizes of glass tubes, and the pewter or glass being inserted, the flask and tube will be ready for use, provided the tube is bent to the proper curve. This is easy enough to perform with the pewter, but not quite so easy with the glass tube, which must be held over the flame of a spirit lamp till soft, and then

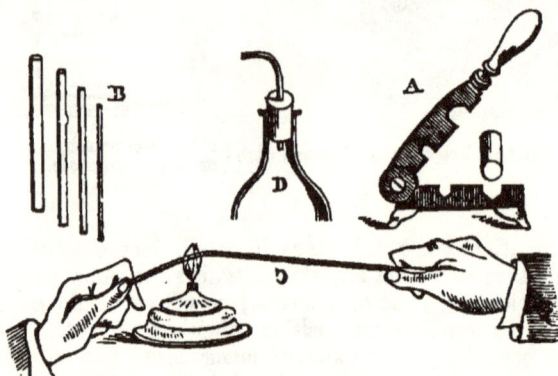

Fig. 96. A. The cork squeezer. B. The cork borers. C. The operation of bending the glass tube over the flame of the spirit-lamp. D. The neck of the flask, with cork and tube bent and fitted complete for use.

bent very gradually to the proper curve. If a short length of the glass tube is heated, it bends too sharply, and the convexity of the glass is flattened, whilst the internal diameter of the tube is lessened, so that at least three inches in length should be warmed, and the heat must not be continued in one place only, but should be maintained in the direction of the bend, the whole manipulation being conducted without any hurry. (Fig. 96.)

Having filled a gas jar with oxygen, it may be removed from the pneumatic trough by sliding it into a plate under the surface of the water, and to prevent the stopper being thrust out accidentally from the jar by the upward pressure of the gas, whilst a little compressed, during the act of passing it into the plate, it is advisable to hold the stopper of the jar firmly but gently, so that it cannot slip out of its place. A number of jars of oxygen may be prepared and arranged in plates, all of which of course must contain a little water, and enough to cover the welt of the jar.

EXPERIMENTS WITH OXYGEN GAS.

This gas was originally discovered by Priestley, in August, 1774, and was first obtained by heating red precipitate—*i.e.*, the red oxide of mercury.

$$HgO = Hg + O.$$

We leave these symbols and figures to be deciphered by the youthful philosopher with the aid of the table of elements, &c., and return to the experiments.

There are certain thin wax tapers like waxed cord, called bougies, which can be bent to any shape, and are very convenient for experiments with the gases. If one of these tapers is bent as in Fig. 97, then lighted and allowed to burn for some minutes, a long snuff is gradually formed, which remains in a state of ignition when the flame of the taper is blown out. On plunging this into a jar of oxygen, it instantly re-lights with a sort of report, and burns with greatly-increased brilliancy, as described by Dr. Priestley in his first experiment with this gas, and so elegantly repeated by Professor Brande in his refined dissertation on the progress of chemical science.

"The 1st of August, 1774, is a *red-letter day* in the annals of chemical philosophy, for it was then that Dr. Priestley discovered dephlogisticated air. Some, sporting in the sunshine of rhetoric, have called this the birthday of pneumatic chemistry; but it was even a more marked and memorable period; it was then (to pursue the metaphor) that this branch of science, having eked out a sickly and infirm infancy in the ill-managed nursery of the early chemists, began to display symptoms of an improving constitution, and to exhibit the most hopeful and unexpected marks of future importance. The first experiment, which led to a very satisfactory result, was concluded as follows :— A glass jar was filled with quicksilver, and inserted in a basin of the same; some red precipitate of quicksilver was then introduced, and floated upon the quicksilver in the jar; heat was applied to it in this situation with a burning-lens, and to use Priestley's own words, *I presently found that air was expelled from it very readily. Having got about three or four times as much as the bulk of my materials, I admitted water into it, and found that it was not imbibed by it. But what surprised me more than I can well express was, that a candle burned in this air with a remarkably vigorous flame, very much like that enlarged flame with which a candle burns in nitrous air exposed to iron or lime of sulphur (i.e., laughing gas); but as I had got nothing like this remarkable appearance from any kind of air besides this peculiar*

Fig. 97.

modification of nitrous air, and I knew no nitrous acid was used in the preparation of mercurius calcinatus, I was utterly at a loss how to account for it." (Fig. 98.)

Second Experiment.

The term oxygen is derived from the Greek (οξυσ, acid, and γενναω, I give rise to), and was originally given to this element by Lavoisier, who also claimed its discovery; and if this honour is denied him, surely he has deserved equal scientific glory in his masterly experiments, through which he discovered that the mixture of forty-two parts by

Fig. 98. A. Glass vessel full of mercury, containing the red precipitate at the top, and standing in the dish B, also containing mercury. C. The burning-glass concentrating the sun's rays on the red precipitate, being Priestley's original experiment.

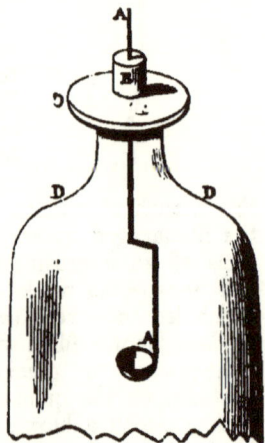

measure of azote, with eight parts by measure of oxygen, produced a compound precisely resembling our atmosphere. The name given to oxygen was founded on a series of experiments, one of which will now be mentioned.

Place some sulphur in a little copper ladle attached to a wire, and called a deflagrating spoon, passed through a round piece of zinc or brass plate and cork, so that the latter acts as an adjusting arrangement to fix the wire at any point required. The combustion of the sulphur, previously feeble, now assumes a remarkable intensity, and a peculiar coloured light is generated, whilst the sulphur unites with the oxygen, and forms sulphurous acid gas. It produces, in fact, the same gas which is formed by burning an ordinary sulphur match. This compound is valuable as a disinfectant, and is a very important bleaching agent, being most extensively employed in the whitening of straw employed in the manufacture of straw bonnets. It is an acid gas, as Lavoisier found, and this property may be detected by pouring a little tincture of litmus into the bottom of the plate in which the gas jar stands. The blue colour of the litmus is rapidly

Fig. 99. A. The deflagrating spoon. B. The cork. C. The zinc, or brass, or tin plate. D D. The gas jar.

changed to red, and it might be thought that no further argument could possibly be required to prove that oxygen was *the* acidifying agent, the more particularly as the result is the same in the next illustration.

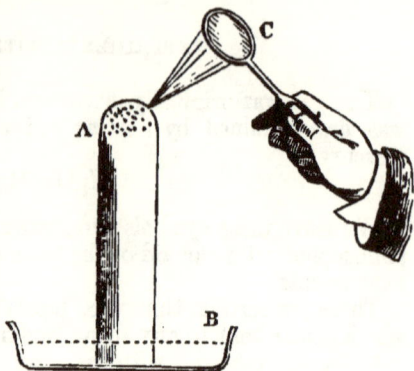

Third Experiment.

Cut a small piece from an ordinary stick of phosphorus under water, take care to dry it properly with a cloth, and after placing it in a deflagrating spoon, remove the stopper from the gas-jar, as there is no fear of the oxygen rushing away, because it is somewhat heavier than atmospheric air; and then, after placing the spoon with the phosphorus in the neck of the jar, apply a heated wire and pass the spoon at once into the middle of the oxygen; in a few seconds a most brilliant light is obtained, and the jar is filled with a white smoke; as this subsides, being phosphoric acid, and perfectly soluble in water, the same litmus test may be applied, when it is in like manner changed to red. The acid obtained is one of the most important constituents of bone.

Fourth Experiment.

A bit of bark-charcoal bound round with wire is set on fire either by holding it in the flame of a spirit-lamp, or by attaching a small piece of waxed cotton to the lower part, and igniting this; the charcoal may then be inserted into a bottle of oxygen, when the most brilliant scintillations occur. After the combustion has ceased and the whole is cool, a little tincture of litmus may also be poured in and shaken about, when it likewise turns red, proving for the third time the generation of an acid body, called carbonic acid—an acid, like the others already mentioned, of great value, and one which Nature employs on a stupendous scale as a means of providing plants, &c., with solid charcoal. Carbonic acid, a virulent poison to animal life, is, when properly diluted, and as contained in atmospheric air, one of the chief alimentary bodies required by growing and healthy plants.

In three experiments acid bodies have been obtained; can we speculate on the result of the next?

Fifth Experiment.

Into a deflagrating spoon place a bit of potassium, set this on fire by holding it in the spoon in the flame of a spirit-lamp, and then rapidly plunge the burning metal into a bottle of oxygen. A brilliant ignition occurs in the deflagrating spoon for a few seconds, and there is little or no smoke in the jar. The product this time is a solid, called potash, and if this be dissolved in water and·filtered, it is found to be clear and bright, and now on the addition of a little tincture of litmus to one half of the solution, it is wholly unaffected, and remains blue; but if with the other half a small quantity of tincture of turmeric is mixed, it immediately changes from a bright yellow solution to a reddish-brown, because turmeric is one of the tests for an alkali; and thus is ascertained by the help of this and other tests that the result of the combustion is not an *acid*, but an *alkali*. The experiment is made still more satisfactory by burning another bit of potassium in oxygen and dissolving the product in water, and if any portion of

the reddened liquid derived from the sulphurous, phosphoric, and carbonic acids taken from the previous experiments, be added to separate
portions of the alkaline solution, they are all restored to their original blue
colour, because an acid is neutralized by an alkali; and the experiment is
made quite conclusive by the restoration of the reddened turmeric to a
bright yellow on the addition of a solution of either of the three acids
already named. Moreover, an acid need not contain a fraction of
oxygen, as there is a numerous class of *hy*dracids, in which the acidifying principle is hydrogen instead of oxygen, such as the hydrochloric,
hydriodic, hydro-bromic, and hydrofluoric acids.

Sixth Experiment.

A piece of watch-spring is softened at one end, by holding it in the
flame of a spirit-lamp, and allowing it to cool. A bit of waxed cotton
is then bound round the softened end, and after being set on fire, is
plunged into a gas jar containing oxygen; the cotton first burns away,
and then the heat communicates to the steel, which gradually takes fire,
and being once well ignited, continues to burn with amazing rapidity, forming drops of liquid dross, which fall to the bottom of the plate—and also
a reddish smoke, which condenses on the sides of the jar; neither the
dross which has dropped into the plate, nor the reddish matter condensed
on the jar, will affect either tincture of litmus or turmeric; they are
neither acid nor alkaline, but *neutral* compounds of iron, called the
sesquioxide of iron (Fe_2O_3), and the magnetic oxide ($Fe_3O_4=FeO.$
Fe_2O_3).

Seventh Experiment.

Some oxygen gas contained in a bladder provided with a proper
jet may be squeezed out, and upon, some liquid phosphorus con-

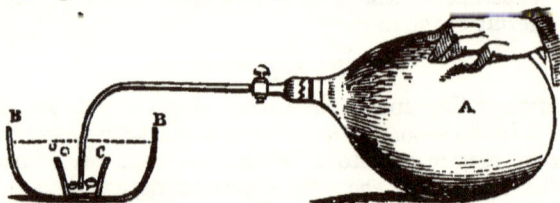

Fig. 100. A. Bladder containing oxygen, provided with a stop-cock and jet leading to,
B, B. Finger glass containing boiling water. c. The cup of melted phosphorus under the
water. The gas escapes from the bladder when pressed.

tained in a cup at the bottom of a finger glass full of boiling water,
when a most brilliant combustion occurs, proving that so long as the
principle is complied with—viz., that of furnishing oxygen to a combustible substance—it will burn under water, provided it is insoluble,
and possesses the remarkable affinity for oxygen which belongs to
phosphorus. The experiment should be performed with boiling water,
to keep the phosphorus in the liquid state; and it is quite as well to hold

a square foot of wire gauze over the finger glass whilst the experiment is being performed. (Fig. 100.)

Eighth Experiment.

Oxygen is available from many substances when they are mixed with combustible substances, and hence the brilliant effects produced by burning a mixture of nitre, meal powder, sulphur, and iron or steel filings; the metal burns with great brilliancy, and is projected from the case in most beautiful sparks, which are long and needle-shaped with steel, and in the form of miniature rosettes with iron filings; it is the oxygen from the nitre that causes the combustion of the metal, the other ingredients only accelerate the heat and rate of ignition of the brilliant iron, which is usually termed a gerb.

Ninth Experiment.

Fig. 101. A. Case of red fire burning downwards, and attached by a copper wire to a bit of leaden pipe B, to sink it. c c. Jar containing water.

A mixture of nitrate of potash, powdered charcoal, sulphur, and nitrate of strontium, driven into a strong paper case about two inches long, and well closed at the end with varnish, being quite waterproof, may be set on fire, and will continue to burn under water until the whole is consumed; the only precaution necessary being to burn the composition from the case with the mouth downward, and if the experiment is tried in a deep glass jar it has a very pleasing effect. (Fig. 101.)

The red-fire composition is made by mixing nitrate of strontia 40 parts by weight, flowers of sulphur 13 parts, chlorate of potash 5 parts, sulphuret of antimony 4 parts. These ingredients must first be well powdered separately, and then mixed carefully on a sheet of paper with a paper-knife. They are liable to explode if rubbed *together* in a mortar, on account of the presence of sulphur and chlorate of potash, and the composition, if kept for any time, is liable to take fire spontaneously.

Tenth Experiment.

Some zinc is melted in an iron ladle, and made quite red hot; if a little dry nitre is thrown upon the surface, and gently stirred into the metal, it takes fire with the production of an intense white light, whilst large quantities of white flakes ascend, and again descend when cold, being the oxide of zinc, and called by the alchemists the "Philosopher's Wool" (ZnO). In this experiment the oxygen from the nitre effects the oxidation of the metal zinc.

Eleventh Experiment.

A mixture of four pounds of nitre with two of sulphur and one and a half of lamp black produces a most beautiful and curious fire, continually projected into the air as sparks having the shape of the rowel of a spur, and one that may be burnt with perfect safety in a room, as the sparks consume away so rapidly, in consequence of the finely divided condition of the charcoal, that they may be received on a handkerchief or the hand without burning them. The difficulty consists in effecting the complete mixture of the charcoal. The other two ingredients must first be thoroughly powdered separately, and again triturated when mixed, and finally the charcoal must be rubbed in carefully, till the whole is of a uniform tint of grey and very nearly black, and as the mixture proceeds portions must be rammed into a paper case, and set on fire; if the stars or pinks come out in clusters, and spread well without other and duller sparks, it is a sign that the whole is well mixed; but if the sparks are accompanied with dross, and are projected out sluggishly, and take some time to burn, the mixture and rubbing in the mortar must be continued; and even that must not be carried too far, or the sparks will be too small. N.B.—If the lamp-black was heated red hot in a close vessel, it would probably answer better when cold and powdered.

Twelfth Experiment.

Into a tall gas jar with a wide neck project some red-hot lamp-black through a tin funnel, when a most brilliant flame-like fire is obtained, showing that finely divided charcoal with pure oxygen would be sufficient to afford light; but as the atmosphere consists of oxygen diluted with nitrogen, compounds of charcoal with hydrogen, are the proper bodies to burn, to produce artificial light.

Thirteenth Experiment. The Bude Light.

This pretty light is obtained by pass - ing a steady current of oxygen gas (escaping at a very low pressure) through and up the centre pipe of an argand oil lamp, which must be supplied with a highly carbonized oil and a very thick wick, as the oxygen has a tendency to burn away the cotton unless the oil is well supplied, and allowed to overflow the wick, as it does in the lamps of the lighthouses. The best whale oil is usually employed, though it would be worth while to test the value of Price's "Belmontine Oil" for the same purpose. (Fig. 102.)

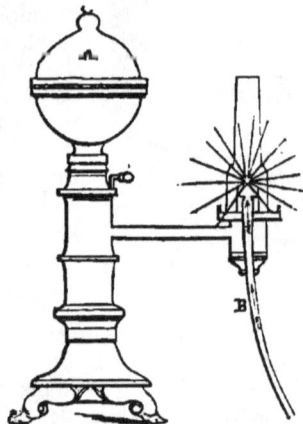

Fig. 102. A. Reservoir of oil. B. The flexible pipe conveying oxygen to centre of the argand lamp.

H

Fourteenth Experiment. A Red Light.

Clear out the oil thoroughly from the Bude light apparatus ; or, what is better, have two lamps, one for oil, and the other for spirit ; fill the apparatus with a solution of nitrate of strontia and chloride of calcium in spirits of wine, and let it burn from the cotton in the same way as the oil, and supply it with oxygen gas.

Fifteenth Experiment. A Green Light.

Dissolve boracic acid and nitrate of baryta in spirits of wine, and supply the Bude lamp with this solution.

Sixteenth Experiment. A Yellow Light.

Dissolve common salt in spirits of wine, and burn it as already described in the Bude light apparatus.

Seventeenth Experiment. The Oxy-calcium Light.

This very convenient light is obtained in a simple manner, either by using a jet of oxygen as a blowpipe to project the flame of a spirit lamp on to a ball of lime ; or common coal-gas is employed instead of the

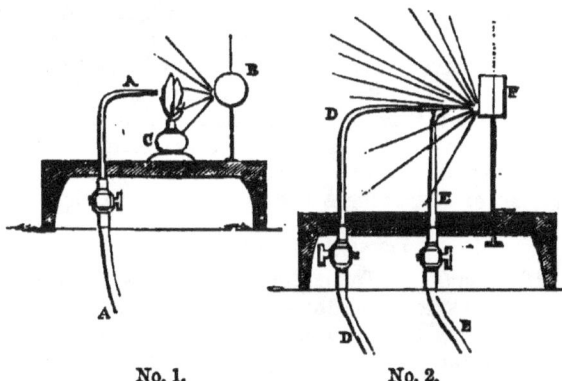

No. 1.   No. 2.

Fig. 103.—No. 1. A. Oxygen jet. B. The ball of lime, suspended by a wire. C. Spirit lamp.
No. 2. D. Oxygen jet. E. Gas (jet connected with the gas-pipe in the rear by flexible pipe) projected on to ball of lime, F.

spirit lamp, being likewise urged against a ball of lime. By this plan one bag containing oxygen suffices for the production of a brilliant light, not equal, however, to the oxy-hydrogen light, which will be explained in the article on hydrogen. (Fig. 103.)

Eighteenth Experiment.

To show the weight of oxygen gas, and that it is heavier than air, the stoppers from two bottles containing it may be removed, one bottle may be left open for some time and then tested by a lighted taper, when it will still indicate the presence of the gas, whilst the other may be suddenly inverted over a little cup in which some ether, mixed with a few drops of turpentine, may be burning—the flame burns with much greater brilliancy at the moment when the oxygen comes in contact with it.

Nineteenth Experiment.

The theory of the effect of oxygen upon the system when inhaled would be an increase in the work of the respiratory organs; and it is stated that after inhaling a gallon or so of this gas, the pulse is raised forty or fifty beats per second: the gas is easily inhaled from a large indiarubber bag through an amber mouthpiece; it must of course be quite pure, and if made from the mixture of chlorate of potash and oxide of manganese, should be purified by being passed through lime and water, or cream of lime.

Twentieth Experiment.

There are certain colouring matters that are weakened or destroyed by the action of light and other causes, which deprive them of oxygen gas or deoxidize them. A weak tincture of litmus, if long kept, often becomes colourless, but if this colourless fluid is shaken in a bottle with oxygen gas it is gradually restored; and if either litmus, turmeric, indigo, orchil, or madder, paper, or certain ribbons dyed with the same colouring matters, have become faded, they may be partially restored by damping and placing them in a bottle of oxygen gas. The effect of the oxygen is to reverse the *deoxidizing* process, *and* to impart oxygen to the colouring matters. By a peculiar process indigo may be obtained quite white, and again restored to its usual blue colour, either by exposure to the air or by passing a stream of oxygen through it.

Twenty-first Experiment.

Messrs. Matheson, of Torrington-street, Russell-square, prepare in the form of wire some of the rarest metals, such as magnesium, lithium, &c. A wire of the metal magnesium burns magnificently in oxygen gas, and forms the alkaline earth magnesia. The metal lithium, to which such a very low combining proportion belongs—viz., 6·5, can also be procured in the state of wire, and burns in oxygen gas with an intense white light into the alkaline lithia, which dissolved in alcohol with a little acetic acid, and burnt, affords a red flame, making a curious contrast between the effects of colour produced by the metallic and oxidized state of lithium.

THE ALLOTROPIC CONDITION OF OXYGEN GAS.

The term allotropy (from αλλοτροπος, of a different nature) was first used by the renowned chemist Berzelius. Dimorphism, or diversity in crystalline form, is therefore a special case of allotropy, and is most amusingly illustrated with the iodide of mercury (HgI), which is made either by rubbing together equal combining proportions of mercury and iodine (both of which are to be found in the Table of Elements, page 86), or by carefully precipitating a solution of corrosive sublimate (chloride of mercury (HgCl)) with one of iodide of potassium, just enough and no more of the latter being added to precipitate the metal, or else the iodide of mercury is redissolved by the excess of the precipitant. It is first of a dirty yellow, and then gradually changes when stirred to a scarlet; if this be collected on a filter, and washed and drained, it is a beautiful scarlet, and when some of this substance is rubbed across a sheet of paper, a bright scarlet is apparent, which may be rapidly changed to a lemon-yellow by heating the paper over the flame of a spirit lamp; and the iodide of mercury is again brought back to a scarlet colour by rubbing down the yellow crystals with the fingers. This experiment may be repeated over and over again with the like results. If some of the scarlet iodide of mercury is sublimed from one bit of glass to another, it forms crystals, derived from the right rhombic prism; when these are scratched with a pin they change again to the scarlet state, the latter when crystallized being in the form of the square-based octohedron.

Other cases of dimorphism may be mentioned—viz., with sulphur, carbonate of lime, and lead, and many others, whilst allotropy is curiously illustrated in the various conditions of charcoal, which, in the more numerous examples, is black and opaque, and in another instance transparent like water. Lamp-black is soft, but the diamond is the hardest natural substance. The allotropic state of sulphur has been already alluded to; phosphorus, again, exists in three modifications: 1st, Common phosphorus, which shines in the dark and emits a white smoke. 2nd, White phosphorus. 3rd, Red or amorphous phosphorus, which does not shine or emit white smoke when exposed to the air, and is so altered in its properties that it may be safely carried in the pocket.

Enough evidence has therefore been offered to show that the allotropic property is not confined to one element or compound, but is discoverable in many bodies, and in no one more so than in the allotropic state of the element oxygen called

OZONE.

The Greek language has again been selected by the discoverer, Schönbein, of Basle, for the title or name of this curious modification of oxygen, and it is so termed from οξεῖν, to smell. The name at once suggests a marked difference between ozone and oxygen, because the latter is perfectly free from odour, whilst the former has that peculiar smell which is called electric, and is distinguishable whenever an electrical machine is at work, or if a Leyden jar is charged by the powerful Rhumkoff, or Hearder coil; it is also apparent when water is decomposed by a current of electricity and resolved into its elements, oxygen and hydrogen. When highly concentrated it smells like chlorine; and the author recollects seeing the first experiments by Schönbein, in England, at Mr. Cooper's laboratory in the Blackfriars-road. Ozone is prepared by taking a clean empty bottle, and pouring therein a very

little distilled water, into
which a piece of clean
scraped phosphorus is
introduced, so as to ex-
pose about one-half of
its diameter to the air in
the bottle, whilst the
other is in contact with
the water. (Fig. 104.)

For the sake of pre-
caution, the bottle may
stand in a basin or soup
plate, so that if the
phosphorus should take
fire, it may be instantly
extinguished by pour-
ing cold water into the
bottle, and should this
crack and break, the
phosphorus is received
into the plate.

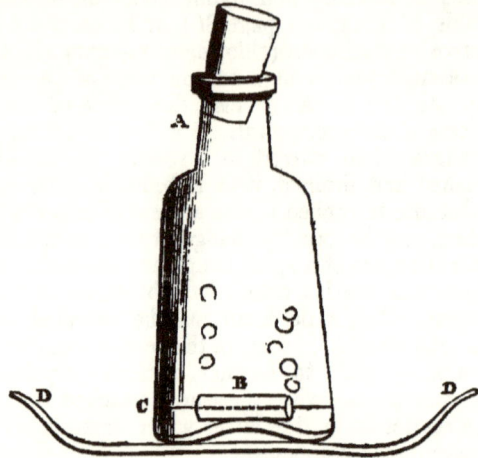

Fig. 104. A. A quart bottle, with the stopper loosely
placed therein. B. The stick of clean phosphorus. C. The
water level just to half the thickness of the phosphorus.
D D. A soup-plate.

When the ozone is
formed the phospho r s can be withdrawn, and the phosphorous-acid
smoke washed out by shaking the bottle; it is distinguishable by its
smell, and also by its action on test paper, prepared by painting with
starch containing iodide of potassium on some Bath post paper; when
this is placed in the bottle containing ozone, it changes the test blue,
or rather a purplish blue.

Ozone is a most energetic body, and a powerful bleaching agent; if
a point is attached to the prime conductor of an electrical machine,
and the electrified air is received into a bottle, it will be found to smell,
and has the power of bleaching a *very* dilute solution of indigo. Ozone

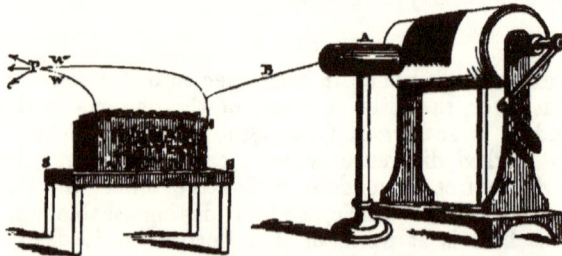

Fig. 105. v. A small voltaic battery standing on the stool with glass legs, s s, and
capable of heating a thin length of platinum wire about two inches long, and bent to form
a point between the conducting wires, w w.—N.B. The voltaic current can be cut off at
pleasure, so as to cool the wire when necessary. A is the prime conductor of an ordinary
cylinder electrical machine. B is the wire conveying the frictional electricity to the
conducting wires of the voltaic battery, where the point P being the sharpest point in the
arrangement, delivers the electrified and ozonized air.

is not a mere creation of fancy, as it can not only be produced by certain methods, but may be destroyed by a red heat.   If a point is prepared with a loop of platinum wire. and this latter, after being connected with a voltaic battery, made red hot, and the whole placed on an insulating stool, and connected with the prime conductor of an electrical machine, it is found that the electrified air no longer smells, the ozone is destroyed; on the other hand, if the voltaic battery is disconnected, and the electrified air again allowed to pass from the cold platinum wire, the smell is again apparent, the air will bleach, and if caused to impinge at once upon the iodide of starch test, changes it in the manner already described. (Fig. 105.)

Ozone is insoluble in water, and oxidizes silver and lead leaf, finely powdered arsenic and antimony; it is a poison when inhaled in a concentrated state, whilst diluted, and generated by natural processes, it is a beneficent and beautiful provision against those numerous smells originating from the decay of animal and vegetable matter, which might produce disease or death : ozone is therefore a powerful disinfectant. The test for ozone is made by boiling together ten parts by weight of starch, one of iodide of potassium, and two hundred of water; it may either be painted on Bath post paper, and used at once, or blotting paper may be saturated with the test and dried, and when required for use it must be damped, either before or after testing for ozone, as it remains colourless when *dry*, but becomes blue after being moistened with water.

Paper prepared with sulphate of manganese is an excellent test for ozone, and changes brown rapidly by the oxidation of the proto-salt of manganese, and its conversion into the binoxide of the metal.

Ozone is also prepared by pouring a little sulphuric ether into a quart bottle, and then, after heating a glass rod in the flame of the spirit lamp, it may be plunged into the bottle, and after remaining there a few minutes ozone may be detected by the ordinary tests.

NITROGEN, OR AZOTE.

Nιτρον, nitre; γενναω, I form; a, privative; ζωη, life.   Symbol, N ; combining proportion, 14.· Also termed by Priestley, *phlogisticated* air.

In the year 1772, Dr. Rutherford, Professor of Botany in the University of Edinburgh, published a thesis in Latin on fixed air, in which he says :—" *By the respiration of animals healthy air is not merely rendered mephitic* (i.e., charged with carbonic acid gas), *but also suffers another change. For after the mephitic portion is absorbed by a caustic alkaline lixivium, the remaining portion is not rendered salubrious; and although it occasions no precipitate in lime-water, it nevertheless extinguishes flame and destroys life.*" Such is the doctor's account of the discovery of nitrogen, which may be separated from the oxygen in the air in a very simple manner. The atmosphere is the great storehouse of nitrogen, and four-fifths of its prodigious volume consist of this element

Composition of Atmospheric Air.

	Bulk.		Weight.
Oxygen	20	......	22·3
Nitrogen	80	......	77·7
	100		100·

The usual mode of procuring nitrogen gas is to abstract or remove the oxygen from a given portion of atmospheric air, and the only point to be attended to, is to select some substance which will continue to burn as long as there is any oxygen left. Thus, if a lighted taper is placed in a bottle of air, it will only burn for a certain period, and is gradually and at last extinguished; not that the whole of the oxygen is removed or changed, because after the taper has gone out, some burning sulphur may be placed in the vessel, and will continue to burn for a limited period; and even after these two combustibles have, as it were, taken their fill of the oxygen, there is yet a little left, which is snapped up by burning phosphorus, whose voracious appetite for oxygen is only appeased by taking the whole. It is for this reason that phosphorus is employed for the purpose of removing the oxygen, and also because the product (phosphoric acid) is perfectly soluble in water, and thus the oxygen is first combined, and then washed out of a given volume of air, leaving the nitrogen behind.

First Experiment.

To prepare nitrogen gas, it is only necessary to place a little dry phosphorus in a Berlin porcelain cup on a wine glass, and to stand them in a soup plate containing water. The phosphorus is set on fire with a hot wire, and a gas jar or cylindrical jar is then carefully placed over it, so that the welt of the jar stands in the water in the soup plate. At first, expansion takes place in consequence of the heat, but this effect is soon reversed, as the oxygen is converted into a solid by union with the phosphorus, forming a white smoke, which gradually disappears. (Fig. 106.)

Supposing two grains of phosphorus had been placed in a platinum tube, and just enough atmospheric air passed over it to convert the whole into phosphoric acid, the weight of the phosphorus would be increased to 4½ grains by the addition of 2½ grains

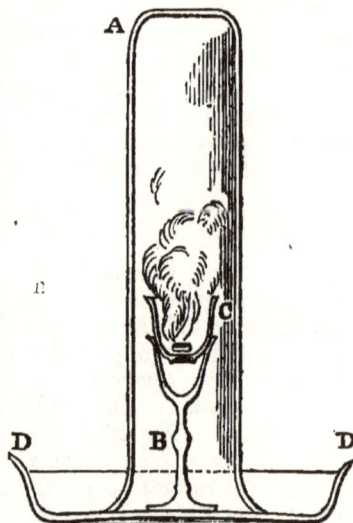

Fig. 106. A. Cylindrical glass vessel, open at one end, and inverted over B, the wine-glass, supporting C, the cup containing the burning phosphorus, and the whole standing in a soup-plate, D D, containing water.

of oxygen; now one cubic inch of oxygen weighs 0·3419, or about ⅓rd
of a grain, hence 7·3 cubic inches of oxygen disappear, which weigh
as nearly as possible 2½ grains, so that as 36·5 cubic inches of air contain 7·3 cubic inches of oxygen, that quantity of air must have passed
over the 2 grains of phosphorus to convert it into 4½ grains of phosphoric acid.

For very delicate purposes, nitrogen is best prepared by passing air
over finely-divided metallic copper heated to redness; this metal absorbs
the whole of the oxygen and leaves the nitrogen. The finely-divided copper
is procured by passing hydrogen gas over pure black oxide of copper.

Second Experiment.

A very instructive experiment is performed by heating a good mass of
tartrate of lead in a glass tube which is herme cally sealed, and being
placed on an iron sup-
port, is then covered
by a capped air jar
with a sliding rod and
stamper, the whole
being arranged in
a plate containing
water. When the
stamper is pushed
down upon the glass
the latter is broken
(Fig. 107), and the air
gradually penetrates
to the finely divided
lead, when ignition oc-
curs, and the oxygen
is absorbed, as demon-
strated by the rise
of the water in the
jar. On the same
principle, if a bottle
is filled about one-
third full with a liquid
amalgam of lead and
mercury, and then
stopped and shaken
for two hours or
more, the finely di-
vided lead absorbs
the oxygen and

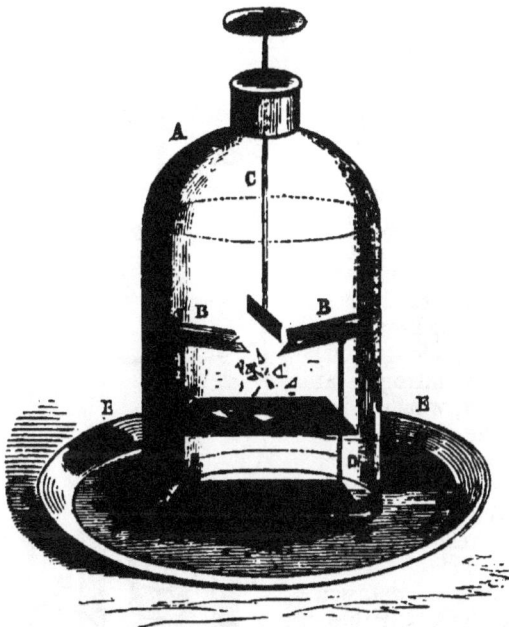

Fig. 107. A. Glass jar, with collar of leather, through which
the stamper, C, works. B B. The tube containing the finely-
divided lead, part of which falls out, and is ignited, and
retained by the little tray just below, being part of the iron
stand, D ·D, with crutches supporting the ends of the glass
tube, and the whole stands in the dish of water, E E.

leaves pure nitrogen. Or if a mixture of equal weights of sulphur
and iron filings, is made into a paste with water in a thin iron cup,
and then warmed and placed under a gas jar full of air standing on the

shelf of the pneumatic trough, or in a dish full of water, the water gradually rises in the jar in about forty-eight hours, in consequence of the absorption of the oxygen gas.

Third Experiment.

Nitrogen is devoid of colour, taste, smell, of alkaline or acid quaiities; and, as we shall have occasion to notice presently, it forms an *acid* when chemically united with oxygen, and an alkali in union with hydrogen. A lighted taper plunged into this gas is immediately extinguished, while its specific gravity, which is lighter than that of oxygen or air, is demonstrated by the rule of proportion.

Weight of 100 cubic inches of air at 60° Fahr., bar. 29·92 in.	Unity.	Weight of 100 cubic inches of nitrogen at 60° Fahr., bar. 29·92 in	Specific gravity of nitrogen.
30·829	: 1 : :	29·952 : :	971

And its levity may be shown very prettily by a simple experiment. Select two gas jars of the same size, and after filling one with oxygen gas and the other with nitrogen gas, slide glass plates over the bottoms of the jars, and proceed to invert the one containing oxygen, placing the neck in a stand formed of a box open at the top; then place the jar containing nitrogen over the mouth of the first, withdrawing the glass plates carefully; and if the table is steady the top gas jar will stand nicely on the lower one. Then (having previously lighted a taper so as to have a long snuff) remove the stopper from the nitrogen jar and insert the lighted taper, which is immediately extinguished, and as quickly relighted by pushing it down to the lower jar containing the oxygen. This experiment may be repeated several times, and is a good illustration of the relative specific gravities of the two gases, and of the importance of the law of universal diffusion already explained at p. 6, by which these gases *mix*, not *combine* together, and the atmosphere remains in one uniform state of composition in spite of the changes going on at the surface of the earth. Omitting the aqueous vapour, or steam, ever present in variable quantities in the atmosphere, ten thousand volumes of dry air contain, according to Graham :—

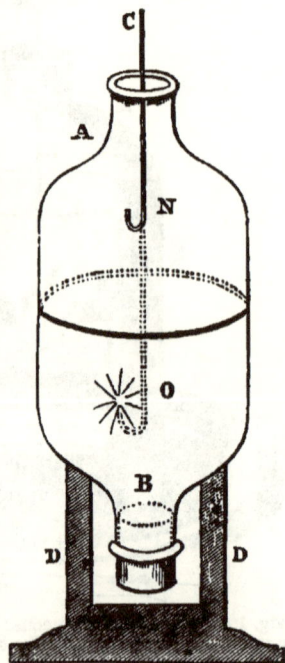

Fig. 108. A. Gas jar containing nitrogen, N, standing on B, another jar full of oxygen, o. The taper, c, is extinguished at N, and relighted at o. D D. Stand supporting the jars.

Nitrogen 7912
Oxygen 2080
Carbonic acid 4
Carburetted hydrogen (CH$_2$) . . . 4
Ammonia . . . . a trace
                                   —————
                                   10,000

Fourth Experiment.

It was the elegant, the accomplished, but ill-fated Lavoisier who discovered, by experimenting with quicksilver and air, the compound
nature of the atmosphere; and it was the same chemist who gave the
name of azote to nitrogen; it should, however, be borne in mind that it
does not necessarily follow because a gas extinguishes flame that it is a *poison*. Nitrogen extinguishes flame, but we inhale enormous quantities of air without any ill effects from the nitrogen or azote that it contains; on the other hand, many gases that extinguish flame are *specific poisons*, such as carbonic acid, carbonic oxide, cyanogen, &c.

Lavoisier's experiment may be repeated by passing into a measured jar, graduated into five equal volumes, four measures of nitrogen and one measure of oxygen; a glass plate should then be slid over the mouth

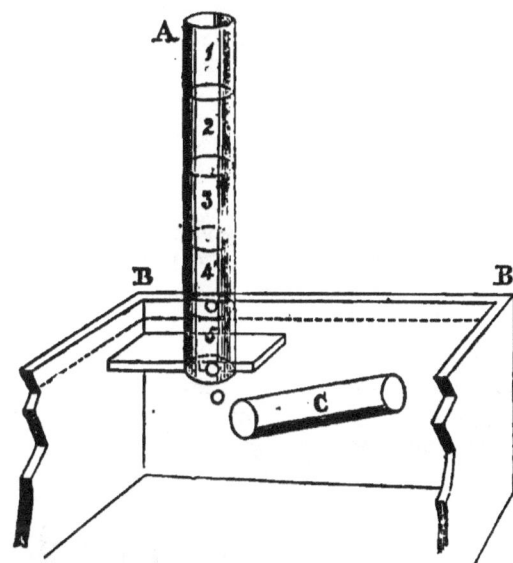

Fig. 109. A. Gas jar divided into five equal parts. B B. Section of pneumatic trough, to show the decantation of gas from one vessel to another. The gas is being passed from o to A, through the water.

of the vessel, and it may be turned up and down gently for. some little
time to mix the two gases, and when the mixture is tested with a lighted
taper, it is found neither to increase nor diminish the illuminating power
and the taper burns as it would do in atmospheric air. (Fig. 109.)

HYDROGEN.

Hydrogen (υδωρ, water; γενναω, I give rise to), so termed by Lavoisier —called by other chemists inflammable air, and phlogiston. Symbol, H ; combining properties, 1.   The lightest known form of matter.

Every 100 parts by weight of water contain 11 parts of hydrogen gas; and as the quantity of water on the surface of the earth represents at least two-thirds of the whole area, the source of this gas, like that of oxygen or nitrogen, is inexhaustible.   Van Helmont, Mayow, and Hales had shown that certain inflammable and peculiar gases could be obtained, but it was reserved for the rigidly philosophic mind of Cavendish to determine the nature of the elements contained in, and giving a speciality to, the inflammable gases of the older chemists.   By acting with dilute acids upon iron, zinc, and tin, Cavendish liberated an inflammable elastic gas ; and he discovered nearly all the properties we shall notice in the succeeding experiments, and especially demonstrated the composition of water in his paper read before the Royal Society in the year 1784.

First Experiment.

Hydrogen is prepared in a very simple manner, by placing some zinc cuttings in a bottle, to which is attached a cork and pewter or bent glass tube, and pouring upon the metal some dilute sulphuric or hydrochloric acid.   Effervescence and ebullition take place, and the gas escapes in large quantities, water being decomposed ; the oxygen passes to the zinc, and forms oxide of zinc, and this uniting with the sulphuric acid forms sulphate of zinc, which may be obtained after the escape of the hydrogen by evaporation and crystallization. (Fig. 110.)

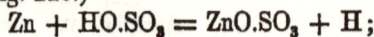

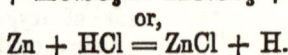

$$Zn + HO.SO_3 = ZnO.SO_3 + H;$$

or,

$$Zn + HCl = ZnCl + H.$$

In nearly all the processes employed for the generation of hydrogen gas, a metal is usually employed, and this fact has suggested the notion that hydrogen may possibly be a metal, although it is the lightest known form of matter ; and it will be observed in all the succeeding experiments that a metallic substance will be employed to take away the oxygen and displace the hydrogen.

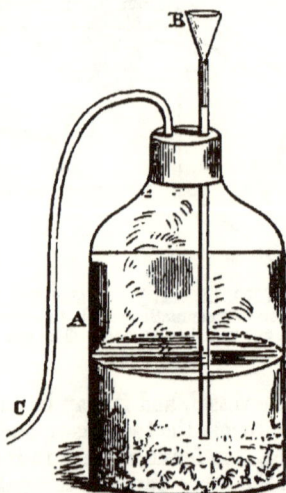

Fig. 110. A. Bottle containing zinc cuttings and water and fitted with a cap and two tubes, the one marked B, containing a funnel, conveys the sulphuric acid to the zinc and water, whilst the gas escapes through the pipe C.

Whenever hydrogen is prepared it should be allowed to escape from the generating vessel for a few minutes before any flame is applied, in order that the atmospheric air may be expelled. The most serious accidents have occurred from carelessness in this respect, as a mixture of hydrogen and air is explosive, and the more dangerous when it takes fire in any close glass bottle.

Second Experiment.

If a piece of potassium is confined in a little coarse wire gauze cage, attached to a rod, and thrust under a small jar full of water, placed on the shelf of the pneumatic trough, hydrogen gas is produced with great rapidity, and is received into the gas jar. The bit of potassium being surrounded with water, is kept cool, whilst the hydrogen escaping under the water is not of course burnt away, as it is whenever the metal is thrown on the *surface* of water.

Third Experiment.

Across a small iron table-furnace is placed about eighteen inches of 1-inch gas-pipe containing iron borings, the whole being red-hot; and attached to one end is a pipe conveying steam from a boiler, or flask, or retort, whilst another pipe is fitted to the opposite end, and passes to the pneumatic trough. Directly the steam passes over the red hot iron borings it is deprived of oxygen, which remains with the iron, forming the rust or oxide of iron, whilst the hydrogen, called in this case *water gas*, escapes with great rapidity. When steam is passed over red-hot charcoal, hydrogen is also produced with carbonic oxide gas, and this in fact is the ordinary process of making *water gas*, which being purified is afterwards saturated with some volatile hydrocarbon and burnt. At first sight, such a mode of making gas would be thought extremely profitable, and in spite of the numerous failures the *discovery* (so called) of *water gas* is reproduced as a sort of *chronic wonder;* but experience and practice have clearly demonstrated that *water gas* is a fallacy, and as long at we can get coal it is not worth while going through the round-about processes of first burning coal to produce steam ; secondly,

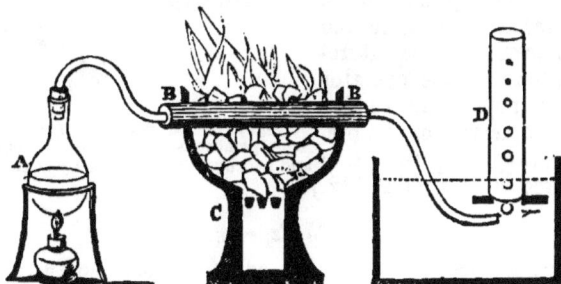

Fig. 111. A. Flask containing water, and producing steam, which passes to the iron tube, B B, containing the iron borings heated red hot in the charcoal stove C. The hydrogen passes to the jar D, standing on the shelf of the pneumatic trough.

of burning coal to heat charcoal, over which the steam is passed to be converted into gas, which has then to be purified and saturated with a cheap hydrocarbon obtained from coal or mineral naphtha; whilst ordinary coal gas is obtained at once by heating coal in iron retorts. (Fig. 111.)

Thus, by the metals zinc, tin, potassium, red-hot iron (and we might add several others), the oxygen of water is removed and hydrogen gas liberated.

Fourth Experiment.

If bottles of hydrogen gas are prepared by all the processes described, they will present the same properties when tested under similar circumstances. A lighted taper applied to the mouths of the bottles of hydrogen, which should be inverted, causes the gas to take fire with a slight noise, in consequence of the mixture of air and hydrogen that invariably takes place when the stopper is removed; on thrusting the lighted taper into the bulk of the gas it is extinguished, showing that hydrogen possesses the opposite quality to oxygen—viz., that it takes fire, but does not support combustion. By keeping the bottles containing the hydrogen upright, when the stopper is removed the gas escapes with great rapidity, and atmospheric air takes its place, so much so that by the time a lighted taper is applied, instead of the gas burning quietly, it frequently astonishes the operator with a loud pop. This sudden attack on the nerves may be prevented by always experimenting with inverted bottles. (Fig. 112.)

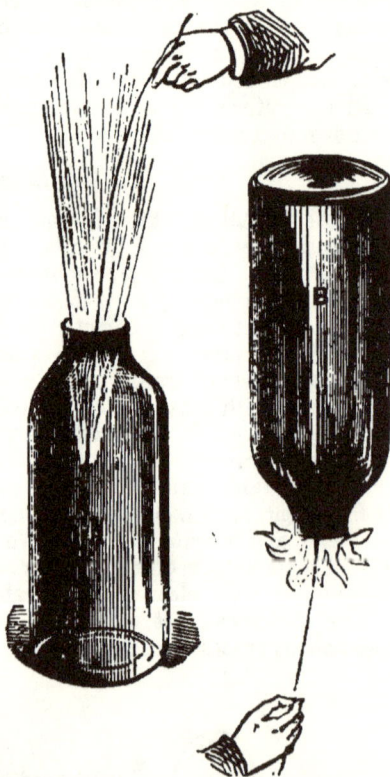

Fig. 112. A. Bottle opened upright, and hydrogen exploding. B. Bottle opened inverted, and hydrogen burning quietly at the mouth.

Fifth Experiment.

Hydrogen is 14·4 lighter than air, and for that reason may be passed into bottles and jars without the assistance of the pneumatic trough. One of the most amusing proofs of its levity is that of filling paper bags or balloons with this gas; and we read, in the accounts of the fêtes at

Paris, of the use of balloons ingeniously constructed to represent animals, so that a regular aerial hunt was exhibited, with this drawback only, that nearly all the animals preferred ascending with their legs upwards, a circumstance which provoked intense mirth amongst the volatile Frenchmen. The lightness of hydrogen may be shown in two ways—first, by filling a little goldbeater's-skin balloon with *pure* hydrogen (prepared by passing the gas made from zinc and dilute pure sulphuric acid through a strong solution of potash, and afterwards through one of nitrate of silver), and allowing the balloon to ascend; and then afterwards, having of course secured the balloon by a thin twine or strong thread, it may be pulled down and the gas inhaled, when a most curious effect is produced on the voice, which is suddenly changed from a manly bass to a ludicrous nasal squeaking sound. The only precautions necessary are to make the gas quite pure, and to avoid flame whilst inhaling the gas. It is related by Chaptal that the intrepid (quære, foolish) but unfortunate aeronaut, Mons. Pilate de Rosio, having on one occasion inhaled hydrogen gas, was rash enough to approach a lighted candle, when an explosion took place in his mouth, which he says "*was so violent that he fancied all his teeth were driven out.*" Of course, if it were possible to change by some extraordinary power the condition of the atmosphere in a concert-room or theatre, all the bass voices would become extremely nasal and highly comic, whilst the sopranos would emulate railway whistles and screech fearfully; and supposing the specific gravity of the air was continually and materially changing, our voices would never be the same, but alter day by day, according to the state of the air, so that the "familiar voice" would be an impossibility.

A bell rung in a gas jar containing air emits a very different sound from that which is produced in one full of hydrogen—a simple experiment is easily performed by passing a jar containing hydrogen over a self-acting bell, such as is used for telegraphic purposes. (Fig. 113.)

Fig. 113. A. Stand and bell. B B. Tin cylinder full of hydrogen, which may be raised or depressed at pleasure, by lifting it with the knob at the top, when the curious changes in the sound of the bell are audible.

Sixth Experiment.

Some of the small pipes from an organ may be made to emit the most curious sounds by passing heavy and light gases through them; in these experiments bags containing the gases should be employed, which may drive air, oxygen, carbonic acid, or hydrogen, through the organ pipes at precisely the same pressure.

Seventh Experiment.

One of those toys called "The Squeaking Toy" affords another and ridiculous example of the effect of hydrogen on sound, when it is used in a jar containing this gas. (Fig. 114.)

Eighth Experiment.

An accordion played in a large receptacle containing hydrogen gas demonstrates still more clearly what would be the effect of an orchestra shut up in a room containing a mixture of a considerable portion of hydrogen with air, as the former, like nitrogen, is not a poison, and only kills in the absence of oxygen gas.

Ninth Experiment.

Some very amusing experiments with balloons have been devised by Mr. Darby, the eminent firework manufacturer, by which they are made to carry signals of three kinds, and thus the motive or ascending power may be utilized to a certain extent.

Fig. 114. The squeaking toy, used in a jar of hydrogen.

Mr. Darby's attention was first directed to the manufacture of a good, serviceable, and cheap balloon, which he made of paper, cut with mathematical precision; the gores or divisions being made equal, and when pasted together, strengthened by the insertion of a string at the juncture; so that the skeleton of the balloon was made of string, the whole terminating in the neck, which was further stiffened with calico, and completed when required by a good coating of boiled oil. These balloons are about nine feet high and five feet in diameter in the widest part, exactly like a pear, and tapering to the neck in the most graceful and elegant manner. They retain the hydrogen gas remarkably well for many hours, and do not leak, in consequence of the paper of which they are made being well selected and all holes stopped, and also from the circumstance of the pressure being so well distributed over the interior by the almost mathematical precision with which they are cut, and the careful preparation of the paper with proper varnish. One of their greatest recommendations is cheapness; for whilst a gold-beater's skin balloon of the same size would cost about 5*l.*, these can be furnished at 5*s.* each in large quantities.

A balloon required to carry one or more persons must be constructed of the best materials, and cannot be too carefully made; it is therefore a somewhat costly affair, and as much as 200*l.*, 500*l.*, and even 1000*l.* have been expended in the construction of these aerial chariots.

The chief points requiring attention are:—first, the quality of the silk; secondly, the precision and scrupulous nicety required in cutting

out and joining the gores ; thirdly, the application of a good varnish to
fill up the pores of the silk, which must be insoluble in water, and sufficiently elastic not to crack.

The usual material is Indian silk (termed Corah silk), at from 2*s.* to
2*s.* 6*d.* per yard.

The *gores* or parts with which the balloon is constructed require, as
before stated, great attention ; it being a common saying amongst
aeronauts, " *that a cobweb will hold the gas if properly shaped,*" the
object being to diffuse the pressure equally over the whole bag or
balloon.

The varnish with which the silk is rendered air-tight can be made
according to the private recipe of Mr. Graham, an aeronaut, who states
that he uses for this purpose two gallons of linseed oil (boiled), two
ditto (raw), and four ounces of beeswax ; the whole being simmered
together for one hour, answers remarkably well, and the varnish is
tough and not liable to crack.

For repairing holes in a balloon, Mr. Graham recommends a cement
composed of two pounds of black resin and one pound of tallow,
melted together, and applied on pieces of varnished silk to the apertures.

The actual cost of a balloon will be understood from information also
derived from Mr. Graham. His celebrated " Victoria Balloon," which
has passed through so many hairbreadth escapes, was sixty-five feet
high, and thirty-eight feet in diameter in the broadest part ; and the
following articles were used in its construction :—

	£	*s.*	*d.*
1400 yards of Corah silk, at 2*s.* 6*d.* per yard . .	175	0	0
The netting weighed 70 lbs.	20	0	0
Extra ropes weighed 20 lbs. at 2*s.* per lb. . . .	2	0	0
The car weighed 25 lbs.	7	0	0
Varnish, wages, &c.	16	0	0
	£220	0	0

Thirty-eight thousand cubic feet of coal gas were required to fill this
balloon, charged by one company 20*l.*, by others from 9*l.* to 10*l.* ; and
eight men were required to hold the inflated baggy monster.

Such a balloon as described above is a mere soap bubble when compared with the " New Aerial Ship" now building in the vicinity of New
York ; the details are so practical and interesting, that we quote nearly
the whole account of this mammoth or Great Eastern amongst balloons,
as given in the *New York Times.*

"An experiment in scientific ballooning, greater than has yet been
undertaken, is about to be tried in this city. The project of crossing
the Atlantic Ocean with an air-ship, long talked of, but never accomplished, has taken a shape so definite that the apparatus is already prepared and the aeronaut ready to undertake his task.

"The work has been conducted quietly, in the immediate vicinity of
New York, since the opening of spring. The new air-ship, which has

been christened the City of New York, is so nearly completed, that but few essentials of detail are wanting to enable the projectors to bring it visibly before the public.

"The aeronaut in charge is Mr. T. S. C. Lowe, a New Hampshire man, who has made thirty-six balloon ascensions.

"The dimensions of the City of New York so far exceed those of any balloon previously constructed, that the bare fact of its existence is notable. Briefly, for so large a subject, the following are the dimensions :—Greatest diameter, 130 feet ; transverse diameter, 104 feet ; height, from valve to boat, 350 feet; weight, with outfit, $3\frac{1}{4}$ tons; lifting power (aggregate), $22\frac{1}{2}$ tons ; capacity of gas envelope, 725,000 cubic feet.

"The City of New York, therefore, is nearly five times larger than the largest balloon ever before built. Its form is that of the usual perpendicular gas-receiver, with basket and lifeboat attached.

"Six thousand yards of twilled cloth have been used in the construction of the envelope. Reduced to feet, the actual measurement of this material is 54,000 feet—or nearly 11 miles. Seventeen of Wheeler and Wilson's sewing machines have been employed to connect the pieces, and the upper extremity of the envelope, intended to receive the gas-valve, is of triple thickness, strengthened with heavy brown linen, and sewed in triple seams. The pressure being greatest at this point, extraordinary power of resistance is requisite. It is asserted that 100 women, sewing constantly for two years, could not have accomplished this work, which measures by miles. The material is stout and the stitching stouter.

"The varnish applied to this envelope is a composition the secret of which rests with Mr. Lowe. Three or four coatings are applied, in order to prevent leakage of the gas.

"The netting which surrounds the envelope is a stout cord, manufactured from flax expressly for the purpose. Its aggregate strength is equal to a resistance of 160 tons, each cord being capable of sustaining a weight of 400 lbs. or 500 lbs.

"The basket which is to be suspended immediately below the balloon is made of rattan, is 20 feet in circumference and 4 feet deep. Its form is circular, and it is surrounded by canvas. This car will carry the aeronauts. It is warmed by a lime-stove, an invention of Mr. O. A. Gager, by whom it was presented to Mr. Lowe. A lime-stove is a new feature in air voyages. It is claimed that it will furnish heat without fire, and is intended for a warming apparatus only. The stove is $1\frac{1}{2}$ feet high, and 2 feet square. Mr. Lowe states that he is so well convinced of the utility of this contrivance, that he conceives it to be possible to ascend to a region where water will freeze, and yet keep himself from freezing. This is to be tested.

"Dropping below the basket is a metallic lifeboat, in which is placed an Ericsson engine. Captain Ericsson's invention is therefore to be tried in mid-air. Its particular purpose is the control of a propeller, rigged upon the principle of the screw, by which it is proposed to obtain

ɪ

a regulating power. The application of the mechanical power is in-geniously devised. The propeller is fixed in the bow of the lifeboat, projecting at an angle of about forty-five degrees. From a wheel at the extremity twenty fans radiate. Each of these fans is 5 feet in length, widening gradually from the point of contact with the screw to the extremity, where the width of each is 1½ feet. Mr. Lowe claims that by the application of these mechanical contrivances his air-ship can be readily raised or lowered, to seek different currents of air; that they will give him ample steerage way, and that they will prevent the rotatory motion of the machine. In applying the principle of the fan, he does not claim any new discovery, but simply a practical development of the theory advanced by other aeronauts, and partially reduced to practice by Charles Green, the celebrated English aeronaut.

"Mr. Lowe contends that the application of machinery to aerial navigation has been long enough a mere theory. He proposes to reduce the theory to practice, and see what will come of it. It is estimated that the raising and lowering power of the machinery will be equal to a weight of 300 lbs., the fans being so adjusted as to admit of very rapid motion upward or downward. As the loss of three or four pounds only is sufficient to enable a balloon to rise rapidly, and as the escape of a very small portion of the gas suffices to reduce its altitude, Mr. Lowe regards this systematic regulator as quite sufficient to enable him to control his movements and to keep at any altitude he desires. It is his intention to ascend to a height of three or four miles at the start, but this altitude will not be permanently sustained. He prefers, he says, to keep within a respectable distance of mundane things, where 'he can see folks.' It is to be hoped his machinery will perform all that he anticipates from it. It is a novel affair throughout, and a variety of new applications remain to be tested. Mr. Lowe, expressing the utmost confidence in all the appointments of his apparatus, assured us that he would certainly go, and, as certainly, would go into the ocean, or deliver a copy of Monday's *Times* in London on the following Wednesday. He proposes to effect a landing in England or France, and will take a course north of east. A due easterly course would land him in Spain, but to that course he objects. He hopes to make the trip from this city to London in forty-eight hours, certainly in sixty-four hours. He scouts the idea of danger, goes about his preparations deliberately, and promises himself a good time. As the upper currents, setting due east, will not permit his return by the same route, he proposes to pack up the City of New York, and take the first steamer for home.

"The air-ship will carry weight. Its cubical contents of 725,000 feet of gas suffice to lift a weight of 22½ tons. With outfit complete its own weight will be 3½ tons. With this weight 19 tons of lifting power remain, and there is accordingly room for as many passengers as will care to take the venture. We understand, however, that the company is limited to eight or ten. Mr. Lowe provides sand for ballast, regards his chances of salvation as exceedingly favourable, places implicit faith in the strength of his netting, the power of his machinery, and the buoyancy of his lifeboat, and altogether considers himself secure from the hazard of disaster. If he accomplish his voyage in safety, he will have done more than any air navigator has yet ventured to undertake. If he fail, the enterprise sinks the snug sum of 20,000 dollars. Wealthy men who are his backers, sharing his own enthusiasm, declare failure impossible, and invite a patient public to wait and see."

A night ascent witnessed at any of the public gardens is certainly a stirring scene, particularly if the wind is rather high. On approaching the balloon, swayed to and fro by the breeze, it seems almost capable of crushing the bold individual who would venture beneath it; seen as a large dark mass in the yet dimly-lighted square, it appears to be incapable of control; when the inflation is completed, the aeronaut, all importance, seats himself in the car, and blue lights, with other fireworks, display the victim who is to make a "last ascent," or perhaps *descent*. Finally the word is given, the ropes are cast off, and the bulky chariot rises majestically to the sound of the National Anthem. The crowd see no more, but the next day's *Times* reports the end of the aerial journey.

Balloons can never be of any permanent value as means of locomotion until they can be steered; and this is a problem, the solution of which is something like *perpetual motion*. In the first place, a balloon of any size exposes an enormous surface to the pressure and force of the winds; and when we consider that they move at the rate of from three to eighty miles per hour, it will be understood that the fabric of the balloon itself must give way in any attempt to tear, work, or pull it against such a force. Secondly and lastly, the power has not yet been created which will do all this without the inconvenience of being so *heavy* that the steering engine fixes the balloon steadily to the earth by its obstinate gravity. When engines of power are constructed without the aeronaut's obstacle of weight—when balloons are made of thin copper or sheet-iron, then we may possibly hear of the voyage of the good ship *Aerial*, bound for any place, and quite independent of dock, port, and the host of dues (*quere*), which the sea-going ships have to disburse. It is, however, gratifying to the zeal and perseverance of those who dream of aerial navigation, to know that a balloon is not quite useless; and here we may return to the consideration of Mr. Darby's signals, which are of various kinds, and intended to appeal to the senses by night as well as by day; and first, by *audible sounds*. Such means have long been recognised, from the ancient float and bell of the "Inchcape Rock," to the painful minute-gun at sea, or the shrill railway whistle and detonating signals employed to prevent the horrors of a collision between two trains. The signal sounds are produced by the explosion of shells capable of yielding a report equal to that of a six-pounder cannon, and they are constructed in a very simple manner. A ball, composed of wood or copper, and made up by screwing together the two hemispheres, is attached to a shaft or tail of cane or lance-wood, properly feathered like an arrow; at the side opposite to that of the arrow—viz., at its antipodes, is placed a slight protuberance containing a minute bulb of glass filled with oil of vitriol, and surrounded with a mixture of chlorate of potash and sugar, the whole being protected with gutta-percha, and communicating by a touch-hole with the interior, which is of course filled with gunpowder. These shells are attached to a circular framework by a strong whipcord, which passes to a central fuse, and are detached one after the other as the slow fuse (made hollow on the principle of the argand lamp) burns steadily away. Directly a shell falls to the ground, the little bulb containing the oil of vitriol breaks, and the acid coming in contact with the chlorate of potash and sugar, causes the mixture to take fire, when the gunpowder explodes. During the siege of Sebastopol many similar mines were prepared by the Russians in the earth, so that when an unfortunate soldier trod upon the spot, the concealed mine blew up and seriously injured him; such petty warfare is as bad as shooting sentries, and a cruel application of science, that unnecessarily increases the miseries of war without producing those grand results for which the truly great captains, Wellington and Napoleon, only warred. (Fig. 115.)

The bill distributor consists of a long piece of wood, to which are

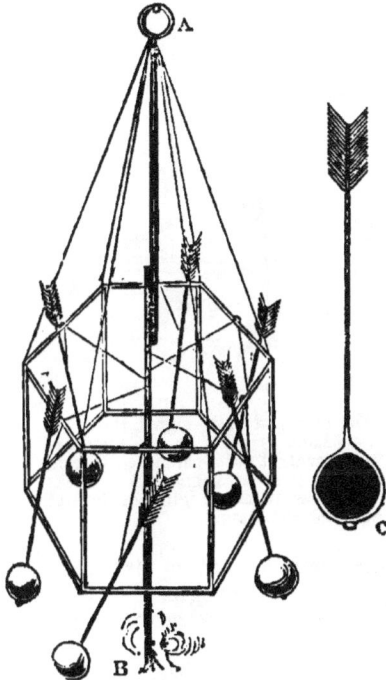

Fig. 115. A. Ring attached to balloon, carrying an hexagonal framework with six shells. B. Hollow fuse, which burns slowly up to the strings, and detaches each shell in succession. C. Section of shell. The shaded portion represents the gunpowder.

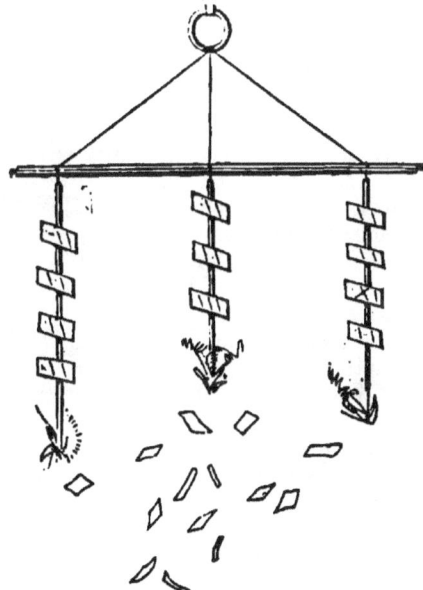

Fig. 116. The bill distributor, consisting of three hollow fuses, with bills attached in packets.

attached a number of hollow fuses, with packets of bills, protected from
being burned or singed by a thin tin plate; 10,000 or 20,000 bills can
thus be delivered, and the wind assists in scattering them, whilst the balloon travels over a distance of many miles. It must be recollected that
in each case the shells and the bills are detached by the string burning
away as the fire creeps up from the fuse. (Fig. 116.)

Another most ingenious arrangement, also prepared by Mr. Darby,
is termed by the inventor, the "Land and Water Signal," and may be
thus described:—A short hollow ball of gutta-percha, or other convenient material, five or six
inches in diameter, and filled
with printed bills, or the information, whatever it may
be, that is required to be
sent, is attached to a cap to
which a red flag, having the
words "*Open the shell,*" and
four cross sticks, canes, or
whalebones with bits of cork
at equal distances, are fitted.
The whole is connected by
a string to the fuse as before
described. These signals
are adapted for land and
water: in either case they
fall upright, and in consequence of the sticks pro-
jecting out they float well
in the water, and can be
seen by a telescope at a distance of three miles. (Fig.
117.) Many of these sig-
nals were sent away by Mr.
Darby from Vauxhall; one
was picked up at Harwich,
another at Brighton, a third
at Croydon; in the latter
case it was found by a cot-
tager, who, fearing gunpow-

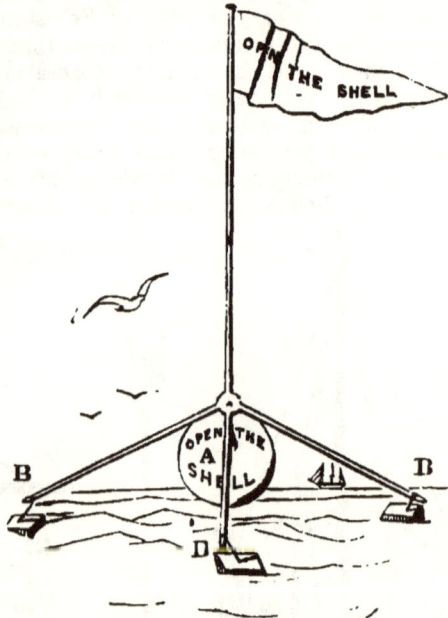

Fig. 117. The land and water signal, which remains upright on land, or floats on the surface of
water. A. The water-tight gutta-percha shell, containing the message or information. B B B. Sticks
of cane to keep the flag in an upright position; at
the ends are attached cork bungs.

der and combustibles, did not examine the shell, but having mentioned
the circumstance to a gentleman living near him, they agreed to cut it
open; and intelligence of their arrival, in this and the other cases, was
politely forwarded to Mr. Darby at Vauxhall Gardens.

Balloons, like a great many other clever inventions, have been despised
by military men as new-fangled expedients, toys, which may do very well
to please the gaping public, but are and must be useless in the field.
Over and over again it has been suggested that a balloon corps for
observation should be attached to the British army, but the scheme has

been rejected, although the expense of a few yards of silk and the generation of hydrogen gas would be a mere bagatelle as compared with the transport and use of a single 32-pounder cannon. The antiquated notions of octogenarian generals have, however, received a great shock in the fact that the Emperor Napoleon III. was enabled, by the assistance of a captive balloon, to watch the movements and dispositions of the Austrian troops; and with the aid of the information so obtained, he made his preparations, and was rewarded by the victory of Solferino; and as soon as the battle was over Napoleon III. occupied at Cavriana the very room and ate the dinner prepared for his adversary, the Emperor Francis Joseph.

Over and over again the most excellent histories have been written of aerostation, but they all tend to one truth, and that is, the great danger and risk of such excursions; and to enable our readers to form their own judgment, a chronological list of some of the most celebrated aeronauts, &c., is appended.

1675. Bernair attempted to fly—*killed*.
1678. Besnier attempted to fly.
1772. L'Abbé Desforges announced an aerial chariot.
1783. Montgolfier constructed the first air balloon.
  „ Roberts *frères*, first gas balloon, destroyed by the peasantry of Geneva, who imagined it to be an evil spirit or the moon.
1784. Madame Thiblé, the first lady who was ever up in the clouds; she ascended 13,500 feet.
  „ Duke de Chartres, afterwards *Egalité* Orleans, travelled 135 miles in five hours in a balloon.
  „ Testu de Brissy, equestrian ascent.
  „ D'Achille, Desgranges, and Chalfour—Montgolfier balloon.
  „ Bacqueville attempted a flight with wings.
  „ Lunardi—gas balloon.
  „ Rambaud—Montgolfier balloon, which was burnt.
  „ Andreani—Montgolfier balloon.
1785. General Money—gas balloon, fell into the water, and not rescued for six hours.
  „ Thompson, in crossing the Irish Channel, was run into with the bowsprit of a ship whilst going at the rate of twenty miles per hour.
  „ Brioschi—gas balloon ascended too high and burst the balloon; the hurt he received ultimately caused his *death*.
  „ A Venetian nobleman and his wife—gas balloon—*killed*.
  „ Pilatre de Rozier and M. Romain—gas balloon took fire—both *killed*.
1806. Mosment—gas balloon—*killed*.
  „ Olivari—Montgolfier balloon—*killed*.
1808. Degher attempted a flight with wings.
1812. Bittorf—Montgolfier balloon—*killed*.
1819. Blanchard, Madame—gas balloon—*killed*.

1819. Gay Lussac—gas balloon, ascended 23,040 feet above the level of the sea. Barometer 12·95 inches; thermometer 14·9 Fah.
   „   Gay Lussac and Biot—gas balloon for the benefit of science. Both philosophers returned safely to the earth.
1824. Sadler—gas balloon—*killed*.
   „   Sheldon—gas balloon.
   „   Harris—gas balloon—*killed*.
1836. Cocking—parachute from gas balloon—*killed*.
1847. Godard—Montgolfier balloon fell into and extricated from the Seine.
1850. Poitevin, a successful French aeronaut.
   „   Gale, Lieut.—gas balloon—*killed*.
   „   Bixio and Barral—gas balloon.
   „   Graham, Mr. and Mrs.—gas balloon.—Serious accident ascending near the Great Exhibition in Hyde Park.
   „   Green, the most successful living aeronaut of the present time.

Of the 41 persons enumerated, 14 were killed, and nearly all the aeronauts met with accidents which might have proved fatal.

Fig. 118. Flying machine (*theoretical*).

Tenth Experiment.

Soap bubbles blown with hydrogen gas ascend with great rapidity, and break against the ceiling; if interrupted in their course with a lighted taper they burn with a slight yellow colour and dull report.

Eleventh Experiment.

By constructing a pewter mould in two halves, of the shape of a tolerably large flask, a balloon of collodion may be made by pouring the collodion *inside* the pewter vessel, and taking care that every part is properly covered; the pewter mould may be warmed by the external application of hot water, so as to drive off the ether of the collodion, and when quite dry the mould is opened and the balloon taken out. Such balloons may be made and inflated with hydrogen by attaching to them a strip of paper, dipped in a solution of wax and phosphorus, and sulphuret of carbon; as the latter evaporates, the phosphorus takes fire and spreads to the balloon; which burns with a slight report. The pewter mould must be very perfectly made, and should be bright inside; and if the balloons are filled with oxygen and hydrogen, allowing a sufficient excess of the latter to give an ascending power, they explode with a loud noise directly the fire reaches the mixed gases.

Twelfth Experiment.

In a soup-plate place some strong soap and water; then blow out a number of bubbles with a mixture of oxygen and hydrogen; a loud report occurs on the application of flame, and if the room is small the window should be placed open, as the concussion of the air is likely to break the glass.

Thirteenth Experiment.

Any noise repeated at least thirty-two times in a second produces a musical sound, and by producing a number of small explosions of hydrogen gas inside glass tubes of various sizes, the most peculiar sounds are obtained. The hydrogen flame should be extremely small, and the glass tubes held over it may be of all lengths and diameters; a trial only will determine whether they are fit for the purpose or not.

Fourteenth Experiment.

Flowers, figures, or other designs, may be drawn upon silk with a solution of nitrate of silver, and the whole being moistened with water, is exposed to the action of hydrogen gas, which removes the oxygen from the silver, and reduces it to the metallic state.

In like manner designs drawn with a solution of chloride of gold are produced in the metallic state by exposure to the action of hydrogen gas. Chloride of tin, usually termed muriate of tin, may also be reduced in a similar manner care being taken in these experiments that the fabric upon which the letters, figures, or designs are painted with the metallic solution be kept quite damp whilst exposed to the hydrogen gas.

Fifteenth Experiment.

A mixture of two volumes of hydrogen with one volume of oxygen explodes with great violence, and produces two volumes of steam, which condense against the sides of the strong glass vessel, in which the experiment may be made, in the form of water. As the apparatus called the Cavendish bottle, by which this experiment only may be safely performed, is somewhat expensive, and requires the use of an air-pump, gas jars with stop-cocks, and an electrical machine and Leyden jar, other and more simple means may be adopted to show the combination of oxygen and hydrogen, and formation of water.

If a little alcohol is placed in a cup and set on fire, whilst an empty cold gas jar is held over the flame, an abundant deposition of moisture takes place from the combustion of the hydrogen of the spirits of wine. Alcohol contains six combining properties of hydrogen, with four of charcoal and two of oxygen. If a lighted candle, or an oil, camphine, Belmontine, or gas flame, is placed under a proper condenser, large quantities of water are obtained by the combustion of these substances (Fig. 119).

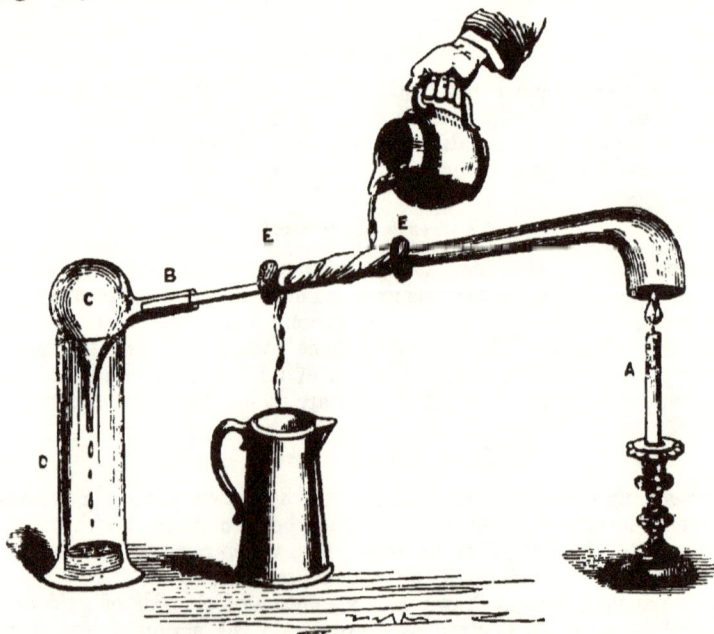

Fig. 119. A. A burning candle, or oil or gas lamp. Copper head and long pipe fitting into B C, the receiver from which the condensed water drops into D. E E. Two corks fitted, between which is folded some wet rag.

Sixteenth Experiment.

During the combustion of a mixture of two volumes of hydrogen with one of oxygen, an enormous amount of heat is produced, which is usefully applied in the arrangement of the oxy-hydrogen blowpipe. The flame of the mixed gases produces little or no light, but when directed on various metals contained in a small hole made in a fire brick, a most intense light is obtained from the combustion of the metals, which is variously coloured, according to the nature of the substances employed. With cast-iron the most vivid scintillations are obtained, particularly if after having fused and boiled the cast-iron with the jet of the two gases, one of them, viz., the hydrogen, is turned off, and the oxygen only directed upon the fused ball of iron, then the carbon of the iron burns with great rapidity, the little globule is enveloped in a shower of sparks, and the whole affords an excellent notion of the principle of Bessemer's patent method of converting cast-iron at once into pure malleable iron, or by stopping short of the full combustion of carbon, into cast-steel.

The apparatus for conducting these experiments is of various kinds, and different jets have been from time to time recommended on account of their alleged safety. It may be asserted that all arrangements proposed for burning any quantity of the *mixed* gases are extremely dangerous : if an explosion takes place it is almost as destructive as gunpowder, and should no particular damage be done to the room, there is still the risk of the sudden vibration of the air producing permanent deafness. If it is desired to burn the *mixed* gases, perhaps the safest apparatus is that of Gurney; in this arrangement the mixed gases bubble up through a little reservoir of water, and thus the gasholder — viz., a bladder, is cut off from the jet when the combustion takes place. (Fig. 120.) This jet is much recommended by Mr. Woodward, the highly respected President of the Islington Literary and Scientific Institution,

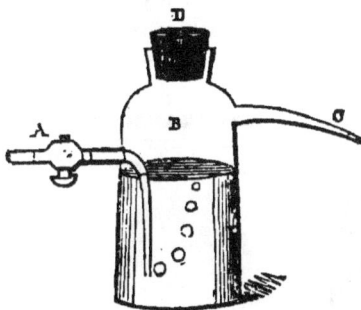

Fig. 120. Gurney's jet. A. Pipe with stop-cock leading from the gas-holder. B. The little reservoir of water through which the mixed gases bubble. C. The jet where the gases burn. D. Cork, which is blown out if the flame recedes in the pipe, c.

and may be fitted up to show the phenomena of polarized light, the microscope, and other interesting optical phenomena.

Mr. Woodward states, that a series of experiments, continued during many years, has proved, that while the bladder containing the mixed gases is under pressure, the flame cannot *be made* to pass the safety chambers, and consequently an explosion is impossible; and even if through extreme carelessness or design, as by the removal of pressure or the contact of a spark with the bladder, an explosion occurs, it can produce no other than the momentary effect of the alarm occasioned by

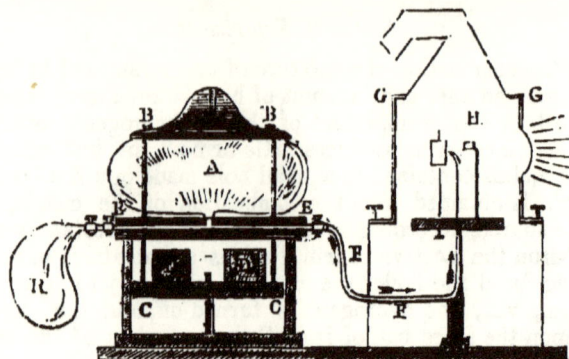

Fig. 121. A. The bladder of mixed gases, pressed by the board, B B, attached by wire supports to another board, O O, which carries the weights, D D. E E. Pipe to which the bladder, A, is screwed, and when A is emptied, it is re-filled from the other bladder, R. F F F. Pipe conveying mixed gases to the lantern, G G, where they are burnt from a Gurney's jet, H.

the report; whereas, when the gases are used in separate bags under a pressure of two or three half hundredweights, if the pressure on one of the bags be accidentally removed or suspended, the gas from the other will be forced into it, and if not discovered in time, will occasion an explosion of a very dangerous character; or if through carelessness one of the partially emptied bags should be filled up with the wrong gas, effects of an equally perilous nature would ensue.

In the oxy-hydrogen blowpipe usually employed, the gases are kept quite separate, either in gasometers or gas bags, and are conveyed by distinct pipes to a jet of very simple construction, devised by the late Professor Daniell, where they

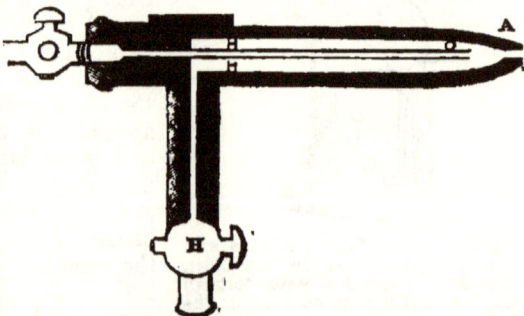

Fig. 122. Daniell's jet. O O. The stop-cock and pipe conveying oxygen, and fitting inside the larger tube H H, to which is attached a stop-cock, H, connected with the hydrogen receiver. A. The orifice near which the gases mix, and where they are burnt.

mix in very small volumes, and are burnt at once at the mouth of the jet. (Fig. 122.)

The gases are stored either in copper gasometers or in air-tight bags of Macintosh cloth, capable of containing from four to six cubic feet of gas, and provided with pressure boards. The boards are loaded with two or three fifty-six pound weights to force out the gas with sufficient pressure, and of course must be equally weighted; if any change of weight is made, the stop-cocks should be turned off and the light put out, as the most disastrous results have occurred from carelessness in this respect. (Fig. 123.)

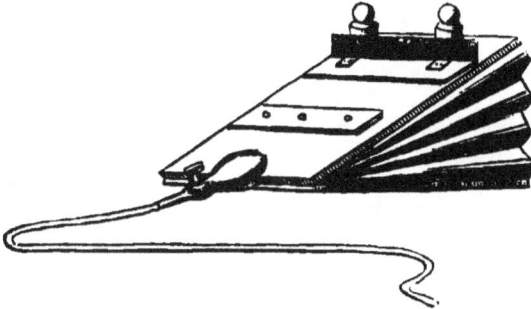

Fig. 123. Gas bag and pressure boards.

The oxy-hydrogen jet is further varied in construction by receiving the gases from separate reservoirs, and allowing them to mix in the upper part of the jet, which is provided with a safety tube filled with

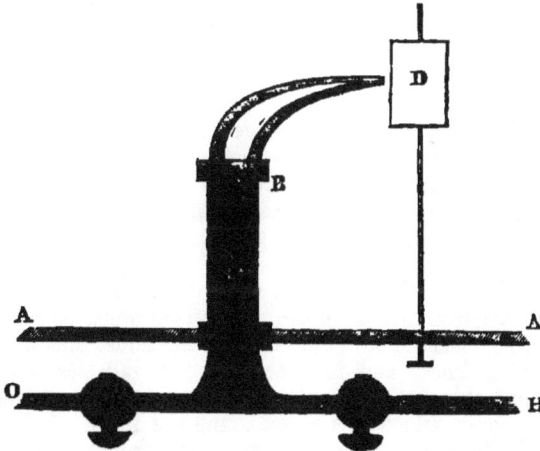

Fig. 124. A A. Board to which B B is fixed. o. Oxygen pipe. H. Hydrogen pipe.
c c. Space filled with wire gauze. D. Lime cylinder.

circular pieces of wire gauze. (Fig. 124.) With this arrangement a most intense light is produced, called the Drummond or lime light, and coal gas is now usually substituted for hydrogen.

Seventeenth Experiment.

There are many circumstances that will cause the union of oxygen and hydrogen, which, if confined by themselves in a glass vessel, may be preserved for any length of time without change; but if some powdered glass, or any other finely-divided substance with sharp points, is introduced into the mixed gases at a temperature not exceeding 660° Fahrenheit, then the gases silently unite and form water.

This curious mode of effecting their combination is shown in a still more interesting manner by perfectly clear platinum foil, which if introduced into the mixed gases gradually begins to glow, and becoming red-hot causes the gases to explode. Or still better, by the method first devised by Dobereiner, in 1824, by which finely prepared spongy platinum—*i.e.*, platinum in a porous state, and exposing a large metallic surface—is almost instantaneously heated red-hot by contact with the mixed gases. When this fact became known, it was further applied to the construction of an instantaneous light, in which hydrogen was made to play upon a little ball of spongy platinum, and immediately kindled. These Dobereiner lamps were possessed by a few of the curious, and would no doubt be extensively used if the discovery of phosphorus had not supplied a cheaper and more convenient fire-giving agent. When the spongy platinum is mixed with some fine pipeclay, and made into little pills, they may (after being slightly warmed) be introduced into a mixture of the two gases, and will silently effect their union. The theory of the combination is somewhat obscure, and perhaps the simplest one is that which supposes the platinum sponge to act as a conductor of electric influences between the two sets of gaseous particles; although, again, it is difficult to reconcile this theory with the fact that powdered glass at 660°, a bad conductor of electricity, should effect the same object. The result appears to be due to some effects of surface by which the gases seem to be condensed and brought into a condition that enables them to abandon their gaseous state and assume that of water.

When Sir H. Davy invented the safety-lamp, he was aware that, in certain explosive conditions of the air in coal mines, the flame of the lamp was extinguished, and in order that the miner should not be left in the dreary darkness and intricacies of the galleries without some means of seeing the way out, he devised an ingenious arrangement with thin platinum wire, which was coiled round the flame of the lamp, and fixed properly, so that it could not be moved from its proper place by any accidental shaking. When the flame of the safety-lamp, having the platinum wire attached, was accidentally extinguished by the explosive atmosphere in which it was burning, the platinum commenced glowing with an intense heat, and continued to emit light as long as it remained in the dangerous part of the mine. Sir H. Davy warned those who might use the platinum to take care that no portion of the thin wire passed *outside* the wire gauze, for the obvious reason that, if ignited outside the wire gauze protector, it would inflame the fire-damp.

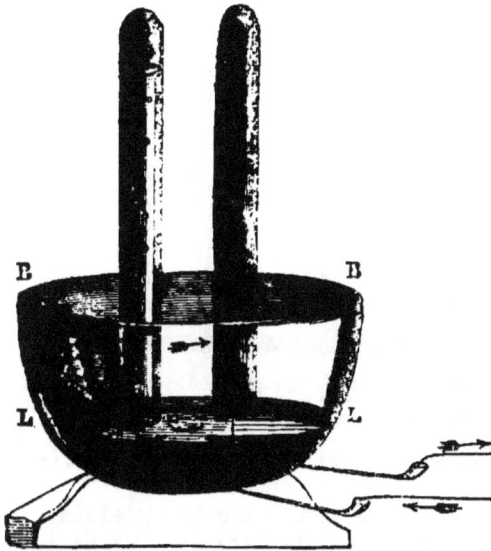

Fig. 125. P P. Two platinum plates connected with wires to the cups. The wires are passed through holes in the finger-glass, B B, and are fixed perfectly steady by pouring in cement composed of resin and tallow to the line L L. Two glass tubes filled with water acidulated with sulphuric acid, and placed over the platinum plates in finger-glass, which also contains dilute sulphuric acid to improve the conducting power of the water. The wires of the battery are placed in the cups, and the arrows show the direction of the current of electricity.

Eighteenth Experiment.

Water is decomposed by passing a current of voltaic electricity through it by means of two platinum plates, which may be connected with a ten-cell Grove's battery. The gases are collected in separate tubes, and the experiment offers one of the most instructive illustrations of the composition of water. (Fig. 125.)

There is a current of electricity passing from and between two platinum plates decomposing water, offering the converse of the Dobereiner experiment, and highly suggestive of the probability of the theory already advanced in explanation of the singular combination of oxygen and hydrogen in the presence of clean platinum foil, and more especially when we consider the operation of Grove's gas battery, in which a current of electricity is produced by pieces of platinum foil covered with finely-divided platinum, called platinum black; each piece is contained in a separate glass tube filled alternately with oxygen and hydrogen, and by connecting a great number of these tubes a current of electricity is obtained, whilst the oxygen and hydrogen are slowly absorbed and disappear, having combined and formed water, although placed in separate glass tubes. (Fig. 126.)

The analysis of water is shown very perfectly on the screen by fitting up some very small tubes and platinum wires in the same manner as shown in fig. 125. The vessel in which the tubes and wires are contained with the dilute sulphuric acid must be small, and arranged so as to pass nicely into the space usually filled by the picture in an ordinary magic lantern, or, still better, in one lighted by the oxy-hydrogen or lime light. If the dilute acid is coloured with a little solution of indigo, the gradual displacement of the fluid by the production of the two gases is very perfectly developed on the screen when the small voltaic battery is attached to the apparatus; and of course a large number of persons may watch the experiment at the same time.

With respect to the application of the light produced from a jet or

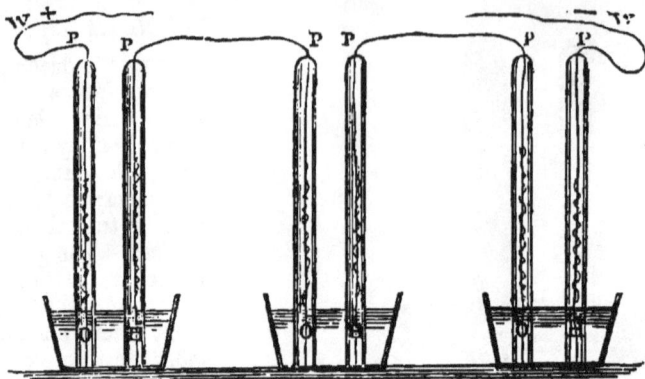

Fig. 126. Grove's gas battery consists of tubes containing oxygen and hydrogen alternately, and having a thin piece of platinum foil, P, inserted by the blowpipe in each glass tube. The foil hangs down the full length of the interior of the glass. Each pair of tubes is contained in a little glass tumbler containing some dilute sulphuric acid, and the hydrogen tube, H, of one pair, is connected with the oxygen tube, O, of the next. w w. The terminal wires of the series.

the mixed gases thrown upon a ball of lime, it may be stated that for many years the dissolving view lanterns and other optical effects have been produced with the assistance of this light; and more lately Major Fitzmaurice has condensed the mixed gases in the old-fashioned oil gas receivers, and projected them on a ball of lime; and it was this light thrown from many similar arrangements that illuminated the British men-of-war when Napoleon III. left her Majesty's yacht at night in the docks at Cherbourg.

Mr. Sykes Ward, of Leeds, has also proposed a most simple and excellent application of the oxy-hydrogen light for illumination *under* the

Fig. 127. Cherbourg.

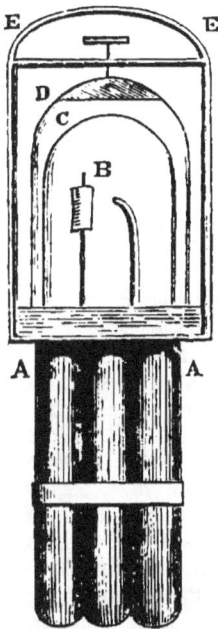

Fig. 128. A A. Tube reservoir to hold the mixed gases. B. The jet and lime ball. D. The first glass shade, held down by a cap and screw. C. The second glass shade. E E. The handle by which it is lowered into the water.

surface of water, and for the convenience of divers, who are frequently obliged to cease their operations in consequence of the want of light. Mr. Ward's submarine lamp consists of a series of very strong copper tubes, which are filled with the mixed gases by means of a force-pump; and in order to prevent the lamp being extinguished, it burns under *double* glass shades, which are desirable in order to prevent the glass immediately next to the light cracking by contact with the cold water.

The author tried this lamp at Ryde, and although the coast-guards objected to the production of a brilliant light at night, which they stated might be mistaken for a signal and would cause some confusion amongst the war vessels in the immediate neighbourhood, enough experiments were made, to show that the Ward lamp would burn for a considerable time under water, and could be kept charged with the gas by means of a process that was easily workable in the boat. The gases were taken out mixed in gas bags, and pumped into the reservoir when required. With a much larger reservoir greater results could be obtained; and if nautilus diving bells are to be used in modern warfare, they will require a powerful light to show them their prey, so that they may attach the explosives which are to blow great holes in the men-of-war.

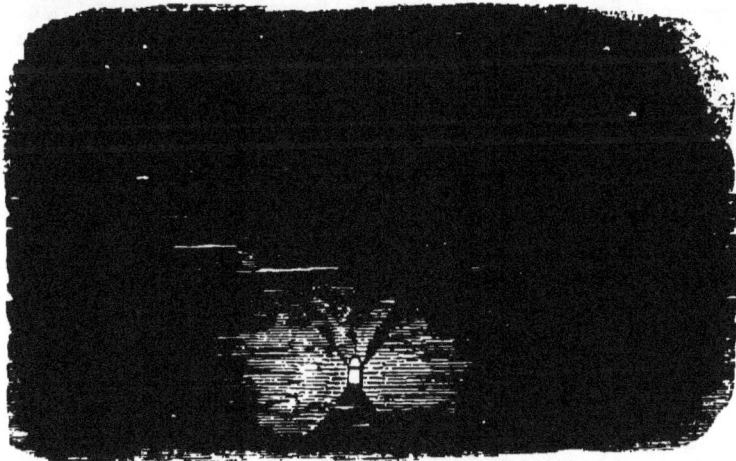

Fig. 129. Submarine lamp.

CHAPTER XI.

CHLORINE, IODINE, BROMINE, FLUORINE.

The four Halogens, or Producers of Substances like Sea Salt.

Chlorine (χλωρος, green). Symbol, Cl. Combining proportion, 35·5. Specific gravity, 2·44. Scheele termed it dephlogisticated muriatic acid; Lavoisier, oxymuriatic acid; Davy, chlorine.

The consideration of the nature of this important element introduces to our notice one of the most original chemists of the eighteenth century—viz., the illustrious Scheele, who was born at Stralsund, in 1742, and in spite of every obstacle, fighting his "battle of life" with sickness and sorrow, he succeeded in making some of the most valuable discoveries in science, and amongst them that of chlorine gas. It was in the examination of a mineral solid—viz., of manganese—that Scheele made the acquaintance of a new gaseous element; and in a highly original dissertation on manganese, in 1774, he describes the mode of procuring what he termed *dephlogisticated muriatic acid*—a name which is certainly to be regretted, from its absurd length, but a title which was strictly in accordance with the then established theory of phlogiston; and if the latter is considered synonymous with hydrogen, quite in accordance with our present views of the nature of this element. Scheele discovered the leading characteristics of chlorine, and especially its power of bleaching, which is alone sufficient to place this gas in a high commercial position, when it is considered that all our linen used formerly to be sent to Holland, where they had acquired great dexterity in the ancient mode of bleaching—viz., by exposure of the fabric to atmospheric air or the action of the damps or dews, assisted greatly by the agency of light. Some idea may be formed of the present value of chlorine, when it is stated that the linen goods were retained by the Dutch bleachers for nine months; and if the spring and summer happened to be favourable, the operation was well conducted; on the other hand, if cold and wet, the goods might be more or less injured by continual exposure to unfavourable atmospheric changes. At the present time, as much bleaching can be done in nine weeks as might formerly have been conducted in the same number of months; and the whole of the process of chlorine bleaching is carried on independent of external atmospheric caprices, whilst the money paid for the process no longer passes to Holland, but remains in the hands of our own diligent bleachers and manufacturers.

First Experiment.

As Scheele first indicated, chlorine is obtained by the action of the black oxide of manganese, on "the Spirit of Salt," or hydrochloric acid; and the most elementary and instructive experiment showing its preparation can be made in the following manner:—

K

Place in a clear Florence oil-flask, to which a cork and bent tube have
been first fitted, some strong fuming hydrochloric acid. Arrange the
flask on a ring-stand, and then pass the bent tube either to a Wolfe's
bottle containing some pumice stone moistened with oil of vitriol, or to
a glass tube containing
either pumice or as-
bestos wetted with the
same acid. . Another
glass tube, bent at
right angles, passes
away from the Wolfe's
bottle into a receiving
bottle. (Fig. 130). On
the application of heat,
the hydrochloric gas is
driven off. from its so-
lution in water, and
any aqueous vapour
carried up is retained
by the asbestos or pu-
mice stone wetted
with oil of vitriol; the
application of the lat-
ter is called *drying the
gas*—i.e., depriving it
of all moisture; some-

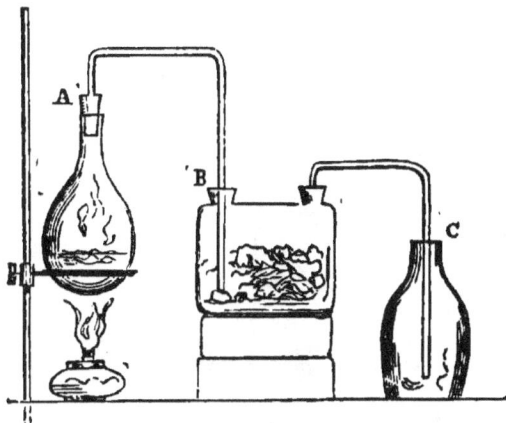

Fig. 130. A. Flask containing the fuming hydrochloric
acid, which is gently boiled by the heat of the spirit lamp.
B. Tube passing to the Wolfe's bottle, containing pumice-
stone or asbestos moistened with sulphuric acid. C.
Second tube passing into a dry empty bottle, which receives
the hydrochloric acid gas.

times the salt called chloride of calcium is used for the same purpose, and
it must be understood by the juvenile chemist that gases are not dried
like towels, by exposure to heat, or *by putting them in bladders before the
fire*, as we once heard was actually recommended, but by causing the gas
charged with invisible steam to pass over some substance having a great
affinity for water. The dry hydrochloric gas falls into the bottle, and displaces the air, being about one-fourth heavier than the latter, and gradually overflowing from the mouth of the vessel, produces a white smoke,
which is found to be acid by litmus paper, but has no power to bleach,
and is not green; it is, in fact, a combination of one combining proportion of chlorine with one of hydrogen, and to detach the latter, and
set the chlorine free, it is necessary to convey the hydrochloric gas to
some body which has an affinity for hydrogen. Such a substance is
provided in the use of the black oxide of manganese, which is placed
either in a small flask or in a tube provided with two bulbs, and when
heated with the lamp it separates the hydrogen from the hydrochloric
gas, and forms water, which partly condenses in the second bulb. And
now the gas that escapes is no longer acid and fuming with a white
smoke on contact with the air; but is green, has a strong odour,
bleaches, and is so powerful in its action on all living tissues, that it
must be carefully avoided and not inhaled; if a small quantity is accidentally inhaled, it produces a violent fit of coughing, which lasts a

considerable time, and is only abated by inhaling the diluted vapour of ammonia, or ether, or alcohol, and swallowing milk and other softening drinks. (Fig. 131).

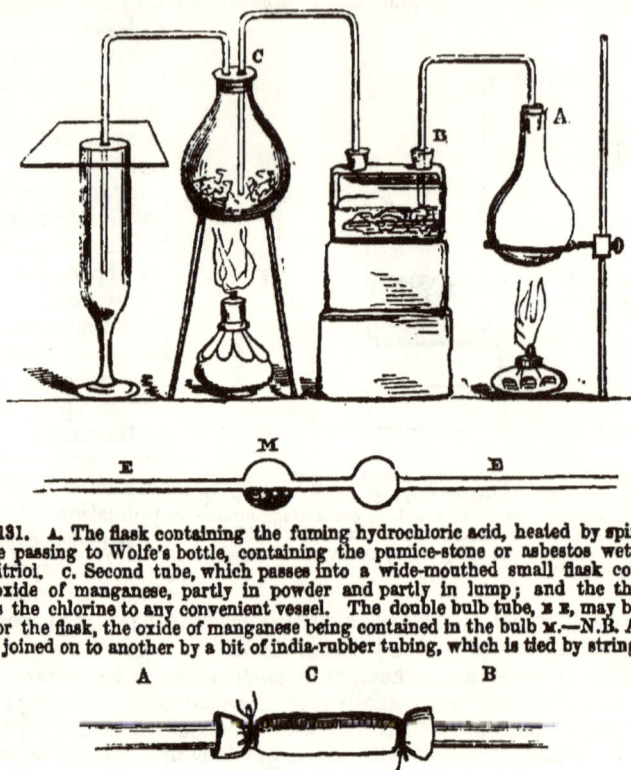

Fig. 131. A. The flask containing the fuming hydrochloric acid, heated by spirit lamp. B. Tube passing to Wolfe's bottle, containing the pumice-stone or asbestos wetted with oil of vitriol. C. Second tube, which passes into a wide-mouthed small flask containing black oxide of manganese, partly in powder and partly in lump; and the third tube conveys the chlorine to any convenient vessel. The double bulb tube, E E, may be substituted for the flask, the oxide of manganese being contained in the bulb M.—N.B. Any tube may be joined on to another by a bit of india-rubber tubing, which is tied by string.

Tube A is joined to tube B by the caoutchouc pipe c, tied with packthread.

Second Experiment.

The mode of preparing chlorine, as already given, though very instructive, is troublesome to perform; a more simple process may therefore be described:—

Pour some strong hydrochloric acid upon powdered black oxide of manganese contained in a Florence oil-flask, taking care that the whole of the black powder is wetted with the acid so that none of it clings to the bottom of the flask in the dry state to cause the glass to crack on the application of heat. A cork and bent glass tube is now attached, and conveyed to the pneumatic trough; on the application of heat to the mixture in the flask the chlorine is evolved, and may be collected in stoppered bottles, the first portion that escapes, although it contains atmospheric air, should be carefully collected in order to prevent any

K 2

132   BOY'S PLAYBOOK OF SCIENCE.

accident from inhaling the gas, and it will do very well to illustrate the bleaching power of the gas, and therefore need not be wasted. The above process may be described in symbols, all of which are easily deciphered by reference to the table of elements, page 86.

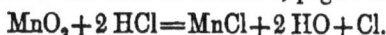

$$Mn O_2 + 2\,HCl = Mn Cl + 2\,HO + Cl.$$

Third Experiment.

Another and still more expeditious mode of preparing a little chlorine, is by placing a small beaker glass, containing half an ounce of chlorinated lime, usually termed chloride of lime or bleaching powder, carefully at the bottom of a deep and large beaker glass, and then, by means of a tube and funnel, conveying to the chloride of lime some dilute oil of vitriol, composed of half acid and half water; effervescence immediately occurs from the escape of chlorine gas, and as it is produced it falls over the sides of the small beaker glass into the large one, when it may be distinguished by its green colour. If a little gas be dipped out with a very small beaker glass arranged as a bucket, and poured into a cylindrical glass containing some dilute solution of indigo, and shaken therewith, the colour disappears almost instantaneously; and if a piece of Dutch metal is thrown into the beaker glass it will take fire if enough chlorine has been generated, or some very finely-powdered antimony will demonstrate the same result. Thus, with a few beaker glasses, some chloride of lime, sulphuric acid, a solution of indigo, and a little Dutch metal, the chief properties of chlorine may be displayed. (Fig. 132.)

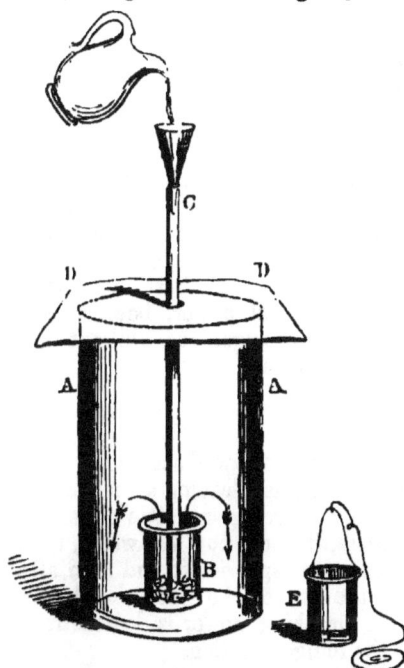

Fig. 132. A A. The large beaker glass. B. The small one, containing the chloride of lime. C. The tube and funnel down which the dilute sulphuric acid is poured. D D. Sheet of paper over top of large glass, with hole in centre to admit the tube. E. The little beaker used as a bucket.

Fourth Experiment.

Into a little platinum spoon place a small pellet of the metal sodium, and after heating it in the flame of a spirit lamp, introduce the metal into a bottle of chlorine, when a most intense and brilliant combustion occurs, throwing out a vivid yellow light, and the heat is frequently so great that the bottle is cracked. After the combustion, and when the bottle is cool, it is usually lined with a white powder, which will be found to taste exactly the same as salt, and, in fact, is that substance, produced by the combination of chlorine, a virulent poison, with the metal sodium, which takes fire on contact with a small quantity of water; and hence the use of salt for the preparation of chlorine gas when it is required on the large scale.

	Parts.
Common salt	4
Black oxide of manganese	1
Sulphuric acid	2
Water	2

Fifth Experiment.

Some Dutch metal, or powdered antimony, or a bit of phosphorus, immediately takes fire when introduced into a bottle containing chlorine gas, forming a series of compounds termed chlorides, and demonstrating by the evolution of heat and light, the energetic character of chlorine, and that oxygen is not the only supporter of combustion; chlorine gas has even, in some cases, greater chemical power, because some time elapses before phosphorus will ignite in oxygen gas, whilst it takes fire directly when placed in a bottle of chlorine.

Sixth Experiment.

The weight and bleaching power of chlorine are well shown by placing a solution of indigo in a tall cylindrical glass, leaving a space at the top of about five inches in depth. By inverting a bottle of chlorine over the mouth of the cylindrical glass, it pours out like water, being about two and a half times heavier than atmospheric air, and then, after placing a ground glass plate over the top of the glass, the chlorine is recognised by its colour, whilst the bleaching power is demonstrated immediately the gas is shaken with the indigo solution.

Seventh Experiment.

As a good contrast to the last experiment, another cylindrical jar of the same size may be provided, containing a solution of iodide of potassium with some starch, obtained by boiling a teaspoonful of arrowroot with some water; any chlorine left in the bottle (sixth experiment) may be inverted into the top of this glass and shaken, when it turns a beautiful purple blue in consequence of the liberation of iodine by the chlorine, whose greater affinity for the base produces this result. The colour is caused by the union of the iodine and the starch, which form together a beautiful purple compound, and thus the apparent anomaly of destroying and producing colour with the same agent is explained.

Eighth Experiment.

Dry chlorine does not bleach, and this fact is easily proved by taking a perfectly dry bottle, and putting into it two or three ounces of fused chloride of calcium broken in small lumps, then if a bottle full of chlorine is inverted over the one containing the chloride of calcium, taking the precaution to arrange a few folds of blotting paper with a hole in the centre on the top of the latter to catch any water that may run out of the chlorine bottle at the moment it is inverted, the gas will be dried by contact with the chloride of calcium, and if a piece of paper, with the word chlorine written on it with indigo, and previously made hot and dry, is placed in the chlorine, no change occurs, but directly the paper is removed, dipped in water, and placed in a bottle of damp chlorine, the colour immediately disappears. (Fig. 133.)

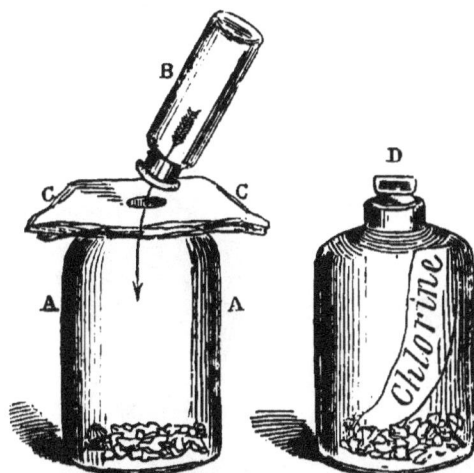

Fig. 133. A A. Dry bottle, containing chloride of calcium. B. Bottle of chlorine. The arrow indicates the gas.   C C. The blotting-paper, to catch any water from the bottle, B. D. The bottle closed, and containing the paper.

This experiment shows that chlorine is only the means to the end, and that it decomposes water, setting free oxygen, which is supposed to exert a high bleaching power it its *nascent* state, a condition which many gases are imagined to assume just before they take the gaseous state, a sort of intermediate link between the solid or fluid and the gaseous condition of matter. The nascent state may possibly be that of ozone, to which we have already alluded as a powerful bleaching agent.

Ninth Experiment.

A piece of paper dipped in oil of turpentine emits a dense black smoke, and frequently a flash of fire is perceptible, directly it is plunged into a bottle containing chlorine gas; here the gas combines only with the hydrogen of the turpentine, and the carbon is deposited as soot.

Tenth Experiment.

If a lighted taper is plunged into a bottle of chlorine it continues to burn, emitting an enormous quantity of smoke, for the reason already explained, and demonstrating the perfection of the atmosphere in which we live and breath, and showing that had oxygen gas possessed the same properties as chlorine, the combustion of compounds of hydrogen and carbon would have been impossible, in consequence of the enormous quantity of soot which would have been produced, so that some other element that would freely enter into combination with it must have been provided to produce both artificial light and heat. Chlorine is a gas which cannot be inhaled, and ozone presents the same features, as a mouse confined for a short time with an excess of ozone soon died; but ozone is the extraordinary condition of oxygen; the element in the ordinary state is harmless, and is the one which enters so largely into the composition of the air we breathe.

Eleventh Experiment.

When one volume of olefiant gas (pre- pared by boiling one measure of alcohol and three of sulphuric acid) is mixed with two volumes of chlorine, and the two gases agitated together in a long glass ves- sel for a few seconds, with a glass plate over the top, which should have a welt ground perfectly flat, they unite on the ap- plication of flame, with the production of a great cloud of black smoke, arising from the deposited carbon, whilst a sort of roaring noise is heard during the time that the flame passes from the top to the foot of the glass. (Fig. 134.)

Twelfth Experiment.

Formerly Bandannah handkerchiefs were in the highest estimation, and no gentleman's toilet was thought complete without one. The pattern was of the simplest kind, consisting only of white spots on a red or other coloured ground. These spots were produced in a very in- genious manner by Messrs. Monteith, of Glasgow, by pressing together many layers of silk with leaden plates perforated with holes; a solution of chlorine was then poured upon the upper plate, and pressure being applied it penetrated the whole mass in the direction of the holes,

Fig. 134. Remarkable deposition of carbon during the combustion of one volume of olefiant gas with two of chlorine.

bleaching out the colour in its passage. This important commercial result may be imitated on the small scale by placing a piece of calico dyed with Turkey red between two thick pieces of board, each of which is perforated with a hole two inches in diameter, and corresponding accurately when one is placed upon the other. The pieces of board may be squeezed together in any convenient way, either by weights, strong vulcanized india-rubber bands or screws, and when a strong solution of chlorine gas or of chloride of lime is poured into the hole and percolates through the cloth, the colour is removed, and the part is bleached almost instantaneously by first wetting the calico with a little weak acid, and then pouring on the solution of chloride of lime. On removing and washing the folded red calico it is found to be bleached in all the places exposed to the solution, and is now covered with white spots. (Fig. 135.)

Fig. 135. A. Circular hole in the upper piece of wood, a similar one being perforated in the lower one. B B. The strong india-rubber bands. The bleaching solution is poured into A.

IODINE.

Iodine (Ἰώδης, violet coloured). Symbol, I; combining proportion, 127·1; specific gravity, 4·948. Specific gravity of iodine vapour, 8·716.

In the previous chapter, devoted to the element chlorine, little or nothing has been said of that inexhaustible storehouse of chlorine, iodine, and bromine—viz., the boundless ocean. Some one has remarked that, as it is possible the air may contain a little of everything capable of assuming the gaseous form, so the ocean may hold in a state of solution a modicum of every soluble substance, in proof of which we have lately read of some very important experiments resulting in the separation of the metal silver from sea water, not certainly in any profitable quantity, but quite enough to prove its presence in the ocean.

No elaborate research is necessary to ascertain the presence of chlorine, when it is remembered that Schafhäutl has calculated, that all the oceans on the globe contain three millions fifty-one thousand three hundred and forty-two cubic geographical miles of salt, or about five times more than the mass of the Alps.

Now, salt contains about 60 per cent. of chlorine gas, and therefore the bleachers can never stand still for want of it; but iodine is not so plentiful, and was discovered by M. Courtois, of Paris, in *kelp*, a substance from which he prepared carbonate of soda, or washing soda; but as this is now more cheaply prepared from common salt, the kelp is at present required only for the iodine salts it contains, as also for the chloride of potassium. Kelp is obtained by burning dried sea-weeds in a shallow pit ; the ashes accumulate and melt together, and this fused mass broken into lumps forms kelp. The ocean bed no doubt has its fertile and barren plains and mountains, and amongst the so-called "oceanic meadows" are to be mentioned the two immense groups and bands of sea-weed called the Sargasso Sea, which occupy altogether a space exceeding six or seven times the area of Germany.

The iodine is contained in the largest proportion in the deep sea plants, such as the long elastic stems of the fucus palmatus, &c. The kelp is lixiviated with water, and after separating all the crystallizable salts, there remains behind a dense oily-looking fluid, called "iodine ley," to which sulphuric acid is added, and after standing a day or two the acid "ley" is placed in a large leaden retort, and heated gently with black oxide of manganese. The chlorine being produced very slowly, liberates the iodine, as already demonstrated in experiment seven, p. 133, and it is collected in glass receivers.

Iodine, when quite pure and well crystallized, has a most beautiful metallic lustre, and presents a bluish-black colour, affording an odour which reminds one at once of the " sea smell."

First Experiment.

A few grains of iodine placed in a flask may be sublimed at a very gentle heat, and afford a magnificent violet vapour, which can be poured out of the flask into a warm bottle. If the bottle is cold the iodine condenses in minute and brilliant crystals. (Fig. 136.)

Second Experiment.

Upon a thin slice of phosphorus place a few small particles of iodine ; the heat produced by the combination of the two elements soon causes the phosphorus to take fire.

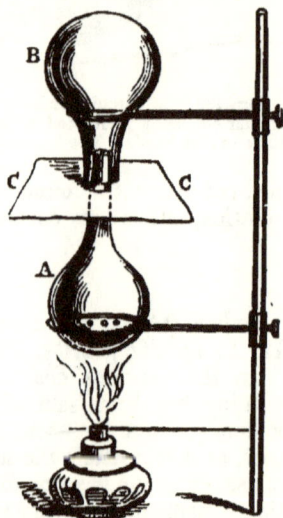

Fig. 136. A. Flask containing iodine heated by spirit lamp. B. Cold flask above to receive the vapour. c c. Sheet of cardboard to cut off the heat from the spirit lamp.

Third Experiment.

Heat a brick, and then throw upon it a few grains of iodine; by holding a sheet of white paper behind, the splendid violet colour of the vapour is seen to great advantage. It was by the discovery of iodine in the ashes of sponge—which had long been used as a remedy for goitre, a remarkable glandular swelling—that this element began to be used for medical purposes, and the important salt called iodide of potassium is now used in large quantities, not only in medicine, but likewise for that most fascinating art, which has made its way steadily, and is now practised so extensively, under the name of *photography.*

THE ART OF PHOTOGRAPHY.

It was the great George Stephenson who asked the late Dean Buckland the posing question, " Can you tell me what is the power that is driving that train?" alluding to a train which happened to be passing at the moment. The learned dean answered, "I suppose it is one of your big engines." "But what drives the engine?" "Oh, very likely a canny Newcastle driver." "What do you say to the light of the sun?" "How can that be?" asked Buckland. "It is nothing else," said Stephenson. "It is light bottled up in the earth for tens of thousands of years; light, absorbed by plants and vegetables, being necessary for the condensation of carbon during the process of their growth, if it be not carbon in another form; and now, after being buried in the earth for long ages in fields of coal, that latent light is again brought forth and liberated, made to work—as in that locomotive—for great human purposes."

Such was the opinion of the most original and practical man that ever reasoned on philosophy; and could he have lived to realize the thorough adaptation and business use of light in the art of photography, he would have said, man is only imitating nature, and in producing photographs he must employ the same agent which in ages past assisted to produce the coal.

In another part of this elementary work we shall have to consider the nature of light; here, however, the chemical part only of the process of photography will be discussed.

Many years ago (in the year 1777) Jenny Lind's most learned countryman, Scheele, discovered that a substance termed chloride of silver, obtained by precipitating a solution of chloride of silver with one of salt, blackened much sooner in the violet rays than in any other part of the spectrum. He says, "Fix a glass prism at the window, and let the refracted sunbeams fall on the floor; in this coloured light put a paper strewed with luna cornua (horn silver or chloride of silver), and you will observe that this horn silver grows sooner black in the violet ray than in any of the other rays."

In 1779, Priestley directed especial attention to the action of light on plants; and the famous Saussure, following up these and other experiments, determined that the carbonic acid of plants was more generally decomposed into carbon and oxygen in the blue rays of the spectrum; these facts probably suggested the bold theory of Stephenson already alluded to. Passing by the intermediate steps of photography, we come to the second year of the present century, and find in the Journal of the Royal Institution a paper by Wedgwood, entitled "An Account of a Method of Copying Paintings upon Glass, and of making Profiles, by the Agency of Light upon Nitrate of Silver; with observations, by H. Davy." Such a paper would lead the reader to suppose that very little remained to be effected, and that mere details would quickly establish the art; but in this case the experimentalists were doomed to disappointment, as, after producing their photographs, they could not make them permanent; they had not yet discovered the means of *fixing* the pictures. Nearly fourteen years elapsed, when the subject was again taken up by Niépcè, of Chalons, with little success, so far as the fixing was concerned; and twenty-seven years had passed away since the experiments of Wedgwood and Davy, when, in 1829, Niépcè and Daguerre executed a deed of co-partnership for mutually investigating the matter. These names would suggest a rapid progress; but, strange to relate, ten years again rolled away, the father Niépcè had in the meantime died, and a new contract was made between the son and M. Daguerre, when, in January, 1839, the famous discovery was made known to the world, and in July of the same year the French Government granted a pension for life of six thousand francs to Daguerre, and four thousand to the son of Niépcè, who had so worthily continued the experiments commenced by his father. The triumph of the industrious French experimentalists was not, however, to be unique; across the Channel another patient and laborious philosopher had completed on paper precisely the same kind of results as those obtained by Daguerre on silver plates. Mr. Fox Talbot, in England, had immortalized himself by a discovery which was at once called the Talbotype, and for which a patent was secured in 1841. Having thus hastily sketched a brief history of the art, we may now proceed to the details of the process.

First Experiment.

A photogenic drawing, so called, but now termed *a positive copy*, is prepared by placing some carefully selected paper, which is free from spots or inequalities (good paper is now made by several English manufacturers, although some kinds of French paper, such as Cansan's, are in high repute), in a square white hard porcelain dish containing a solution of common salt in distilled water, 109 grains of salt to the pint. The paper is steeped in this solution for ten minutes, and then taken out and pressed in a *clean* wooden press, or it should be dabbed dry on a *clean* flat surface with a *clean* piece of white calico, which may be kept specially for this duty and not used for anything else, and it is well that all would-be photographers should understand that neatness and cleanliness are perfectly indispensable in conducting these processes. If a design were required for the armorial bearings of the art of photography, it might certainly be most fanciful, but the motto must be *cleanliness* and *neatness*, and in preparing paper it should not be unnecessarily handled, but lifted by the corners only. The object of dabbing the paper is to prevent the salt accumulating in large quantities in one part of the paper and the reverse in another, and to distribute the salt equally through the whole. The paper being now dried, is called salted paper, and is rendered sensitive when required by laying it down on a solution of ammonio-nitrate of silver, prepared by adding ammonia to a solution containing sixty grains of nitrate of silver to the ounce of distilled water, until the whole of the oxide of silver is re-dissolved, except a very small portion. A few drops of nitric acid are also recommended to be added, and after allowing the solution to stand, it may be poured off quite clear, and is ready for use either in the bath, or if economy must be rigidly adhered to, the salted paper may be laid flat on a board, and held in its place with four pins at the corners, and then just enough to wet the surface of the paper may be run along the side of a glass spreader, and the liquid gently drawn over the surface of the

Fig. 137. A. The glass spreader with cork handle. B. The silver solution clinging to rod and paper by capillary attraction. c c c c. Four pins holding down the paper on a board.—N.B. The spreader is made of glass rod three-eighths thick.

salted paper, which is allowed to dry on a flat surface for a few minutes, and afterwards hung up by one corner to a piece of tape stretched across the room, until quite dry, and then placed in a blotting-book fitting into a case which completely excludes the light. Copying-paper should be made at night, as the day is then free for all photographic operations requiring an abundance of light. It will not keep long, and should be used the next day.

A piece of lace, a skeleton leaf, a sharp engraving on thin paper, and above all things, a negative photograph on glass or paper, is easily copied by placing the prepared paper with the prepared side (carefully protected from the light) upwards on any flat surface, such as plate glass; upon this is arranged the bit of lace or the negative photograph with the face or picture downwards, another bit of plate glass is then placed over it, and weights arranged at the corners; after exposure to the sun's rays for thirty minutes, more or less (according to the dullness or bright aspect of the day), the picture is brought into a dark room and examined by the light of a candle or by the light from a window covered with yellow calico, and after placing a paper weight on one corner of the lace, or negative picture, or copying paper, it may be carefully lifted in one part, and if the copy is sufficiently dark, is ready for fixing, but if it is faint the lifted corner is carefully replaced, the upper glass is laid on, and the picture again exposed to the light. Should the position of the lace or negative be changed during the examination, re-exposure is useless, and would only produce a double and confused picture, as it would be impossible to lay the lace or the negative exactly in the same place again on the copying paper.

The manipulations just described are much facilitated by using a copying-frame or press, which consists of a square woodenframe with a thick plate-glass window; upon this are placed the negative picture and the copying paper, and the two are brought in close contact by means of a board at the back pressed by a hand-screw. (Fig. 138.) After the photogenic drawing or positive copy is taken, it is fixed by being placed in a solution of hyposulphite of soda, consisting of one fluid ounce of saturated solution to eight of water. The saturated

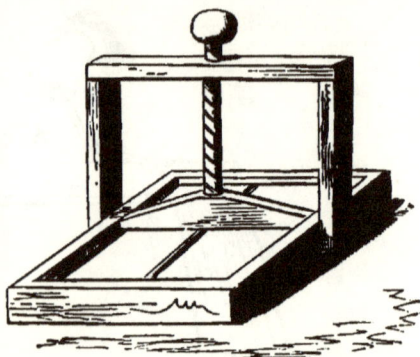

Fig. 138. The back of the copying-frame, showing the hand-screw and pressure-board. The plate glass inside is set in the base of the frame, and is of course the part exposed to the light.

solution of hyposulphite of soda is conveniently kept in a large bottle for use, and in order to improve the colour a very little chloride of gold is added to the fixing solution, the picture must now be thoroughly washed, dried, and pressed.

Second Experiment.

Another mode of preparing the copying paper, called albumen paper, is to take the whites of four eggs, and four ounces of distilled water containing one hundred and sixty grains of chloride of ammonium; these are beaten up with a fork or a bundle of feathers, and as the froth is produced it is skimmed off by a silver spoon into another basin, or a beaker glass, and being allowed to settle for twelve hours it is strained through fine muslin, and is ready for use. The best paper is floated on the surface of this liquid for three minutes, taken out, and dried at once on a hot plate.

In floating paper one corner is first laid down, and care taken not to enclose any air bubbles, which would prevent the fluid wetting the paper, whilst the remainder of the paper is slowly laid upon the surface of the fluid.

The albumen paper is excited by laying it for five minutes on a solution of nitrate of silver, seventy-two grains to the ounce of water, and when dry it will keep for three days. This copying paper is used in the same manner as the last, and fresh eggs only must be used in its preparation, because stale ones soon cause the copy to change and blacken all over from the liberation of sulphur, which unites with the silver. The colour of the copy is sometimes improved by a solution of hot potash, and by dipping the well-washed picture, after the use of the hyposulphite of soda, in a very dilute solution of hydrosulphuret of ammonia.

Third Experiment.

In the Daguerreotype process, a silver plate, after being thoroughly cleaned and polished, is exposed to the vapour of iodine, and is thus rendered so sensitive that it may be at once exposed in the camera. In the Talbotype process, the same principle is apparent, and paper is prepared by first covering its surface with iodide of silver, which is afterwards rendered sensitive to the action of light by means of an excess of nitrate of silver, as follows :—

One side of a sheet of selected Cansan's paper is first covered (by means of a spreader) with a solution of nitrate of silver (thirty grains to the ounce of water), hung up in a dark room and dried; it is then immersed in a solution of iodide of potassium of five hundred grains to a pint of distilled water, for five or ten minutes, and immediately changes to a yellow colour in consequence of the pre-cipitation of the yellow iodide of silver; it is then well washed with plenty of water, and being dried, may be kept for any length of time, and is called "iodized paper." Light has no action whatever upon it. To render the paper sensitive, three solutions are prepared in separate bottles, and marked 1, 2, 3.

No. 1, contains a solution of nitrate of silver, fifty grains to the ounce of water.

No. 2, glacial acetic acid.

No. 3, a saturated solution of gallic acid.

With respect to No. 3, Mr. William Crookes has shown, that when a saturated solution of gallic acid is required in large quantities, that it is better to dissolve at once two ounces of gallic acid in six ounces of alcohol (60° over proof); to hasten solution, the flask may be con-veniently heated by immersion in hot water; when cold it should be filtered, mixed with half a drachm of glacial acetic acid, and preserved in a stoppered bottle for use; so prepared it will keep unaltered for a considerable length of time. The gallic acid is not precipitated from this solution by the addition of water; consequently, if in any case desirable, the development of a picture may be effected with a much stronger bath than the one usually employed. To obtain a solution of about the same strength as a saturated aqueous solution, such as No. 3, half a drachm of the alcoholic solution is mixed with two ounces of water; but for my particular purpose, says Mr. Crookes, referring to the wax-paper process, " I prefer a weaker bath, which is prepared by mixing half a drachm with ten ounces of water." In either case it will be found necessary to add solution of nitrate of silver in small quantities, as the developing picture seems to require it.

Returning again to the solutions marked 1, 2, 3, the numbers will assist the memory in mixing the proportions of each. If the paper is required to be used *at once*, a drachm of each may be mixed together and spread over the iodized paper (of course, in a dark room), which is then transferred to a clean blotting-book of white bibulous paper, and being placed in the paper-holder may be taken to the camera and exposed at once. If the paper is not required to be used immediately, the solutions are mixed in the proportions of the numbers—viz., one of No. 1, two of No. 2, three of No. 3; and in making the mixture, it is advisable to keep a measure specially for No. 3, the gallic acid, or else the measure, if used for the three solutions, will have to be washed out every time, which is very troublesome, particularly where water is not plentiful.

If the excited paper is required to be kept some hours before use, No. 3 must be added in still larger proportion, as much as ten or even twenty measures of No. 3 to two of No. 2, and one of No. 1, being used, and even this large dilution is frequently insufficient to prevent the paper spoiling in hot weather; therefore if the temperature is high, too much reliance must not be placed on this paper, as it is peculiarly disappointing, after walking some miles to romantic and beautiful scenery, to find, when developing the pictures in the evening, that the paper used was all spoilt before exposure; and it will be seen presently that when the excited paper is to be carried about for use, it is better to adopt the wax-paper process.

After the excited iodized paper is exposed in the camera—and the time of exposure cannot be taught, as that speciality is only acquired by experience, and may vary from five to thirty minutes, or even more—the invisible picture is developed and rendered visible, not by exposure to the vapour of mercury, as in Daguerre's process with silver plates, but by a mixture of one of No. 1 with four of No. 3. The development is carefully watched by looking through the negative placed before a lighted candle, and the time of development may vary from ten to thirty minutes, and all the time the picture must be kept wet with the solution, so that it is better perhaps to make a bath of the solution and lay the picture on its surface than to pour the liquid over the picture. After the development is matured, the picture is now washed in clean water, and fixed temporarily, if required, by immersion in a bath containing 200 grains of bromide of potassium in one pint of water, or permanently by the hyposulphite of soda, made by mixing one part of a saturated solution with five or ten of water, or one ounce of the salt to six or twelve of water; but, as before mentioned, it is better to keep a Winchester quart full of a saturated solution of hyposulphite of soda, and then it is always ready for use instead of employing the weights and scales, and continually weighing out portions of the salt. The picture after fixing is thoroughly washed with water, and being dried is now placed between the folds of a wax book—*i.e.*, some leaves of blotting-paper are kept saturated with white wax, and when a picture is placed between them, and a hot iron passed over the outside sheet, the wax enters the pores of the paper, and after removing any excess of wax by passing the picture through a book of bibulous paper, over which the hot flat iron is passed, the negative picture at last is ready for use, and any number of positive copies may be taken from it, as already described in the first experiment, page 139.

This mode of manipulation is called the Talbotype, and before dismissing the subject another process of iodizing the paper may be explained.

To a solution of nitrate of silver of twenty, thirty, or fifty grains to the ounce of water, a sufficient number of the crystals of iodide of potassium is added, first to produce the yellow iodide of silver, and then to dissolve it, so that the yellow precipitate appears with a small quantity, and disappears with an excess of the iodide. If this solution is spread over sheets of paper, and these latter then placed in a bath of water, the iodide of silver is precipitated on the surface, and after plenty of washing to remove the excess of iodide of potassium, the paper may be dried, and will keep for any length of time without change. This paper may be excited, exposed, developed, fixed, and waxed, as already explained.

Fourth Experiment. The Wax-paper Process.

This mode of taking negative photographs begins where the talbotype ends—viz., by *first* waxing the paper perfectly and evenly, as already explained, Cansan's negative paper being preferred. The wax paper is now well soaked in a bath, made by dissolving one hundred grains of iodide of potassium, six grains of cyanide of potassium, four grains of fluoride of potassium, ten grains of bromide of potassium, ten grains of chloride of sodium, in one pint of fresh whey, with the addition of a little alcohol and a few grains of iodine. When soaked in this solution for about one hour, the paper is taken out and hung up to dry.

N.B. With respect to iodizing the wax paper, it is almost better to obtain it ready prepared, and then every sheet may be relied on. Mr. Melhuish, of Blackheath and Holborn, supplies it in any quantity, and his paper never fails; the operator has then only to perform the sensitizing and developing processes. To render the iodized paper sensitive it is immersed for about six minutes in a bath containing a solution of nitrate of silver (thirty-five grains to the ounce of water, with forty drops of glacial acetic acid); the paper is now removed, and washed in two trays of common clear rain-water or distilled water, and is then dried off between folds of blotting-paper.

This process may be performed on the previous evening by the light of a candle, or by day in a room lit by one window covered with four thicknesses of yellow calico, and after the paper is dry it will keep for three weeks or a month, and may be exposed in a camera with a three-inch lens of eighteen-inch focus, with the inch diaphragm, on a bright day from five to fifteen minutes; in bad weather the exposure must be longer. The picture may be carried home and rendered visible or developed by immersion in a bath containing a saturated solution of gallic acid, and as the developing continues, a few drops of the sensitizing solution of nitrate of silver and glacial acetic acid may be added. Finally, the picture is fixed by immersion for a quarter of an hour in a solution of hyposulphite of soda (four ounces of the crystal to one pint of water, or one part of the saturated solution to eight of water), and being well washed, is then dried, hung before the fire to melt the wax, and is now ready to print from.

Fifth Experiment. Albumen on Glass Process.

Albumen is the scientific name for the white of egg, of which four ounces by measure are mixed with one ounce and a half of distilled water, and after being whisked to a froth, are removed by a spoon into another basin or a beaker glass, and allowed to stand for several hours and then filtered. Mr. Crookes has recommended a very ingenious, simple, and useful filter. (Fig. 139.) He says: "This simple and inexpensive piece of apparatus, which any instrument maker or glass-blower can supply at a few hours' notice, will be found invaluable in almost every photographic process on glass. The sponge has this great advantage over all other kinds of filters, that thick gelatinous liquids— e.g., honey, albumen, gelatine, meta-gelatine, or the various preservative syrups —flow through it with the utmost readiness; whilst at the same time dust, air bubbles, or froth, and dried particles

Fig. 100. A B. Glass tube, bent as in picture. C. Piece of damp sponge squeezed into the head of the tube. Any liquid poured in at B will flow through the sponge until it has attained the same level in A.

floating in the liquid, are effectually kept back, and if fitted with stoppers, collodion might be filtered in it; or if the ends were fitted together with a bit of flexible pipe, the stoppers might be dispensed with altogether.

Having poured the albumen on a perfectly clean glass plate, taking care to have sufficient to run freely over the surface of the glass, the excess is then gently drained off and the plate turned so as to have the coated side downwards; it is then fixed in a sling made by taking a stout bit of string about three feet long, which is doubled and knotted at the fold, leaving the two ends free; two small triangles or stirrups of silver wire looped at one corner are now tied on to the ends of the string, and these form a support for the opposite edges of the glass plate to rest on; the two strings are knotted together at a

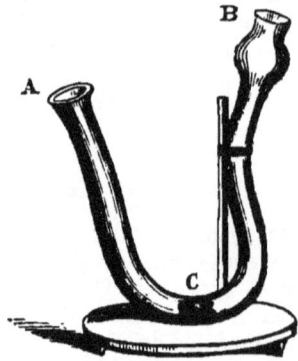

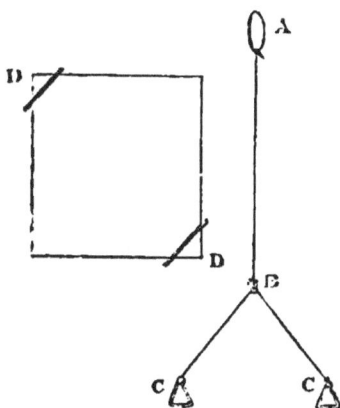

Fig. 140. A. Loop for finger. B. The knot which prevents the stirrups of silver wire, C C, slipping off the corners of the glass plate. D D. The opposite corners of the glass plate on which the stirrups are placed.

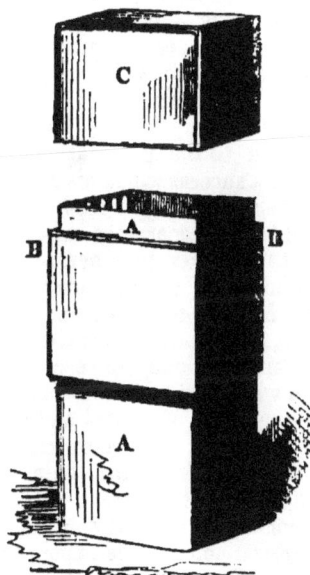

Fig. 141. A A. Tin box, with partitions to hold glass plates. B B. The outer jacket, between which and the box, A, the lid or cover, C, slides.

convenient distance from the stirrups to prevent the glass slipping out, and the plate is now rotated rapidly over a heated metallic surface, such as an iron box containing some burning charcoal or the *warming pan*, care being taken to avoid dust as much as possible, and to use only the whites of new-laid eggs. (Fig. 140.) The glass plate, covered with dry albumen, is now iodized to a straw colour by exposure over a box containing iodine, as in the Daguerreotype process, and is sensitized by immersion for three or four minutes in a bath containing a solution of nitrate of silver (twenty-five grains to an ounce of water); the plate is afterwards washed in distilled water and left to dry spontaneously, of course in a darkened room. The plates may then be placed ready for use in a very ingenious tin box devised by Mr. Crookes, which keeps them perfectly light-tight even in the sun, and at the same time is less bulky than the ordinary wooden ones. It is made of tin plate, the cover sliding tight over the top, and more than half way down the sides; light is further excluded by means of an outer jacket of tin, which is soldered to the box a little below the centre. The cover thus slides between the case and the jacket, and renders injury to the plates by the entrance of light an impossibility. (Fig. 141.)

The sensitive albumenized glass plate is exposed in the camera from fifteen to thirty minutes, and developed (much in the same way as the paper pictures) with one ounce of a saturated solution of gallic acid containing ten or fifteen drops of the sensitizing solution. The plate is usually placed on a levelling stand, and the solution poured on the glass plate; the development is slow, and may be quickened sometimes by the application of heat.

The picture is fixed by immersion for a short time in a bath containing one part of a saturated solution of hyposulphite of soda in eight of water. The pictures produced by this process are exquisitely defined, provided always the camera is well focussed, and to assist.this operation a magnifying glass may be employed. After removal from the hyposulphite of soda the plate is well washed with water, and being allowed to dry spontaneously, is now ready to print from.

Sixth Experiment. *The Collodion on Glass Process.*

The glass plates for this, as well as the albumen on glass process, should be cleaned by rubbing them over first with a mixture of Tripoli powder and ammonia, which is washed off under a tap, and the glass being drained is rubbed dry and polished with a clean calico duster kept exclusively for this purpose.

The iodized collodion is now poured on, and the excess returned to the bottle. Collodion can be made very easily, but if prepared without due precautions, it cannot be used afterwards, and reminds one of the old story of the enthusiastic son, who, when asking his father's permission to espouse the beloved, enumerated amongst her other accomplishments, the fact that she *could* make a pudding, and was answered by the bluff question, "But can you eat it afterwards?" So it is with collodion: a great deal of messing and loss of time is saved by purchasing it of the various makers, amongst whom may be specially noticed Mr. Richard Thomas, of 10, Pall Mall, who has devoted the whole of his attention to the preparation of this important photographic chemical, and with a success which his numerous patrons can well testify. The collodion is sold either mixed with the iodizing solution, or the two can be obtained separately, with directions on the bottles as to the quantities to be mixed together.

The plate covered with the iodized collodion is quickly transferred to a bath containing a solution prepared in the following manner:— Dissolve four ounces of nitrate of silver in eight ounces of water, and to this add twenty grains of iodide of potassium in one ounce of water; shake them together, and then pour the whole into fifty-six ounces of distilled water, and in half an hour add one ounce of alcohol and half an ounce of ether; agitate the whole and filter the next morning. The collodion plate is kept in this solution for a certain period, only learnt by experience, and should be occasionally lifted out to see if a uniform transparency is obtained; say that the immersion may be continued for five minutes, it is now ready for the camera, and may be exposed from about one to two minutes, or more if the light is deficient; the time of exposure is also a matter of *practice*, mere directions can be of no use in this stage of the process.

The picture is developed on a levelled stand, with a solution of three

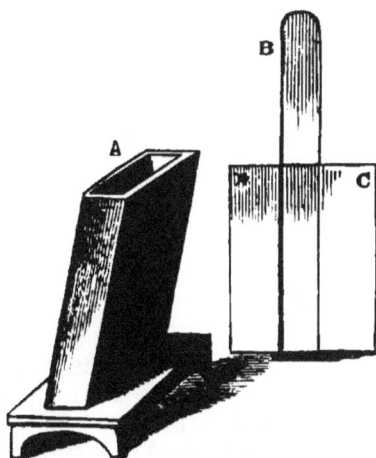

Fig. 142. A. Glass or gutta-percha bath to hold the sensitizing solution. B. Glass, with piece cemented on the end to hold the prepared glass plate, C, whilst dipped in the bath, A. The plate C has a cross in one corner to show prepared side.

grains of pyrogallic acid in three ounces of water, to which sixty drops of glacial acetic acid have been added. When fully developed the plate is washed with water and fixed with a solution of hyposulphite of soda, consisting of one part of the saturated solution to eight of water, again thoroughly but gently washed, so as not to endanger the separation of the film from the glass; it is allowed to dry spontaneously, and being coated with amber varnish (a solution of amber in chloroform) is now ready to print from. (Fig. 123.) It is, perhaps, hardly necessary to add, that the sensitizing and developing processes must be performed in a dark room.

Fig. 143. First effect of peripatetic photography on the rural population.

BROMINE.

Bromine (βρῶμος, a bad odour). Symbol, Br. Combining proportion, 80. Specific gravity, 2·966.

In a previous portion of this work, the connexion between chlorine, iodine, and bromine has been pointed out; and as we have to notice the colour of the element bromine, the chromatic union of the triad may be alluded to. These elements present very nearly all the colours of the spectrum:

Bromine . . . . red to orange.
Chlorine . . . . yellow to green.
Iodine . . . . blue, indigo, violet.

These three elements also furnish examples of the three conditions of matter; iodine being a solid, bromine a fluid, chlorine a gas; the relation of their combining proportions is also curious: as might be expected, the fluid bromine takes an intermediate position, and (according to the axiom that half the sum of the extremes is equal to the mean) by dividing the combining proportions of iodine and chlorine, and adding them together, we have, as nearly as possible, the combining proportion of bromine:

Chlorine . . . . $35 \div 2 = 17\cdot75$
Iodine . . . . $126 \div 2 = 63$

$$80\cdot75$$

The combining proportion of bromine is 80, but 80·75 is so near, that it may reasonably be conjectured future experiments will reduce the number of the three elements, and may prove that they are only modifications of a single one. This is the only kind of alchemy which is tolerated in the nineteenth century, and any philosopher who will reduce the number of elements, and prove that some of them are only modifications of others, will achieve a renown that must transcend the *éclat* of all previous discoverers.

Bromine was discovered by Balard, in 1826, and, like chlorine and iodine, is a constituent of sea water. The chief source of bromine is a mineral spring at Kreutznach, in Germany. The process by which it is obtained offers a good example of chemical affinity; the water of the mineral spring is evaporated, all crystallizable salts removed, and a current of chlorine gas passed through the remaining solution, which changes to a yellow colour, in consequence of the liberation of the bromine by the combinations of chlorine with the bases previously united with the former; the liquid is then shaken with ether, which dissolves out the bromine. In the next place, the etherial solution is agitated with strong solution of potassa, and is thus obliged to part with the bromine, which is converted into bromate of potassa; this is ultimately changed by fusion to bromide of potassium; and by distillation with black oxide of manganese and sulphuric acid, the bromine is finally obtained. Six processes are therefore necessary before the small quantity of bromine contained in the mineral spring-water, is separated.

First Experiment.

Bromine is a very heavy fluid, which should be preserved by keeping it in a bottle covered with water; when required, a few drops may be removed by means of a small tube, and dropped into a warm bottle, which is quickly filled with the orange-red vapour. If some phosphorus is placed in a deflagrating spoon, and exposed to the action of bromine vapour, it takes fire spontaneously.

Second Experiment.

Powdered antimony sprinkled into the vapour of bromine immediately takes fire.

Third Experiment.

A burning taper immersed in a bottle containing the vapour of bromine is gradually extinguished.

Fourth Experiment.

Liquid bromine exposed to a freezing mixture of ice and salt, or reduced to a temperature of about eight degrees below zero, solidifies into a yellowish-brown, brittle, crystalline mass.

Fifth Experiment.

A solution of indigo shaken with a small quantity of the vapour of bromine is quickly bleached. Many substances, when brought in contact with liquid bromine, combine with explosive violence, and therefore experiments with liquid bromine are not recommended, as all the most instructive and conclusive results can be obtained by the use of the vapour of bromine, which is easily procured by allowing a few drops to fall into a warm, dry bottle.

Bromine, as already mentioned, is used in the art of photography.

FLUORINE.

Symbol, F.   Combining proportion, 19.

This singular element seems almost to embody the ancient idea of the alchemists, being a sort of *alkahest*, or universal solvent; or in plainer language, its affinities for other bodies are so powerful, that it attacks every substance (not even excepting gold), at the moment of its liberation, and combines therewith, so that its isolation has not yet been effected. Chemists who assert that they have been able to obtain fluorine in the elementary condition, pronounce it to be a gas which possesses the colour of chlorine; but the experiments, as hitherto conducted, render that statement extremely doubtful.

The only interesting fact connected with fluorine, is the remarkable property of attacking glass and other silicious bodies, belonging to its combination with hydrogen gas, called hydrofluoric acid. This acid is easily obtained and used by placing some powdered fluorspar in a leaden tray six inches square and two inches deep. If sulphuric acid is now mixed with the powdered spar, so as to form a thin paste, and heat applied, the vapour of the hydrofluoric acid quickly rises, and can be employed to etch a glass plate upon which a drawing may have been previously traced by scratching away the wax, with which it is first coated. By heating the glass plate before a fire, a sufficient quantity of wax is soon melted on to it by merely rubbing the wax against the glass plate; any excess should be avoided, if a well-executed drawing is required to be etched on its surface. (Fig. 144.)

Fig. 144. ᴀ ᴀ ᴀ. The glass plate, with the waxed side downwards, placed on the leaden tray containing the fluorspar and sulphuric acid. ᴅ. Spirit lamp.

The wax plate must not remain too long over the leaden tray, as the heat is apt to melt the wax, when the acid not only attacks those parts from which the wax has been removed by the etching needle, but also the surface of the glass generally, and thus the clearness of the design is spoilt. After exposure—and it is as well to prepare two or three glass plates for the experiment—the wax is quickly removed by rubbing and washing with oil of turpentine, and the design (beautifully etched into the glass) is then apparent.

CHAPTER XII.

CARBON, BORON, SILICON, SELENIUM, SULPHUR, PHOSPHORUS.

THIS group of non-metallic elements has been frequently styled "Metalloids," meaning substances allied to, but not possessing, all the properties belonging to a metallic substance; and therefore perhaps the expression, non-metallic solids, is the best that can be adopted. They may be subdivided into two classes of three each, which have properties more or less allied to each other—viz.,

Carbon, Boron, Silicon; and
Selenium, Sulphur, Phosphorus.

CARBON.

Symbol, C; Combining Proportion, 6.

This element has almost the property of ubiquity, and is to be found not only in all animal and vegetable substances, in common air, sea, and fresh water, but also in various stones and minerals, and especially in chalk and limestone.

There is, perhaps, no element which offers a greater variety of amusing experiments and elementary facts than carbon, whether it be considered either in its simple or combined state.

A piece of carbon, in the shape of the Koh-i-Noor, was one of the chief attractions at the first Exhibition in Hyde Park. The diamond is the hardest and most beautiful form of charcoal; how it was made in the great laboratory of nature, or how its particles came together, seems to be a mystery which up to the present time has not yet been solved, at all events no artificial process has yet produced the diamond.

Sir D. Brewster, speaking of the Koh-i-Noor, remarks that on placing it under a microscope, he observed several minute cavities surrounded with sectors of polarized light, which could only have been produced by the expansive action of a *compressed gas or fluid*, that had existed in the cavities when the diamond was in the *soft* state.

Now it is known that bamboo, which is of a highly silicious nature, has the property of depositing in its joints a peculiar form of silica, called tabasheer. Silicon is one of the triad with carbon—*i.e.*, it is allied to carbon on account of certain analogies; may it not then be supposed that, in times gone by, ages past, when the atmosphere was known to be highly charged with carbonic acid gas, there might possibly have existed some peculiar tree which had not only the power of decomposing carbonic acid (possessed by all plants at the present period), but was enabled, like the bamboo, to deposit, not silica, which is the oxide of silicium, but carbon, the purest form of charcoal—viz., the diamond? Speculation in these matters is ever more rife than stern proof, and it may be stated, that all attempts to manufacture this precious gem (like those of the alchemists with gold and silver) have most signally failed.

First Experiment.

Box and various woods, dried bones, and different organic matters, placed in a nearly close iron or other vessel, and heated red hot, so that all volatile matter may escape, leave behind a solid black substance called charcoal. If that kind obtained from bones, and termed bone black or ivory black, is roughly powdered, and placed in a flask with some solution of indigo or some vinegar, or syrup obtained by dissolving common moist sugar in water, and boiled for a short period, the colour is removed, and on filtering the liquid it is found to be as clear and colourless as water, provided sufficient ivory black has been employed.

Second Experiment.

Charcoal is a disinfectant, and is used for respirators; it has even been recommended medically, and charcoal lozenges can be bought at various chemists' shops. If a few drops of a strong solution of hydrosulphuret of ammonia (which has the agreeable odour belonging to putrid eggs) is mixed with half a pint of water, it will of course smell strongly, and likewise precipitate Goulard water, or a solution of acetate of lead black; but on shaking the water with a few ounces of charcoal, it no longer smells of sulphuretted hydrogen, and if filtered and poured into a solution of lead does not turn it black. This chemical action of charcoal, independent of its seeming mechanical attraction for colouring matter, would appear to show that the pores of charcoal contain oxygen, which in that peculiar condensed state destroys colouring matter, and oxidizes other bodies.

Third Experiment.

A very satisfactory experiment, proving that the diamond and plumbago or black lead are identical with charcoal, although differing in outward form and purity, can be made at a little cost, by purchasing a fragment of refuse diamond, called "*boart*," of Mr. Tennant of the Strand. A small piece costs about five shillings. The fragment should be carefully supported by winding some *thin* platinum wire round it, as, if the wire is too thick, it cools down the heat of the bit of diamond and prevents it kindling in the oxygen gas. A difficulty may arise in preparing the fragment, in consequence of the wire continually slipping off. The "boart" should therefore be grasped by the thumb and first finger, and the wire wound round; then it must be carefully turned and again wound across with the platinum wire, as in the sketch below. (Fig. 145.)

A piece of black lead (so called) may now be taken from a lead pencil and also supported by platinum wire; likewise a bit of common bark charcoal or hard coke. Three bottles of oxygen should now be prepared from chlorate of potash and oxide of manganese, an extra bottle being provided for the diamond in case there should be any failure in its ignition.

Fig. 145. A. The platinum wire. B. The fragment of "boart" or refuse diamond.

The bark charcoal can be first ignited by holding a corner in the spirit lamp for a few seconds; when plunged into oxygen it immediately kindles and burns with rapidity, and if the cork is well fitted, the product of combustion—viz., carbonic acid gas—is retained for future examination. The small piece of black lead is next heated red hot in the flame of the spirit lamp, and being attached by its platinum support to a stiff copper wire thrust through a cork, which fits the bottle of oxygen, is placed whilst red hot in the gas, and continues to glow until consumed. The fragment of diamond is by no means, however so easily ignited, the flame of the spirit lamp must be urged upon it with the blowpipe; when quite red hot, an assistant may remove the stopper from the bottle of oxygen, and the person heating the diamond should plunge it instantly into the gas; if this is dexterously managed, the fragment of *boart* glows like a little star, and the combustion frequently continues till the piece diminishes so much that it falls out of its platinum support.

Sometimes the diamond cools down without igniting, the same process must therefore be repeated, and a few extra bottles of oxygen will prevent disappointment, as every failure destroys the purity of the gas by admixture with atmospheric air when the stopper is removed. (Fig. 146.)

Fig. 146. A. Bottle containing bark charcoal. B. Ditto the plumbago or black lead. C. Ditto the diamond.

The combustion having ceased in the three bottles, the corks are removed, and the glass stoppers again fitted for the purpose of testing the *products*, which offer no apparent indication of any change, as oxygen and carbonic acid gas are both invisible. In each bottle a new combination has been produced; the charcoal, the black lead, the diamond have united with the oxygen, in the proportion of six parts of carbon to sixteen parts of oxygen, to form twenty-two parts of carbonic acid gas, which may be easily detected by pouring into each bottle a small quantity of a solution of slacked lime in water, called lime water. This test is easily made by shaking up common slacked lime with rain or distilled water for about an hour, and then passing it through a calico or paper filter. The test, though perfectly clear when poured in, becomes immediately clouded with a white precipitate, usually termed a *milkiness*, no doubt in allusion to the London milk, which is supposed to contain a notable proportion of chalk and water, for in this case the precipitate is chalk, the carbonic acid from the diamond and the charcoal having united with the lime held in solution by the water and formed carbonate of lime, or chalk, a substance similar in composition to marble, limestone, Iceland or double refracting spar, these three being nearly similar in composition, and differing only, like carbon and the diamond, in external appearance.

The milkiness, however, must not be held as conclusive of the presence of carbonic acid gas until a little vinegar or other acid, such as hydrochloric or nitric, has been finally added; if it now disappears with effervescence (like the admixture of tartaric acid, water, and carbonate of soda), the little bubbles of carbonic acid gas again escaping slowly upwards, leaving the liquid in the three bottles quite clear, then the experimentalist may sum up his labours with these effects, which prove in the most decisive manner that common charcoal, black lead, and the diamond, are formed of one and the same element—viz., carbon.

Fourth Experiment.

Having effected the synthesis (or combining together) of the diamond and oxygen, it is no longer possible to recover it in its brilliant and beautiful form. If the product of combustion is retained in a flask made of thin, hard glass, and two or three pellets of the metal potassium are placed in directly after the diamond has ceased to burn, and the flame of a spirit lamp applied till the potassium ignites, then the metal, by its great affinity for oxygen, takes away and separates it again from that which was formerly the diamond; but instead of the jewel being deposited, there is nothing but *black*, shapeless, and minute particles of carbon obtained, if the potash produced is dissolved in water, and the charcoal separated by a filter.

Fifth Experiment.

Chalk is made by uniting carbonic acid gas with lime; it may therefore be employed as a source of the gas, by placing a few lumps of chalk, or marble, or limestone, in a bottle such as was used in the generation of hydrogen gas; on the addition of some water and hydrochloric acid, effervescence takes place from the escape of carbonic acid gas, and the cork and pewter pipe being adapted, it may be conveyed by its own gravity into glasses, jugs, or any other vessels, and a pneumatic trough will not be required. Carbonic acid gas has a specific gravity of 1·529, and is therefore rather more than half as heavy again as atmospheric air.

Sixth Experiment.

In order to satisfy the mind of the operator that the gas obtained from chalk is similar to the *product of combustion from the diamond*, some lime-water may be placed in a glass, and the gas from the bottle allowed to bubble through it; instantly the same milkiness is apparent, which again vanishes on the addition of acid. And this experiment is rendered still more striking if a lighted taper be placed in the glass just after the addition of the acid, when it will be immediately extinguished.

Seventh Experiment.

If a lady's muff-box, supported by threads or chains, is hung on one end of a scale-beam, and counterbalanced by a scale pan and a few shot, it is

immediately depressed on pouring into the muff-box a quantity of carbonic acid gas, which may have been previously collected in a large tin
vessel. After showing the weight of the gas, the box is detached
from the scale-beam, and the contents poured upon a series of lighted
candles, which are all extinguished in succession. (Fig. 147.)

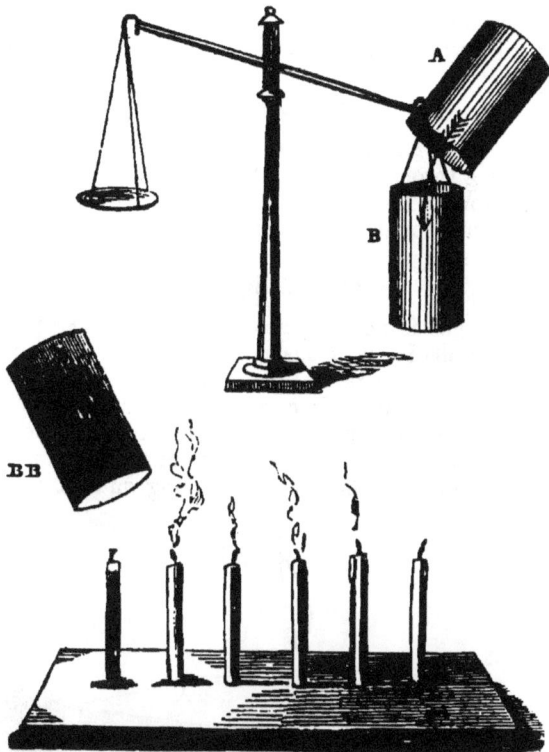

Fig. 147. A. Carbonic acid gas poured out of the tin box into B, the muff-box.
B B. Detached muff-box, and candles extinguished by the carbonic acid gas poured from it.

Eighth Experiment.

The property of carbonic acid gas of extinguishing flame, as compared with the contrary property of oxygen, is nicely shown by first
passing into a large and tall gas jar one half of its volume of oxygen
gas; a large cork perforated with holes may be introduced, so as
to float upon the surface of the water in the gas jar, and is usefully
employed to break the violence with which the carbonic acid enters the
gas jar, as it is passed in to fill up the remaining half volume of the gas
jar, which now contains oxygen at the top, and carbonic acid gas at the
bottom. On testing the contents of the jar with a lighted taper, it
burns fiercely in the oxygen, but is immediately extinguished in the

carbonic acid gas, being alternately lighted and put out as it is raised or depressed in the gas jar.

Ninth Experiment.

A little treacle, water, and a minute portion of size, may be placed with some yeast in a quart bottle, to which a cork and pewter or glass pipe is attached; directly the fermentation begins, quantities of carbonic acid gas may be collected, and tested either with lime-water or the lighted taper.

Tenth Experiment.

Some clear lime-water placed in a convenient glass is quickly rendered milky on passing through it the air from the lungs by means of a glass tube; thus proving that respiration and (as shown by the ninth experiment) fermentation, as well as the combustion of charcoal, produce carbonic acid gas.

Eleventh Experiment.

Carbonic acid gas is not only generated by the above processes, but is liberated naturally in enormous quantities from volcanoes, and from certain soils: hence the peculiar nature of the air in the Grotto del Cane. Dogs thrust into this cave drop down immediately, and are immediately revived by the tender mercies of the guides, who throw them into the adjoining lake. This natural phenomenon is well imitated by taking a box, open at the top, and nailing on to it a frame of card-

Fig. 148. A A. The box model of the Grotto del Cane. B B. Cardboard fixed in front of box, and painted to imitate rocks. c. Carbonic acid gas bottle, with bent tube passing through hole in the side of the box. A taper introduced at D burns in the upper, and is extinguished in the lower, part of the model.

board, which may be painted to represent rocks, taking care that a portion (about three inches deep) at the lower part is well pasted to the box at the edges, so that the gas may be retained; a hole is perforated at the top side to admit a lighted taper, and another at the side for the pipe from the carbonic acid bottle; when the bottom is filled with gas, a taper is applied, which is found to burn in the upper part, but is immediately extinguished when it reaches the lower division, where the three inches of pasteboard prevent it falling out : thus showing in a simple manner why a guide may enter the cave with impunity, whilst the dog is rendered insensible because immersed in the gas. (Fig. 148.)

Twelfth Experiment.

Many fatal accidents have occurred in consequence of the air in deep pits, graves, &c., becoming unfit for respiration by the accumulation of carbonic acid gas, which may arise either from cavities in the soil, where animal matter has undergone decomposition, or it may happen from the depth and narrowness of the hole or well preventing a proper draught or current of air, so that it becomes foul by the breathing of the man who is digging the pit. Air which contains one or two per cent. of carbonic acid will support the respiration of man, or maintain the flame of a candle; but it produces the most serious results if inhaled for any length of time; a lighted candle let down into a well (suspected to contain foul air) before the descent of the person who is to work in it, may burn, but does not indicate the presence of the small percentage of the poison, carbonic acid. Frequently no trouble is taken to test the air with a lighted candle; a man is lowered by his companions, who see him suddenly become insensible, another is then lowered quickly to rescue him, and he shares the same fate; and indeed cases have occurred where even a third and a fourth have blindly and ignorantly rushed to their death in the humane attempt to rescue their fellow creatures. What is to be done in these cases? Are the living to remain idle whilst the unfortunate man is suffocating rapidly at the bottom of the pit? No; provided they do not venture themselves into the pit, they may try every known expedient to alter the condition of the foul air, so as to enable them to descend to the rescue. One should be despatched to any neighbouring house or cottage for a pan of burning coals; if any slacked lime is to be had, it may be rapidly mixed with water, and poured down the side of the pit; a bundle of shavings set on fire and let down, keeping it to one side, so as to establish a current; or even the empty buckets constantly let down empty and pulled up full of the noxious air, may appear a somewhat absurd step to take, but under the circumstances any plan that will change the air sufficiently to enable another person to descend must be adopted; in proof of which the following experiments may be adduced :

Fill a deep glass jar with carbonic acid, and ascertain its presence with a lighted taper; if a beaker glass to which a string is attached is let down into the vessel and drawn up, and then inverted over a lighted taper, the utility of this simple plan is at once rendered apparent ; the beaker glass represents the empty bucket, and can be let down and pulled up full of carbonic acid until a sensible change in the condition of the atmosphere is produced. The best plan, however, is to set the air in motion by heat obtained from burning matter, or even a kettle of boiling water, lowered by a cord, and this fact is well shown by putting a small flask full of boiling water, and corked, at the bottom of the deep glass jar containing the carbonic acid gas, which rises like other gases when sufficiently heated, and passing away, mixes with the surrounding air. (Fig. 149.)

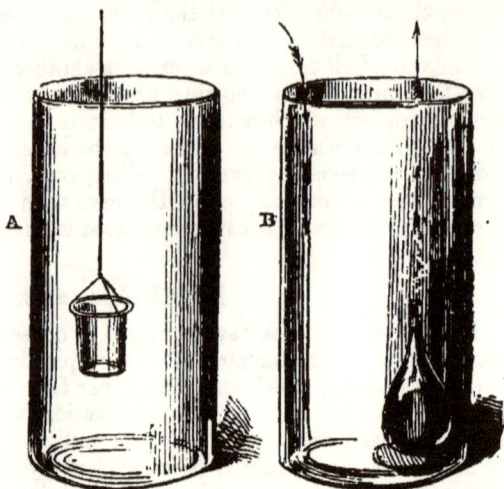

Fig. 149. A. Deep jar containing carbonic acid gas, which is being removed by the little glass bucket. B. Jar containing corked flask of boiling water on a pad ; the heated gas rises and the cold air descends to take its place.

Thirteenth Experiment.

Carbonic acid gas dissolved in water under considerable pressure, forms that most agreeable drink called soda-water ; the gas is not only useful in this respect, but has been applied most successfully by Mr. Gurney to extinguish a fire on a gigantic scale, which had been burning for years in the waste of a coal mine in Scotland. The same gas, generated suddenly by the combustion of a mixture of nitre, coke dust, and clay, or plaster of Paris, in vessels of a peculiar construction, has formed the subject of a patent by Phillips, since merged into the Fire Annihilator Company. The instrument is peculiarly adapted for shipping, and might, if properly used, be the means of saving many ships and valuable lives. (Fig. 150.)

Its practical value is established by the test of actual use : in the streets, by the Leeds Fire Brigade, and by firemen of the Fire Annihilator Company, temporarily stationed at Liverpool and Manchester.

The Fire Annihilator has been formally recognised by the Government Emigration Commissioners, who introduced into the Passengers' Act, 1852, in § 24, the alternative, " *Or other apparatus for extinguishing fire,*" with distinct reference to this invention, and subsequently by formal order authorized their officers to pass ships carrying Fire Annihilators.

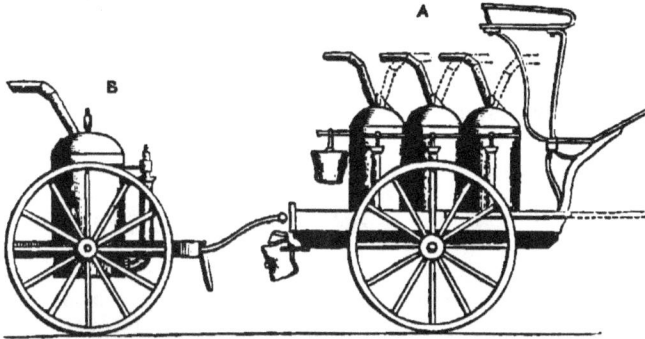

Fig. 150. A. A carriage with six fire annihilators, No. 5 size, fitted with moveal
The body of the carriage forms a tank for forty gallons of water; the tank is :
bunghole in the platform; a patent tap is fitted to the rear of the carriage; a
placed near the end upright of the rail; a hand-pump is placed in the box
carriage; a leather bucket with foot-holds and three canvas buckets are hun;
carriage; a hammer for removing and driving on the cover of the fire annihilat
nut wrench for the No. 10 truck, are placed in the box. B. A fire annihilator, No
with moveable pipe, on a spring truck, is attached to the carriage.
  The battery is fitted with shafts for one horse. A pole is also provided to fix a
shafts, so that the battery may be drawn by hand.

Monsieur Adolphe Girard has proposed that all houses sho
provided with an apparatus for the generation of carbonic ac

Fig. 151. A. Tank containing acid, communicating by a pipe with B, half fi
chalk and water. c c c c. Pipes conveying carbonic acid from the generator 1
ceiling, where it is discharged from numerous holes on the fire beneath.

placed outside the building, which is to be conveyed along the ceiling by means of pipes perforated with numerous holes, and to be put in operation directly a fire breaks out. This plan, however ingenious, could hardly supply the carbonic acid gas with sufficient rapidity, and it is to be feared would utterly fail in practice. (Fig. 151.)

BORON.

Symbol, B; combining proportion, 10·9.

Discovered by Homberg, in 1702, in borax, which is a biborate of soda ($NaO,2BO_3$), and is used very extensively in the manufacture of glass; also for glazing stoneware and soldering metals; it is also a valuable flux in various crucible operations, whilst in testing minerals with the blowpipe it is invaluable. Borax is made either from tincal, a substance that occurs naturally in some parts of India, China, and Persia, or by the addition of carbonate of soda to boracic acid, a substance obtained from the volcanic districts of Tuscany, whence it is imported to this country, and used in the manufacture of borax.

The element boron may be obtained by placing some pure boracic acid and some small bits of potassium in a tube together, and applying the flame of a spirit-lamp, a glow of heat takes place, and when the tube is cold the potash may be washed away, and the boron remains as a dark brownish powder somewhat resembling carbon. M. ·St. Claire Deville and Wöhler have lately made some important discoveries with respect to this element, and disproved the statement that it is uncrystallizable. Their researches prove it to be producible under three forms and of various colours, such as honey-yellow and garnet-red, the crystals in some cases being like diamonds of the purest water—*i.e.*, limpid and transparent. A new combination of aluminium and boron is stated to possess the most remarkable properties. It is harder than the diamond, and in the state of powder will cut and drill rubies, and even the diamond itself, with more facility than diamond powder. Deville and Wöhler incline to the belief that the diamond is dimorphous, and capable (in conditions yet to be described) of assuming the same forms as boron. At a high temperature, boron, like titanium, absorbs *nitrogen* only from the atmosphere, and rejects the oxygen. (Query, may not some of those remarkably hard black diamonds prove to be boron?)

SILICON.

Symbol, Si; combining proportion, 21·3.

The great Berzelius was the first to obtain this element in 1823. Silicon in the pure state is a dark brown powder; if ignited at a very high temperature it assumes a chocolate colour, which is supposed to be the allotropic condition, because it no longer burns when heated moderately in oxygen or air, and is not attacked by hydrofluoric acid.

M

The most interesting combination of silicon is the teroxide called silicic acid, silica (SiO_3). Silicon is next to oxygen so far as regards its plentifulness, and is found in the state of silica in nearly every mineral, but especially in rock crystal, quartz, flint, sand, jasper, agate, and tripoli. It is largely used in the manufacture of glass, and a most useful "soluble glass" is obtained by melting together in a crucible fifteen parts of sand, ten parts of carbonate of potash, and one part of charcoal.

Cold water merely washes away the excess of alkali, and after this is done the powdered soluble glass may be boiled with water in the proportion of one of the former with five of the latter, when it gradually dissolves; the solution may be evaporated to a thick pasty fluid, which looks like jelly when cool, and on exposure to the air in thin films changes to a transparent, colourless, brittle, but not hard glass. Wood, cotton, and linen fabrics are rendered less combustible when coated with this glass, which excludes the oxygen of the air, and it has lately been employed to fill up the porous and capillary openings in stone exposed to the atmosphere, and is stated to be very efficacious as a preservative of the stone in some cases.

SULPHUR.

Symbol, S; combining proportion, 16.

Sulphur, like charcoal, is of common occurrence in nature, and is chiefly supplied from the volcanic districts of Tuscany and Sicily: there is an abundance of this element in the United Kingdom, but then it is locked up in combination with iron, copper, and lead, under the name of iron pyrites, copper pyrites, galena; and whilst Sicily and Tuscany supply thousands of tons weight in the uncombined state, it is not, of course, worth while to go through expensive operations at home for the separation of sulphur from the ores. During the dispute between Sicily and England, several patents were secured for new and economical processes by which sulphur was obtained from various minerals; and had this country been excluded from a supply of native sulphur, no doubt some of these patents would now be in active operation.

It is almost possible to estimate the commercial prosperity of a country by the sulphur it consumes, not, happily, by their warlike operations, but in the manufacture of oil of vitriol or sulphuric acid, which is the starting point of a great number of useful arts and manufactures.

First Experiment.

Some very curious results may be obtained by heating sulphur at certain temperatures; in the ordinary state it is a pale yellow solid, and when subjected to a temperature of 226° Fahr. it melts to a brownish-yellow, transparent, thin fluid; according to all preconceived notions of the properties of substances which liquify by an increase of heat, it might be imagined that every additional degree of heat would only render the melted sulphur still more liquid, but strange to say, when it reaches a temperature of about 320° Fahr. it changes red, and thick like treacle; and as the heat rises it becomes so tenacious, that the ladle in which it is contained may be inverted, and the sulphur will hardly flow out: at about 482° Fahr. it again becomes liquid, but not so fluid as at the lower temperature. If allowed to cool from 482° Fahr., the above results are simply inverted; the sulphur becomes thick, again liquid, and finally crystallizes in long, thin, rhombic prisms, which are seen most perfectly by first allowing a crust of sulphur to form on the liquid portion, and then having made two holes in this crust, the sulphur is poured out, when the remainder is found in the interior of the crucible crystallized in the form already mentioned. Sulphur takes fire in the air when exposed to a heat of about 560° Fahr., and burns with a pale blue flame; and, as already stated, it may be poured from a considerable height on a still dark night, and produces a continuous column of blue fire, just like an unbroken current of electricity. If the melted and burning sulphur is received into a vessel containing boiling water, it is no longer yellow, but assumes a curious *allotropic* state, in which it is a reddish-brown, transparent, shapeless mass, that may be easily kneaded and used for the purpose of taking casts of seals, which become yellow in a few days, and are found then to be hard and crystallized.

Second Experiment.

Sulphur vapour, in one sense, may be regarded as a supporter of combustion: if a clean Florence oil-flask is filled with copper turnings, and a little roughly-powdered sulphur sprinkled in, and heat applied, the copper glows with an intense heat, and burning in the vapour of the sulphur, produces a sulphuret of copper; from this compound the sulphur may be again obtained by boiling the powdered sulphuret with weak nitric acid, which oxidizes and dissolves the copper, leaving the greater part of the sulphur behind, which may be collected, melted, and burnt, and will be found to display all the properties belonging to that element. This experiment is a very good example of simple analysis; and if the copper is weighed and likewise the combined sulphur, a good notion may be formed of the principles of combining proportions.

Third Experiment.

A little sulphur burnt under a gas jar, or in any convenient box (a hat-box, for instance), produces sulphurous acid (SO_2), which will bleach a wetted red rose or dahlia, and many other flowers. This gas is employed most extensively in bleaching straw, and sundry woollen goods, such as blankets and flannel, and likewise silk, and is perhaps one of the best disinfectants that can be employed; when fever has been raging in the dwellings of the poor, as in cottages, &c., all metallic substances should be removed, the doors and windows closed, the bedding, &c., well exposed, and then a quantity of sulphur should be burnt in an old frying-pan placed on a brick, taking care to avoid the chance of setting the place on fire; after a few hours the doors and windows may be opened, and the disinfectant will be found to have done its work cheaply and surely.

Fourth Experiment.

The presence of sulphur in various organic substances, such as hair, the white of egg, and fibrine, is easily detected by heating them in a solution of potash, and adding acetate of lead as long as the precipitate formed is redissolved; finally the solution must be heated to the boiling point, when it instantly becomes black by the separation of sulphuret of lead.

Fifth Experiment.

Sulphuric acid, HO,SO_3, or oil of vitriol, is made in such enormous quantities that it is never worth while to attempt its preparation on a small scale. In consequence of its great affinity for water, many energetic changes are produced by its action. Oil of vitriol poured on some loaf sugar placed in a breakfast-cup with the addition of a dessert-spoonful of boiling water, rapidly boils and deposits an enormous quantity of black charcoal. If a word be written on a piece of white calico with dilute sulphuric acid, and then rapidly and thoroughly washed out, no visible change occurs; but if the calico is exposed to heat, so that the excess of water is driven off, the remaining and now concentrated oil of vitriol attacks the calico, and the word is indelibly printed in black by the decomposition of the fabric of cotton. A very remarkable process has lately been introduced by Mr. Warren de la Rue, by which paper is converted into a sort of tough parchment-like material. called ametastine, by the action of oil of vitriol and water of a certain fixed strength; and any departure from the exact proportions destroys the toughness of the paper. After the paper has been acted upon by the acid, it becomes extremely tenacious, and will support a considerable weight without breaking. Mr. Smee has used this ametastine in the construction of an hygrometer, and states that it may save many a traveller from catching a severe rheumatism in a damp bed.

Sixth Experiment.

When the vapour of sulphur is passed over red-hot charcoal and the product carefully condensed, a peculiar liquid is obtained, called bisulphide of carbon (CS_2), which possesses a peculiar odour, is extremely transparent and brilliant-looking, and enjoys a high refractive power. This liquid is used as a solvent for phosphorus and other substances, and is extremely volatile and combustible, and burns silently with a pale blue flame. The combustion of its vapour, mixed with certain gases, offers a good example of the fact that slow burning may be a peaceful experiment, whilst very rapid combustion often resolves itself into an explosion. Thus, if a few drops of bisulphide of carbon are dropped into a narrow-mouthed dry quart bottle containing common air, and flame applied, the combustion takes place with rapidity, a rushing or roaring sound being audible, in consequence of the diffused vapour being supplied with more oxygen, and burning more rapidly than it would do if simply consumed from a stick or glass rod wetted with the fluid.   A still greater rapidity of combustion is ensured by dropping some bisulphide of carbon into a long stout cylindrical jar, fifteen inches long and three inches in diameter, containing nitric oxide gas (NO_2); when flame is applied the mixture burns with a bright flash and some noise, and if burnt in a narrow mouthed bottle would most likely blow it to atoms.

The greatest rapidity of combustion, and of course the loudest noise, is obtained by shaking some bisulphide of carbon in a similar stout and strong cylindrical jar filled with oxygen gas, but in this case the jar must be protected with a double cylinder of stout wire gauze; it does not always break, but if it is blown to fragments each particle becomes a lancet-shaped piece of glass, which is capable of producing the most dangerous wounds. (Fig. 152.)

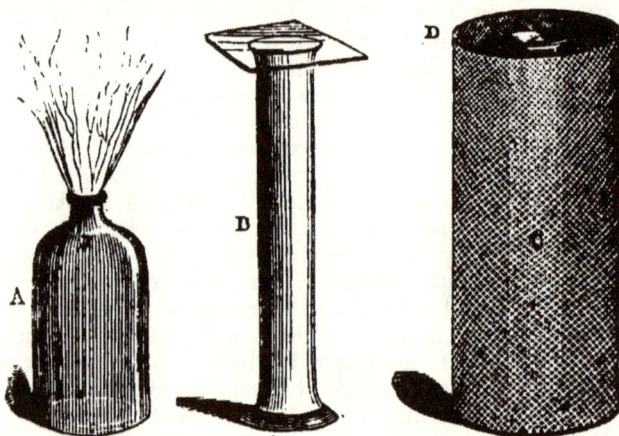

Fig. 152. A. Air and bisulphide of carbon. B. Nitric oxide and ditto. C. Oxygen and ditto. D D. Stout cylinder of double wire gauze, open top and bottom.

SELENIUM.

Selenium ($\sigma\epsilon\lambda\eta\nu\acute{\eta}$, the Moon*); symbol, Se; combining proportion, 39·5.

This new metallic element is allied to sulphur, and is a species of chemical curiosity, being found in minute quantities in various minerals; it may be melted and cast into any form. Medallions of the discoverer (Berzelius) of selenium, in little cases, are imported from Germany, for the cabinets of the curious.

* Called selenium on account of its strong analogy to the metal tellurium (*tellus*, the earth).

PHOSPHORUS.

Phosphorus ($\phi\tilde{\omega}s$, light; $\phi\acute{\epsilon}\rho\epsilon\iota\nu$, to bear; symbol, P; combining proportion, 32.)

Monsieur Salverte, in his work on the Occult Sciences of the Ancients, quotes a remarkable story respecting the probable discovery of the nature of phosphorus in 1761:—"A Prince San Severo, at Naples, cultivated chemistry with some success; he had, for example, the secret of penetrating marble with colour, so that each slab sawed from the block presented a repetition of the figure imprinted on its external surface. In 1761, he exposed some human skulls to the action of different reagents, and then to the heat of a glass furnace, but paying so little attention to his manner of proceeding, that he acknowledged he did not expect to arrive a second time at the same result. From the product he obtained a vapour, or rather a gas was evolved, which kindling at the approach of a light, burned for several months without the matter appearing to die or diminish in weight. San Severo thought he had found the impossible secret of the inextinguishable lamp, but he would not divulge his process, for fear that the vault in which were interred the princes of his family should lose the unique privilege with which he expected to enrich it, of being illuminated with a *perpetual lamp.*" Had he acted like a philosopher of the present day, San Severo would have attached his name to the important discovery of the existence of *phosphorus* in the *bones*, and made public the process by which it might be obtained.

This element, formerly sold at four or five shillings the ounce, has now fallen so much in price, from the greater demand and larger production, that it may be bought for a few shillings the pound, and is imported in tin cases in large quantities from Germany. It was discovered about two hundred years ago by Brandt, a merchant of Hamburg, and may be prepared on a small scale by distilling at a red heat phosphoric acid previously fused with one-fourth. of its weight of powdered charcoal.

First Experiment.

Phosphorus, when pure, is without taste or colour, but generally of a very pale buff-colour, and semi-transparent; it is extremely combustible, and is usually preserved under the surface of water; when perfectly dry, a thin slice will take fire at 60° Fah., and burns with great brilliancy. Considering the heat produced during the combustion of phosphorus, it might be thought that it would infallibly set fire to any ordinary combustible, such as paper or wood, but this is not the case when phosphorus is employed by itself, as may be proved by the following experiment.

Cut five very small pieces of phosphorus, and place them like the five of diamonds on a sheet of cartridge-paper laid upon the table, set the bits of phosphorus on fire, when they will be rapidly burnt away leaving only five black spots, but not firing the paper, as would be the case if some red-hot coals or charcoal were placed in the same position. The cause is very simple. Phosphorus in burning produces phosphoric acid, which is an anti-combustible, and coats the surface of the paper round the spot where the combustion occurs, and acting as a kind of glaze or glass, excludes the oxygen of the air, and prevents the fire spreading.

If some powdered sulphur is sprinkled round the spot where the bit of phosphorus is to be burnt, the case is very different; the heat melts and sets fire to the sulphur, which being uncoated with the phosphoric acid, communicates to the paper; and it is on this principle that lucifer-matches can be used as instantaneous lights. The tip of the wood of which they are composed is first dipped in sulphur, and then the phosphorus composition made of gum, chlorate of potash, vermilion, and phosphorus, is placed over it; and if the latter were used alone without the sulphur, not one match in a hundred would take fire properly.

Second Experiment.

Common phosphorus is perfectly and rapidly dissolved by bisulphide of carbon. The solution must be carefully preserved, as it is a liquid combustible, which takes fire spontaneously after the bisulphide of carbon evaporates; so that wherever it is dropped, a flame, arising from the spontaneous combustion of the finely-divided phosphorus, is sure to be produced. This liquid was recommended many years ago to the Government for the purpose of setting sails of ships or other combustible matter on fire. The solution of phosphorus alone did not answer the purpose, as already explained in the first experiment; but when wax was dissolved with the phosphorus, it then became a most dangerous fluid, which it was recommended should be used in shells, and discharged from a mortar or howitzer in the ordinary manner. Dr. Lyon Playfair was the first to make this proposed application of the solution, and it has since, we believe, been recommended by Captain Norton in his liquid-fire shells.

Third Experiment.

One of the most curious facts in connexion with phosphorus, is its assumption of the allotropic state in what is termed *amorphous* (shapeless) or red phosphorus. This substance, when handled for the first time, might be mistaken for a lump of badly-made Venetian red. There is no risk of its taking fire like the common phosphorus, and it does not (according to Schrötter, of Berlin, who discovered this peculiar condition) exhale those fumes which are so prejudicial to the lucifer-match makers. When the vapour of common phosphorus is continually inhaled, it is said to cause a peculiar and disgusting disease, which terminates in the destruction of the jaw-bone; whilst the bones in other parts of the body become brittle, and arm-bones thus affected are fractured with the slightest blow.

The difference between common and red phosphorus is well shown—

first, by placing a few small pieces of both kinds in separate bottles or vials containing bisulphide of carbon; the common phosphorus, as already explained, quickly dissolves in the liquid, and if poured on a sheet of paper, and hung up, is soon on fire; whilst the red variety is wholly unaffected, and if the bisulphide of carbon is poured off on to paper, it merely evaporates, and no combustion occurs.

The similarity in composition, though not in outward form, is further shown by filling two jars with oxygen gas, and having provided two deflagrating spoons, some common phosphorus is placed in one, and red phosphorus in the other; a wire, gently heated by dipping it into some boiling water, is now applied to the former, which immediately takes fire, and may be plunged into the jar of oxygen gas, when it burns with the usual brilliancy. The red phosphorus, however, must be brought to a much higher temperature (500° Fah.) before it will even shine in the dark, and then with a still further increase of heat it takes fire, and on being placed in the other jar of oxygen burns up much more slowly than the yellow phosphorus, but at last exhibits that brilliant flash of light which is so characteristic of the combustion of phosphorus in oxygen.

The amorphous or red phosphorus is employed in the manufacture of *safety chemical matches*, and M. A. Meunons has secured a patent in England for an improvement in lucifer matches, with a view to obviate the risks of accidental ignition. To attain this end the matches are first cut by a machine from cubes of wood, the cut being stopped at a short distance from the end of each cube, so as to leave the lower extremities adherent. The upper or free extremity of each packet of splints thus formed being coated with wax or sulphur, is dipped in one of the following preparations:—Chlorate of potash, two parts; pulverized charcoal, one part; umber, one part; or, chlorate of potash, sulphur, and umber, in equal parts, thoroughly mixed with glue. The opposite extremity or "cut" of each packet is then painted over with amorphous phosphorus blended with size, so that on separating the matches the phosphorus is only found on the top of each. The matches thus prepared are ignited by breaking off a small piece of the phosphorised end and rubbing it on the opposite extremity covered with the inflammable preparation.

Loud exploding and dangerous lucifers were formerly made by dipping bundles of matches, previously coated with sulphur at the tips, into a thick solution of gum, at a temperature of 104° Fahr., coloured with smalt or red lead, in which was dissolved a certain proportion of chlorate of potash, and also containing finely divided particles of phosphorus obtained by the constant stirring and rubbing of the materials in a mortar. When dry the matches exploded if rubbed against a gritty surface, and there was always a risk of a fragment flying off and entering the eye. To obviate this danger, *silent* or *noiseless lucifer matches* were invented, and the composition used (according to Böttger) is as follows :—Gum arabic, 16 parts by weight; phosphorus, 9 parts; nitre, 14 parts; powdered black oxide of manganese, 16 parts. The above ingredients are worked up in a mortar with water, at 104° Fahr., and the matches previously tipped with sulphur are dipped therein and afterwards dried.

Fourth Experiment.

The combustion of phosphorus under water is easily demonstrated by placing some ordinary stick phosphorus in a metallic cup, and then plunging it rapidly under the surface of boiling water. If a jet of oxygen gas is now directed upon the liquid phosphorus, it burns with great brilliancy.

Fig. 153. A A. Finger-glass of boiling water containing a metallic cup with melted phosphorus. C. Jet of oxygen gas. D D. Sheet of wire gauze.

When the oxygen escapes too rapidly from the jet, it causes some small particles to be thrown out of the water, so that it is advisable to defend the face with a sheet of wire gauze held a few inches above the glass whilst the experiment is being conducted. (Fig. 153.)

Fifth Experiment.

Phosphorus burns and emits beautiful flashes of light in the presence of the gas called peroxide of chlorine (ClO_4), which must be very carefully generated under the surface of water by first placing some cut phosphorus and chlorate of potash at the bottom of a long and stout cylindrical glass nearly full of water; sulphuric acid is then conveyed to the chlorate of potash by means of a syphon, the end of which must be drawn out to a small opening, or else the oil of vitriol will descend too rapidly, and the glass will be cracked by the heat. Immediately the peroxide of chlorine comes in contact with the phosphorus it explodes, and passes again to its original elements, oxygen and chlorine. These bubbles envelope minute particles of phosphorus, which rapidly ascend, like water-spiders, to the surface, and burn as they pass upwards, producing a continual series of sparks of fire, which have an extremely pretty effect. (Fig. 154.) The syphon is of course first filled with water, and as that is displaced, the oil of vitriol takes its place.

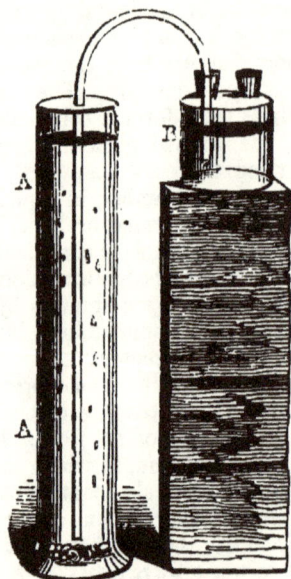

Fig. 154. A A. Tall glass nearly full of water; at the bottom are the chlorate of potash and phosphorus. B. Wolfe's bottle and syphon, conveying the oil of vitriol to bottom of A A.

Sixth Experiment.

If a little phosphorus is placed in a small copper boiler, and the steam allowed to escape from a jet, it is observed to be luminous, in consequence of a minute portion of phosphorus being carried up mechanically with the steam. The same fact is shown very prettily by boiling water in a flask containing some phosphorus.

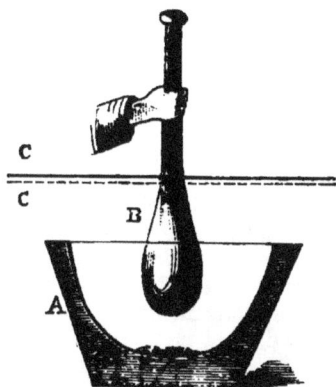

Seventh Experiment.

Phosphorus explodes violently when rubbed with a little chlorate of potash, and in order to perform this experiment safely, it should be made in a strong iron mortar, the pestle of which must be surrounded with a large circle of cardboard and wire gauze; so that when it is brought down upon the phosphorus and chlorate of potash, any particles that may fly out are detained by the shield. Without this precaution the experiment is one of the most dangerous that can be made. (Fig. 155.)

Fig. 155. A. The iron mortar containing the phosphorus and chlorate of potash. B. The pestle, with the shield, c c, composed of a circle of wire gauze, covered with one of cardboard.

Eighth Experiment.

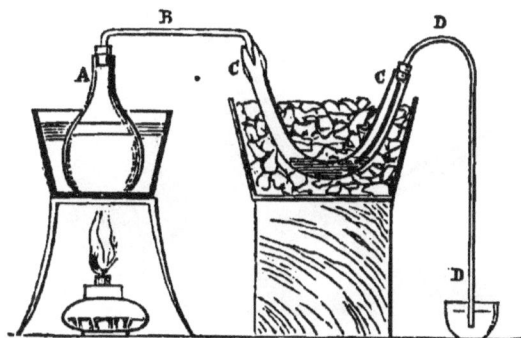

Phosphuretted hydrogen owes its property of spontaneous combustion to the presence of the vapour of a liquid, phosphide of hydrogen (PH_2), which may be prepared by placing some phosphide of calcium into a flask with water heated to a temperature of 140° Fah., and conveying the gas into a U-shaped tube surrounded with a mixture of ice and

Fig. 156. A. The flask containing the phosphide of calcium and water, and placed in a water-bath heated to 140° Fah. B. Bent tube conveying the gas to c c, the U-shaped tube, to which it is attached by india-rubber tubing. c c. The U-shaped tube, surrounded with a freezing mixture. D D. Bent tube, passing into a cup of water to prevent contact with air.

salt. The liquid obtained is colourless, and must be preserved from contact with air, as it takes fire spontaneously directly it is exposed to the atmosphere. (Fig. 156.)

Ninth Experiment.

Phosphide of calcium is quickly prepared by placing some small pieces of lime in a crucible and making them red-hot; if lumps of dry phosphorus are thrown into the crucible, and the cover placed on quickly, and immediately after the phosphorus, the latter unites with the calcium, and forms a brown substance which produces gaseous phosphide of hydrogen (PH_3) when placed in water, and the gas takes fire spontaneously when it comes in contact with the air.

Tenth Experiment.

Phosphorus placed in a retort with a tolerably strong solution of potash, and a small quantity of ether, affords a large quantity of phosphide of hydrogen (commonly called phosphuretted hydrogen) when boiled. The neck of the retort must dip into a basin of water, and the object of the ether is to prevent the combustion of the first bubbles of gas *inside* the retort, which by their explosion would probably break the glass. If the neck of the retort is kept under water in which potash is dissolved, the gas may be generated for many days at pleasure, although it is not a desirable experiment to renew too often, on account of the disagreeable odour produced. (Fig. 157.)

Fig. 157. A retort containing the phosphorus, water, potash, and ether. D. Neck dipping into a basin of water. c. The gas burning, and producing beautiful rings of smoke.

Eleventh Experiment.

When a jar of oxygen is held over the neck of the retort generating the phosphuretted hydrogen, a bright flash of light and explosion are observed; and if the experiment is performed in a darkened room, it is just like a sudden flash of lightning. A bottle of chlorine held over the neck of the retort, and dipping of course in the water of the basin, produces a green flame every time the bubble of gas passes into it. That curious appearance of light, sometimes seen in marshy districts, called will-o'-the-wisp, is supposed to be due to the escape, from decomposing matter, of bubbles of hydrogen, nitrogen, &c., through which the spontaneously inflammable phosphide of hydrogen is diffused.

At a place called Dead Man's Island, near Sheerness, magnificent effects of this kind are sometimes apparent when the mud banks are accidentally stirred at night by a boat-hook. A credible observer says, he once saw there a flash of yellowish-green light, accompanied with noise, about thirty feet in height. The apparent height might be due to the duration of the impression of the flash on the eye, as the light from the burning phosphuretted hydrogen ascended rapidly upwards. The source of this gas appears to be due to the fact, that during the time some Russian ships were watched by the Brest fleet, a number of the sailors died of cholera, and were buried in the banks; the decomposition of the bone containing phosphorus would account for the appearance of light already described.

With the discussion of some of the most interesting properties of the thirteen non-metallic elements we take leave of the subject of chemistry, reserving the consideration of the metals for another popular juvenile work, of which they will form the subject.

In answer to the oft-repeated question, "Where can I get the *things* for the experiments?" it may be stated that every kind of glass vessel and the chemicals mentioned in this chapter, can be procured either of Messrs. Simpson, Maule, and Co., Kennington, or of Griffin and Co., Bunhill-row, or Bolton and Co., High Holborn.

Fig. 159. Will-o'-the-wisp.

Fig. 159. Franklin and his kite.

CHAPTER XIII.

FRICTIONAL ELECTRICITY.

Of all the agents with which man is acquainted, not one can afford a greater source of wonderment to the ignorant, of meditation to the learned, than the effects of that marvellous force pervading all matter called electricity. We look at matter endowed with life, and matter wanting this divine gift, with some degree of interest, depending on our various tastes and occupations ; we know at a glance a bird, a beast, or a fish ; we observe with pleasure and admiration the wonderful changes of nature, and know that a few seeds thrown into the broken clods and well-tilled earth may become either the waving, golden corn-field or in time may grow from the tender little shrub.to the stately forest-tree ; we know all these things because they belong to the visible world, and are continually passing before our eyes : but in looking at the visible, we must not forget and ignore the invisible. It may with truth be stated that the greatest powers of nature are all concealed, and if any truth would lead us from Nature to Nature's God, it is the fact that no visible, solid, tangible agent can work with so much force and power as invisible electricity. Many centuries passed away since the commencement of the Christian era, before the human mind was prepared to appreciate this great power of nature; other forces had claimed attention, and the difference in the presence or absence of two of the imponderable agents, heat and light, as derived from the sun, in the effects of the change of the seasons, and other common facts, had led philosophers to speculate early upon their nature; but electricity, from its peculiar properties, long escaped observation, and it was not until the beginning of the eighteenth century (about 1730) that any material facts had been discovered in this science, when Mr. Stephen Grey, a pensioner of the Charterhouse, discovered what he termed *electrics* and *non-electrics*, and also the use of insulating materials, such as silk, resin, glass, hair, &c.; and it is obvious that, until the latter fact was discovered, the science would remain in abeyance, because there would be no mode of preserving the electrical excitement in the absence of non-conductors of this force.

The year 1750 was remarkable for Volta's discoveries and Dr. Franklin's identification of the electricity of the machine with the stupendous effects of the thunderstorm. Sir Humphry Davy, in 1800, with his commanding genius, threw fresh light upon the already numerous electrical effects discovered. In 1821, Faraday commenced his studies in this branch of philosophy; which he has since so diligently followed up, that he has been for some years, and is still the first electrician of the age. From the commencement of the present century, discoveries have succeeded each other in regular order and with the most amazing results; and now electricity is regularly employed as a money-getting agent in the process of the electrotype and electro-silvering and gilding; also in the electric telegraph; and in a few years we may possibly see it commonly employed as a source of artificial light.

The nature of electricity, says Turner, like that of heat, is at present involved in obscurity. Both these principles, if really material, are so light, subtle, and diffuse, that it has hitherto been found impossible to recognise in them the ordinary characteristics of matter; and therefore electric phenomena may be referred, not to the agency of a specific substance, but to some property or state of common matter, just as sound and light are produced by a vibrating medium. But the effects of electricity are so similar to those of a mechanical agent, it appears so distinctly to emanate from substances which contain it in excess, and rends asunder all obstacles in its course so exactly like a body in rapid motion, that the impression of its existence as a distinct material substance *sui generis* forces itself irresistibly on the mind. All nations, accordingly, have spontaneously concurred in regarding electricity as a material principle; and scientific men give a preference to the same view, because it offers an easy explanation of phenomena, and suggests a natural language intelligible to all.

There are five well-ascertained sources of electricity, and three which are considered to be uncertain. The five sources are friction, chemical action, heat, magnetism, peculiar animal organisms. The three uncertain sources are contact, evaporation, and the solar rays.

First Experiment.

A stick of sealing-wax or a bit of glass tube, perfectly dry, rubbed against a warm piece of flannel, has elicited upon its surface a new power, which will attract bits of paper, straw, or other light materials; and after these substances are endowed with the same force, a repellent action takes place, and they fly off. One of the most convenient arrangements for making experiments with the attractive and repellent powers of electricity is to fix with shell-lac varnish round discs of gilt paper, of the size of a half-crown, at each end of a long straw that is supported about the centre with a silk thread, which may hang from the ceiling or any other convenient support. (Fig. 160.)

The varnish is easily prepared by placing four or eight ounces of shell-lac in a bottle, and pouring enough pyroxylic spirit (commonly termed wood naphtha) upon the lac to cover it. After a short time, and by agitation, solution takes place. In a variety of ways friction is proved to be a source of electricity, and forms a distinct branch of the science, under the name of *frictional* electricity.

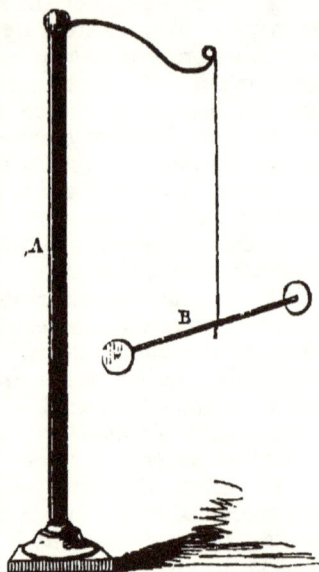

Fig. 160. A. The glass pillar support. B. Straw with discs, hanging by a silk thread.

Second Experiment.

The nature of chemical action has been already explained, and is alluded to here as a source of electricity of which the proof is very simple. A piece of copper and a similar-sized plate of zinc have attached to them copper wires; these plates are placed opposite to, but do not touch each other, in a vessel containing water acidulated with a small quantity of sulphuric acid. When the wires are brought in contact, a current of electricity circulates through the arrangement, but has no power to attract bits of paper, straw, &c. In order to ascertain whether the current of electricity passes or not, a piece of covered copper wire is bent several times round a magnetic needle, so that it has freedom of motion inside the core or hollow formed by twisting the copper wire. This arrangement, properly constructed, is called the galvanometer needle, and is invaluable as a means of ascertaining the passage of electricity derived from chemical action. (Fig. 161.)

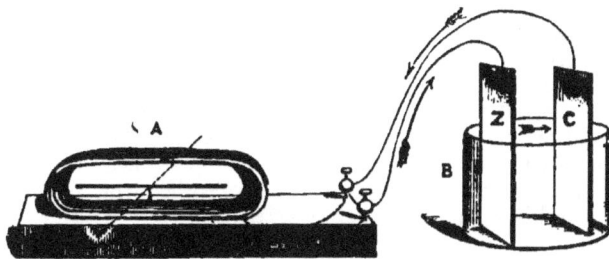

Fig. 161. A. The galvanometer needle. B. Vessel containing weak acid and the zinc and copper plates. The arrows show the path of the electric current.

When the wires leading from the metal plates are connected with the extremities of the coil in the galvanometer, the needle is deflected or pushed aside to the right hand or to the left, according to the direction of the current.

Third Experiment.

The third source of electricity is heat, and the effect of this agent is well shown by twisting together a piece of platinum and silver wire, so as to form one length. If the silver end is attached to any screw of the galvanometer, and the platinum end to the second screw, no movement of the magnetic needle takes place until the heat of a spirit-lamp is applied for a moment to the point of juncture between the silver and platinum wires, when the magnetic needle is immediately deflected.

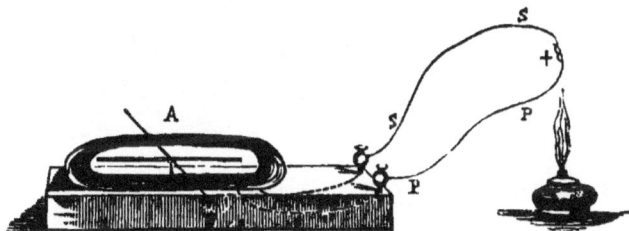

Fig. 162. A. The galvanometer needle, with wires attached. s, s. Silver wire joined to P, P, the platinum wire. The heat of the spirit-lamp is applied at the point of juncture, +.

Fourth Experiment.

The fourth source of electricity—viz., magnetism—requires a some-what more complicated arrangement; and a most delicate galvanometer needle must be provided, to which is attached the extremities of a long spiral coil of copper wire covered with cotton or silk. Every time a bar magnet is introduced inside the coil, so that the conducting wire cuts the magnetic curves, a deflection of the galvanometer needle takes place, and the same effect is produced on the withdrawal of the magnet, the needle being deflected in the opposite direction.

The magnetic spark can be obtained by employing a magnet of sufficient power; and the arrangement for this purpose is very simple. A cylinder of soft iron is provided, and round its centre are wound a few feet of covered thin copper wire, one end of which is terminated with a copper disc well amalgamated, and the other end, after being properly cleaned and coated with mercury, is brought into contact with the disc. Directly this cylinder is laid across the poles of the magnet, and as quickly removed, the point and disc, from the elasticity of the former, separate for the moment, the contact is broken between the point and disc, and a brilliant but tiny spark is apparent.

Fig. 163. A B. Horse-shoe magnet. c. Cylinder of soft iron. D. Coil of copper wire and contact breaker.

Fifth Experiment.

The fifth mode of procuring electricity would require the assistance of an electrical eel, a fine specimen of which (forty inches in length) was exhibited at the Adelaide Gallery some years ago. Various experiments were made with this animal, and the author had the pleasure of witnessing all the ordinary phenomena of frictional electricity, illustrated by Dr. Faraday, with the assistance of the animal electricity derived from this curious creature. Recent experiments have, however, proved that the electric current is induced through the agency of the nervous

N

system. This important fact has been communicated by M. Dubois-Raymond, whose experiment is thus recorded:—A cylinder of wood is firmly fixed against the edge of a table; two vessels filled with salt and water are placed on the table, in such a position that a person grasping the cylinder may, at the same time, insert the fore-finger of each hand in the water. Each vessel contains a metallic plate, and communicates, by two wires, with an extremely sensitive galvanometer. In the instrument employed by M. Dubois-Raymond, the wire is about $3\frac{1}{4}$ miles in length. The apparatus being thus arranged, the experimenter grasps the cylinder of wood firmly with both hands, at the same time dipping the fore-finger of each hand in the saline water. The needle of the galvanometer remains undisturbed; the electric currents passing by the nerves of each arm, and being of the same force, neutralize each other. Now, if the experimenter grasp with energy the cylinder of wood with the right hand, the left hand remaining relaxed and free, immediately the needle will move from west to south, and describe an angle of 30°, 40°, and even 50°; on relaxing the grasp, the needle will return to its original position. The experiment may be reversed by employing the left arm, and leaving the right arm free: the needle will, in this case, be deflected from west to north. The reversing of the action of the needle proves the influence of the nervous force. The conditions, it may be added, essential to the success of the experiment are: 1st, Great muscular and nervous energy; 2nd, The contraction of only one arm at a time; 3rd, Dryness and cleanliness of skin; and 4th, Freedom from any kind of wound on the immersed part.

Sixth Experiment.

In making electrical experiments of the simplest kind, it soon becomes apparent that certain substances, such as glass, sealing-wax, &c., retain the condition of electrical excitement; whilst other bodies, and especially the metals, seem wholly incapable of electrical excitation: hence the classification of bodies into conductors and non-conductors of electricity. This arrangement is not strictly correct, because no substance can be regarded as absolutely a conductor, or *vice versâ*. It is better to consider these terms as meaning the two extremes of a long chain of intermediate links, which pass by insensible gradations the one into the other. In the manufacture of electrical apparatus, glass is of course largely employed, and this substance, with brass and wood, constitute the usual materials. One of the most instructive pieces of apparatus is the electroscope, which can be made with a gas jar, a cork, a piece of glass tube, brass wire and ball, or a flat disc of brass, with some Dutch metal, or still better, gold leaf. The latter is first cut into strips by retaining the leaf between a sheet of well-glazed paper and cutting through the paper and the copper or gold leaf, otherwise it would be impossible to cut the metal, on account of its excessive thinness, except with a gilder's knife and cushion. The cork is next fitted to the gas jar, and perforated with a hole to admit the glass tube, which must be thoroughly dry, and is best coated both inside and out with the shell-lac varnish described at page 175. Some dry silk is wound round the brass wire, so that it remains fixed and upright in the glass tube, the end outside the jar having a ball, or still better, a flat disc of brass attached, and the other extremity being split so as to act like a pair of forceps, to retain a piece of card to which the gold leaves are attached. By removing the cork, tube, and brass wire bodily from the neck of the gas jar, and then in a perfectly still atmosphere carefully bringing the card, slightly wetted with gum at the extremity, on two of the cut gold leaves, they may be stuck on, and the whole is again arranged inside the dry gas jar, and forms the important instrument called the electroscope. (Fig. 164.) With the help of this arrangement, a number of highly instructive experiments are performed.

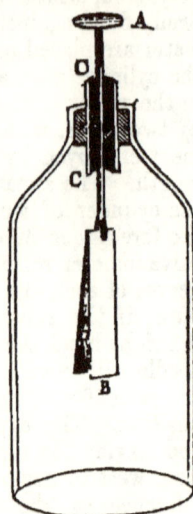

Seventh Experiment.

First, the difference between conductors and non-conductors is admirably shown by rubbing a bit of sealing-wax against a piece of woollen cloth or flannel; on bringing the wax to the brass disc of the electroscope the gold leaves no longer hang quietly side by side, but stand out and repel each other, in obedience to the law *"that bodies similarly electrified repel each other."* If the brass cap is touched whilst the leaves are in this electrical state, they fall again to their original position, showing that sealing-wax, after being excited, retains its electrical condition, as also the gold leaves, because they are supported on glass, or what is termed *insulated*—i.e., cut off from conducting communication with surrounding objects. When, however, the sealing-wax is passed through a damp hand, or the brass disc of the electroscope touched, the electricity is conveyed away to the earth, because the human body is a conductor of electricity.

Fig. 164. A. The brass wire, with flat disc outside, and forceps holding gold leaf B inside the jar. C C. The glass tube.

Eighth Experiment.

When a brass wire is rubbed and brought to the electroscope, the leaves do not move, in consequence of the electricity passing away to the earth through the body as fast as it is generated: it is just like pouring water into a leaky cistern; but if the brass wire is tied to a long stick of sealing-wax, and this latter held in the hand whilst the wire is rubbed with a bit of flannel, then the gold leaves of the electroscope are affected, on account of the insulation of the metal, as every substance which can be rubbed (even fluids, as water) produces electricity.

Ninth Experiment.

An insulating stool is merely a piece of strong square board, supported on glass legs, which should be well varnished. If the assistant stands on this stool and touches the disc of the electroscope, no movement of the leaves takes place until his coat is briskly struck with a piece of dry silk or skin, when the usual repulsion occurs.

Fig. 165. Assistant standing on the insulating stool and touching the disc of the electroscope whilst being struck with a dry handkerchief.

Tenth Experiment.

If a little powdered chalk is placed inside a pair of bellows, and then forcibly ejected on to the disc of the electroscope, the friction of the particles of chalk against the inside of the nozzle of the bellows and against the disc of the instrument soon liberates sufficient electricity to cause the gold leaves to stand out and repel each other.

Eleventh Experiment.

Whilst the leaves of the electroscope are repelled from each other by the application of a bit of rubbed sealing-wax, they may be again caused to approach each other on bringing a dry glass tube previously rubbed with a silk-handkerchief; because the electricity obtained from sealing-wax is different from that procured from glass: the former is called *resinous* or *negative* electricity, the latter *positive* or *vitreous* electricity. Either, separately, is *repulsive* of its own particles, but *attractive* of the other. No electrical excitation can occur without the separation of these two curious states of electricity, and electrical quiescence takes place when the two electricities are brought together; hence the fall of the gold leaves repelled by rubbed wax when the excited glass is brought towards the disc of the electroscope. This experiment may be reversed by repelling the leaves first with the excited glass, and then bringing the rubbed wax, when the same effect takes place.

Twelfth Experiment.

To show the important elementary truth, that in all cases of electrical excitation the two kinds of electricity are generated, take a dry roll of flannel, and holding it as lightly as possible, rub it against a bit of wax. If the flannel is brought to the electroscope, the leaves repel each other, and they immediately fall when the wax is now approached, because the flannel is in the positive or vitreous state of electricity, whilst the sealing-wax is in the negative or resinous condition.

Thirteenth Experiment.

Any kind of friction generates electricity. A little roll brimstone placed in a dry mortar and powdered, and then thrown on to the electroscope, quickly causes the repulsion of the leaves.

Fourteenth Experiment.

A sheet of dry brown paper laid on a flat surface, and vigorously rubbed with a piece of india-rubber, produces so much electricity that sparks and flashes of light are apparent in a dark room when it is lifted from the table; and it affects the leaves of the electroscope very powerfully, so much so that care must be taken to apply it very carefully to the disc, or the violence of the repulsion may cause the fracture of the gold leaves, and then a great deal of time is wasted before they can be put on again.

Fifteenth Experiment.

A dry wig or bunch of horse-hair when combed becomes electrical, and likewise affects the leaves of the electroscope.

Sixteenth Experiment.

Two dry silk ribbons, the one white and the other black, passed rapidly together through the fingers, exhibit sparks and flashes of light when drawn asunder, and also cause the gold leaves to repel each other.

Seventeenth Experiment.

Much instructive amusement is afforded by testing the gold leaves when separated from each other during either of the former experiments, with an excited piece of sealing-wax. If the electricity produced is negative, they repel each other further when the excited wax is approached; if positive, they fall when the excited wax is brought near them.

Eighteenth Experiment.

When fresh, dry, ground coffee is received on to the disc of the electroscope, as it falls from the mill, powerful electrical excitation is displayed, and this is sometimes so apparent, that the particles cling around the lower part of the mill or to the sides of the cup or basin held to catch it.

Nineteenth Experiment.

After playing a tune on a violin, hold the bow (well rosined) to the electroscope, when the usual divergence of the leaves will be apparent.

Twentieth Experiment.

Cut some chips from a piece of wood with a knife attached to a glass handle, and as they fall on to the electroscope the leaves are repelled.

Twenty-first Experiment.

Warm a piece of bombazine by the fire and then draw out some of the threads (which are of two kinds—viz., silk and wool), and place them on the electroscope, when divergence of the leaves immediately takes place.

Twenty-second Experiment.

Put upon the same leg a worsted stocking and over that a silk one, if the latter is now quickly rubbed all over with a dry hand and near the fire, and then suddenly slipped off, the sides repel each other, and the silk stocking retains very much the same shape as if the leg still remained in it, and of course collapses as the electricity passes away.

Twenty-third Experiment.

Electrical machines consist only in the better arrangement of larger pieces of glass and a more convenient mechanical contrivance for rubbing them, and are of two kinds—viz., the cylinder and plate machines; it is usual to give directions for the manufacture of an electrical machine from a common bottle, and doubtless such rude instruments have been made, but as Messrs. Elliott Brothers, of 30, Strand, now supply excellent small machines at a very low cost, it is hardly worth while to incur even a small expense for an instrument that must at the best be a very imperfect one and frequently out of order. (Fig. 166.)

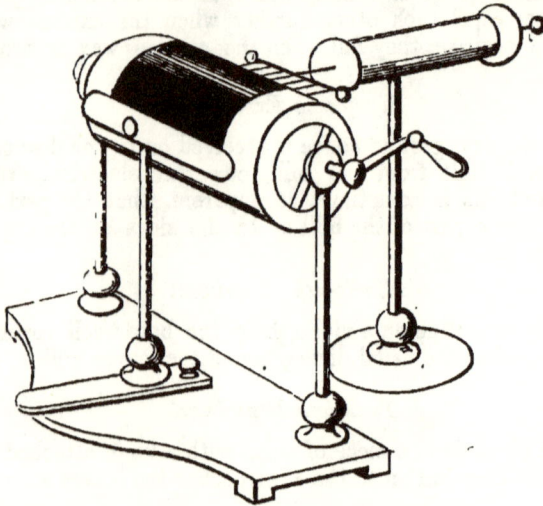

Fig. 166. A cylinder electrical machine.

Plate machines are somewhat more expensive than cylinder ones, but at the same time are more quickly prepared for experiments, and Mr. Hearder, of Plymouth, states, that the *secret* in obtaining the greatest amount of electricity from a cylinder machine, is to keep the inside of the glass absolutely clean, dry, and free from dust. Sometimes the glass of which electrical machines are made is wholly unfit for elec-

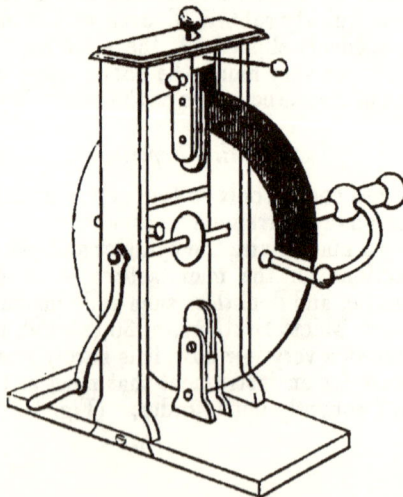

Fig. 167. The ordinary plate electrical machine.

trical purposes, in consequence of the decomposition of the surface from imperfect manufacture and the liberation of the alkali. (Figs. 167, 168.)

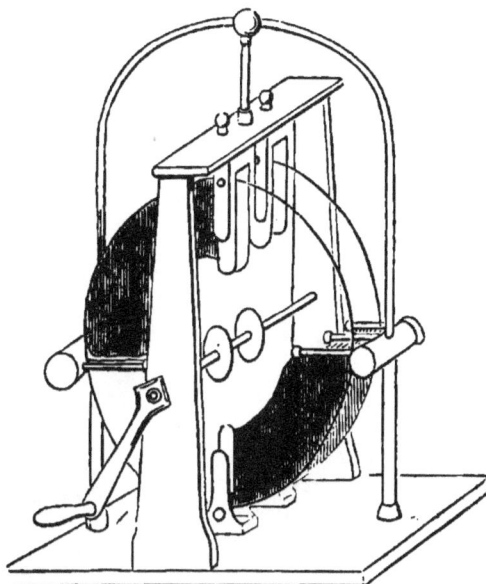

Fig. 168.   Woodward's double plate electrical machine, giving a much la⁻g⁻r
quantity of electricity than Fig. 167.

Twenty-fourth Experiment.

Cylinder and plate machines are furnished with proper rubbers, and before using the instrument it is usual to remove them, and after carefully cleaning the glass with a dry silk handkerchief before a fire, the rubbers are scraped with a paper-knife to remove the old amalgam, and fresh applied by first melting the end of a tallow candle slightly, and after passing this over the rubber, the finely powdered amalgam is now dusted on to it. Electrical amalgam is prepared by fusing one part of zinc with one of tin, and then agitating the liquid mass with two parts of hot mercury placed in a wooden box; when cold it should be carefully powdered and kept in a well-stoppered bottle for use. When the amalgam has been applied, the rubbers are again screwed in their places, and the machine when turned (if the atmosphere is tolerably dry) will emit an abundance of bright sparks.

Twenty-fifth Experiment.

Attraction and repulsion are shown on a larger scale, with the assistance of electrical machines, by placing a fishing rod (the last joint of which is made of glass) in an erect position, and attaching to the extremity a long tassel of paper from which a thin wire passes to the prime conductor of the electrical machine; on turning the instrument, the strips of paper all stand out and repel each other. (Fig. 169.)

Twenty-sixth Experiment.

Suspend from the prime conductor by a chain a circular brass plate

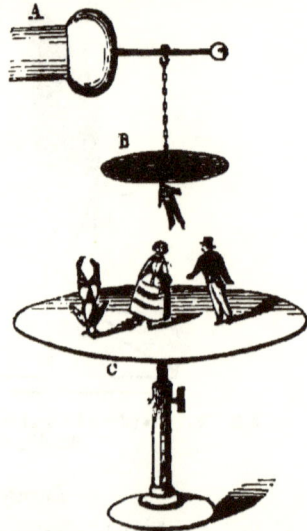

Fig. 169. A A. The glass joint of the fishing-rod, from which the last joint, carrying the paper tassel, B, projects. C. The electrical machine.

Fig. 170. A. Prime conductor. B. Upper brass-plate. C. Lower ditto. The figures are seen between B and C.

and under this place another supported by a brass adjusting stand. If pith figures of men and women are placed on the lower plate, they rise directly the machine is turned, although sometimes, in consequence of irregularity in the adjustment of the centre of gravity, they perversely dance on their heads instead of the usual position; out of half a dozen figures, one only perhaps will be found to dance well, by alternately jumping to the upper plate and falling to the lower one to discharge the excess of electricity; and indeed the experiment will be found to succeed better with one or two only on the plate instead of a number, as they cling together and impede each other's movements. (Fig. 170.)

Twenty-seventh Experiment.

An assistant provided with a wig of well-combed hair presents a most ridiculous appearance when standing on the insulating stool and connected by a wire with the prime conductor of the electrical machine, every hair, when not matted together, standing out in the most absurd manner, when the machine is put in motion.

Twenty-eighth Experiment.

Whilst standing on the stool, sparks may be obtained from his body, and if some tow is tied over a brass ball, and moistened with a little ether, and presented to the tip of his finger, a spark flies off which quickly sets fire to the inflammable liquid.

Twenty-ninth Experiment.

If small discs of tinfoil, cut out with a proper stamp, are pasted in continuous lines over plate glass, or spirally round glass tubes, a very

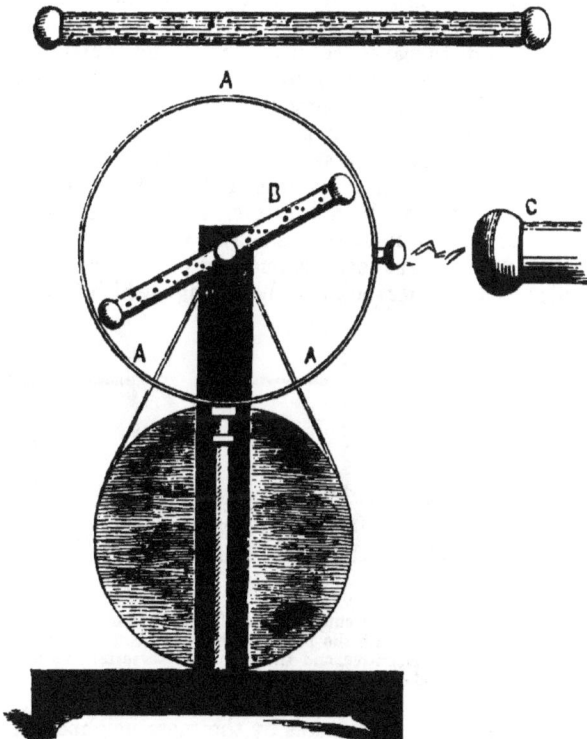

Fig. 171. ᴀ ᴀ ᴀ. A ring of brass wire supported on a glass pillar inside which the spiral tube, ʙ, revolves, and produces beautiful and ever-changing circles of light, when connected with the conductor, c, of the electrical machine.

pretty effect is produced when they receive the sparks from the electrical machine, and the passage of the electricity from one disc to the other produces a vivid spiral or other line of light. When the tube is mounted in a proper apparatus, so as to revolve whilst the sparks pass down the spiral tube, the effect of the continuous electric sparks is much heightened. (Fig. 171.)

Thirtieth Experiment.

A great variety of experiments, depending on the proper arrangement of discs of tinfoil on various tubes of coloured glass are manufactured, and some in the form of windmills, the sails being made luminous by the passage of the electricity. The names of illustrious electricians, beautiful crescents, stars, and even profile portraits, have been produced in continuous streams of electric sparks.

Thirty-first Experiment.

When an electrified body is brought towards another which is not electrical, the latter is thrown into the opposite state of electricity as long as the excited body remains in its neighbourhood; and this condition of electrical disturbance, set up without any contact or supply of electricity, is called *induction*, and involves a vast number of interesting facts, which are thoroughly discussed in Dr. Noad's excellent work on electricity, but can only be briefly alluded to here.

If a number of lengths of brass wire, supplied with balls at the extremities, are supported on glass legs and arranged in a line, with a little pith ball attached to a thread hanging from each end of the length of brass wire, the effect of induction is shown very nicely; and when an excited glass rod is brought towards one end of the series, the rising of the pith balls to each other betrays the change which has occurred in

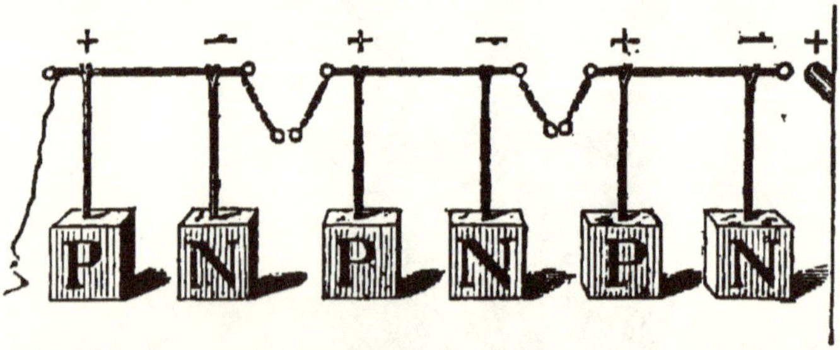

Fig. 172. The lengths of brass wire supported on glass rod pillars indented by blowpipe, so as to retain the brass wires with the pith balls hanging from each series, the letters P and N mean Positive and Negative, and the signs for these terms are placed above. The letters P and N are painted on the blocks which support the glass rods.

the electrical state of the brass wires by the mere neighbourhood of the excited glass tube. The glass tube is electrified positively, and attracts the negative electricity from the brass wire towards the end nearest to it; the other extremity of the brass wire is found to be in the positive state, and this re-acting on the next, and so on throughout the lengths, completes the electrical disturbance in the whole series. (Fig. 172.)

Thirty-second Experiment.

If an insulated brass rod (such as has been described in the last experiment) is touched by the finger whilst under induction, it remains permanently electrified on the removal of the disturbing electrified body; and it is on this principle that the useful electrical machine called the Electrophorus is constructed. This *constant* electrical machine—for it will remain in action during weeks and months if kept sufficiently dry— was invented by Volta in the year 1774, and has been brought to great perfection by Mr. Lewis M. Stuart, of the City of London School; so that with a little additional apparatus the whole of the fundamental prin- ciples of electricity can be demonstrated. It consists of a flat brass or tin circular dish about two feet in diameter and half an inch deep, which is filled with a composition of equal parts of black rosin, shell-lac, and Venice turpentine; the rosin and the Venice turpentine being first melted together, and the shell-lac added afterwards, care of course being taken that the materials do not boil over and catch fire, in which case the pot must be removed from the heat, and a piece of wet baize or other woollen material thrown over it. Another tin or brass circular plate of twelve inches diameter, and supported in the centre with a varnished glass handle nine inches long, is also provided, and the resinous plate being first excited by several smart blows with a warm roll of flannel, the plate held by the glass handle is now laid upon the centre of the resinous one, and if removed directly afterwards, does not afford the electric spark; but if, whilst standing upon the excited resinous plate, it is touched, and then removed by the glass handle, a powerful electric spark is obtained; and this may be repeated over and over again with the like results, provided the plate with the glass handle is touched with the finger just before lifting it from the resinous plate. (Fig. 173.)

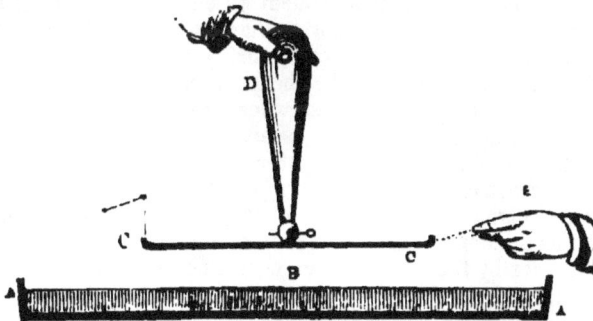

Fig. 173. A A. Large circular tin or brass disc with turned-up edge half an inch deep, and containing the resinous mixture B, which is rubbed with the warm flannel. C C. The upper plate supported by the glass handle D, a pith ball attached to a wire shows the electrical excitation, and the spark is supposed to be passing to the hand E.

The electricity excited on the resinous plate is not lost, and by induction sets up the opposite condition in the plate with the glass handle. The resinous plate, being excited with negative electricity, disturbs the electrical quiescence of the upper plate, and positive electricity is found on the surface touching the resinous plate, and negative electricity on the upper one, so that when it is removed without being touched, the two electricities come together again, and no spark is obtained; but if, as already described, the upper plate is touched whilst under induction, then positive electricity appears to pass from the finger to the negative electricity on the upper side of the plate, when the two temporarily neutralize each other, and then, when the plate is removed, the excess of electricity derived from the earth through the finger becomes apparent. Induction requires no sensible thickness in the conductors, and can be just as well produced on a leaf of gold as on the thickest plate of metal; and it should be remembered that non-conductors do not retain their state of electrical excitation when the disturbing cause is removed, whereas conductors possess this power, and this fact brings us to the consideration of the Leyden jar.

Thirty-third Experiment.

If one side of a dry glass plate is held before and touches a brass ball proceeding from the prime conductor of an electrical machine whilst in action, the other side is soon found to be electrical; this does not arise from the conduction of the electricity through the particles of the glass, but is produced by induction, the side nearest the ball being in the positive state, and the other side negative: as glass is a non-conductor of electricity, the effect is much increased by coating each side with tinfoil, leaving a margin of about two inches of uncovered glass round the covered portion, then, if one side of such a plate is held to the prime conductor of the electrical machine, and the other connected with the ground, a powerful charge is accumulated; and if the opposite sides are brought in contact with a bent brass wire, a loud snapping noise is heard, and the two electricities resident on either side of the glass come together with the production of a brilliant spark, or if the hands are substituted for the bent brass wire, that most disagreeable result is obtained—viz., an *electric shock;* hence these glass plates are sometimes fitted up as pictures, and when charged and handed to the unsuspecting recipient, he or she receives the electric discharge to the great discomfort of their nervous system.

Mica is sometimes substituted for glass, and the late Mr. Crosse, the celebrated electrician, constructed a powerful combination of coated plates of this mineral. It consisted of seventeen plates of thin mica, each five inches by four, coated on both sides with tinfoil within half an inch of the edge. They were arranged in a box with a glass plate between each mica plate, all the upper sides were connected by strips of tinfoil to one side of the box, and all the under surfaces in the same manner with the opposite extremity of the box. They were charged like an ordinary Leyden battery.

Thirty-fourth Experiment.

If the glass plate coated with tin-foil is charged, and then placed up-right on a stand, it may be slowly discharged by placing a bent wire on the edge with the extremities covered with pith balls. The wire balances itself, and continues to oscillate with noise until the electricities of the two surfaces neutralize each other. (Fig. 174.)

Thirty-fifth Experiment.

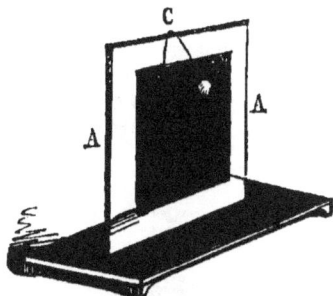

Fig. 174. A A. Glass plate or stand coated with tinfoil on each side, B. C. Wire with pith balls oscillating during the discharge of the glass plate.

It is easy to imagine the glass plate of the last experiment rolled up into the more convenient form of the Leyden jar, which consists of a glass vessel lined both inside and out with tinfoil, leaving some two or three inches of the glass round the mouth uncovered and varnished with shell-lac; a piece of dry wood is fitted into the mouth of the jar, through which a brass wire and chain are passed, and the end outside is fitted with a ball. The Leyden jar is charged by holding the ball to the prime conductor of the electrical machine until a sort of whizzing noise is heard, caused by the excess of electricity passing round the un-covered part of the jar and not through it, as the smallest crack in the glass of the Leyden jar would render it useless. Electricity is sometimes called a fluid, and the fact of collecting it like water in a jar, helps us to understand this analogy.

Fig. 175. The Leyden jar and brass wire discharger.

The noise, the bright spark, or the shock are obtained by grasping the outside with one hand and touching the ball with a brass wire held in the other. (Fig. 175.)

Thirty-sixth Experiment.

The jar is silently discharged if the balls are removed from the discharger and points used instead; so, also, the whole of the electricity produced by an electrical machine in full action may be readily drawn off by a pointed conductor, such as a needle, placed at the end of a brass wire. Electricity passes much more rapidly through points than rounded surfaces, hence the reason why all parts of electrical apparatus are free from sharp points and rough asperities.

Thirty-seventh Experiment.

Extremely thin wires may be burnt by passing the charge of a large Leyden jar through them. The show jars, called specie jars, usually decorated and placed in the windows of chemists' shops, make excellent Leyden jars, when not too thick; and with two of the largest, all

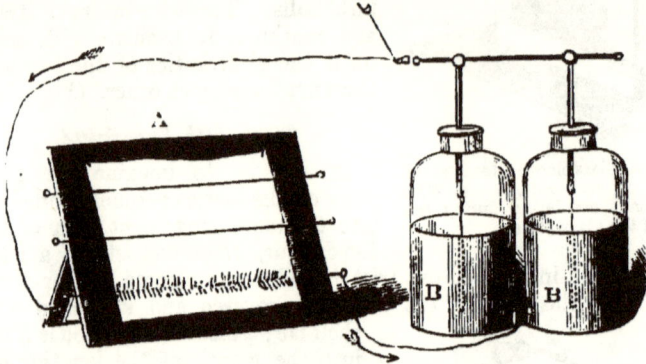

Fig. 176. A. Mahogany board with a sheet of white paper and three pairs of brass wires and balls fixed in the wire, three on each side. The thin wires are stretched between the balls, and the lower one is in course of deflagration. B B. Charged large Leyden battery of two jars; the arrows indicate the path of the electricity.

the interesting effects produced by accumulated electricity may be displayed. To pass the discharge through wires, nothing more is required than to strain them across a dry mahogany board, between two brass wires and balls, and if a sheet of white paper is placed under them, most curious markings are produced by the fine particles of the deflagrated metal blown into the surface of the paper. An arrangement of two or more Leyden jars is usually called a Leyden Battery, just as a single cannon is spoken of as a gun, whilst two or more constitute a battery. (Fig. 176.)

Thirty-eighth Experiment.

Little models of houses, masts of ships, trees, and towers are sold by the instrument makers, and by placing a long balanced wire on the top of the pointed wire of a large Leyden jar, having one end furnished with wool to represent a cloud, a most excellent imitation of the effects of a charged thunder-cloud is produced. The mechanical effect of a flash of lightning has been analysed, and it has been stated, in one instance, that the power developed through fifty feet was equal to a 12,220 horse-power engine, or about the power of the engines of the *Great Eastern*, and that the explosive power was equal to a pressure of three hundred millions of tons. (Fig. 177.)

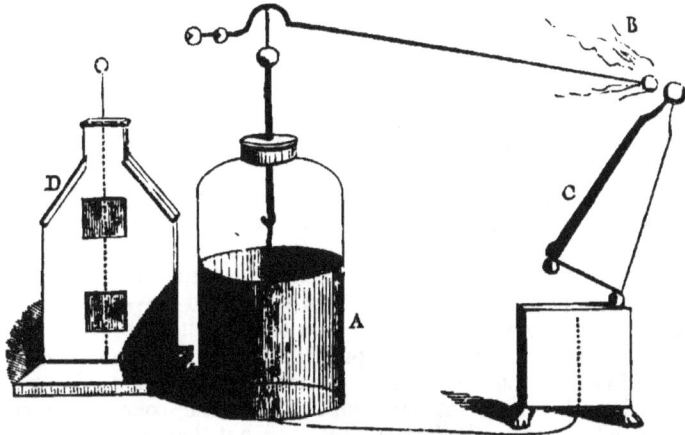

Fig. 177. A. Charged Leyden jar with balanced wire and wool at B, representing a thunder-cloud.   c. The obelisk overturned with the discharge.   D. Another model of the gable end of a house; the square pieces of wood fly out when the continuity of the conductor is broken.

It was the learned but humble minded Dr. Franklin who established the identity between the mimic effects of the electrical machines (such as have been described), and the awe-inspiring thunder and lightning of nature.   A copper rod, half an inch thick, pointed and gilt at the extremity, and carried to the highest point of a building, will protect a circle with a radius of twice its length.   The bottom of the rod must be passed into the earth till it touches a damp stratum.

Fig. 178. A storm.

CHAPTER XIV.

VOLTAIC ELECTRICITY.

IN describing the various means by which electricity may be obtained, it was stated that "Chemical Action" was a most important source of this remarkable agent; at the same time it must be understood that it is not every kind of chemical action which is adapted for the purpose; there are certain principles to be rigidly adhered to—first, in the generation of the force; and secondly, in carrying it by wires so as to be applicable either for telegraphic purposes, or for the highly valuable processes of electrotyping and electro-silvering, plating, and gilding.

A lighted candle, or an intense combustion of coal, coke, or charcoal, no doubt involves the production of electricity, but there are no means at present known by which it may be collected and conducted; when that problem is solved, the cheapest voltaic battery will have been constructed, in which the element decomposed is charcoal, and not a metal, such as iron or zinc. The first and most simple experiment that can be adduced in proof of electrical excitation by chemical means, is to take a bit of clean zinc and a clean half-crown, and placing one on the tongue and the other below it, as long as they remain separate no effect is observed, but directly they are made to touch each other, whilst in that position, a peculiar thrill is rendered evident by the nerves of the tongue, which in this case answers the same purpose as the electroscope already described, and in a short time a peculiar metallic taste is perceptible.

It has been stated over and over again that it was to a somewhat similar circumstance we owe the discovery of voltaic electricity, and the story of the skinned frogs agitated and convulsed by an accidental communication with two different metals, or, as some say, with the electricity from an ordinary machine, has been repeated in nearly every work on the science. Professor Silliman, however, asserts that the galvanic story is doubtful, and is a fabrication of Alibert, an Italian writer of no repute, and that greater merit is due to Galvani than that of being merely the accidental discoverer of this kind of electricity, because he had been engaged for *eleven* years in electro-physiological experiments, using frogs' legs as electroscopes. It was whilst experimenting on animal irritability, Galvani noticed the important fact that when the nerve of a dead frog, recently killed, was touched with a steel needle, and the muscle with a silver one, no convulsions of the limb were produced until the two different metals were brought in contact, and he explained the cause of these singular after-death contortions by supposing that the nerves and muscles of all animals were in opposite states of electricity, and that these nervous contractions were caused by the annihilation, for the time, of this condition, by the interposition of a good conductor between them. This theory of Galvani had several opponents, one of whom, the cele-

O

brated Volta, succeeded in pointing out its fallacy; he maintained that the electrical excitement was due entirely to the metals, and that the muscular contractions were caused by the electricity thus developed passing along the nerves and muscles of the dead animal.

To Volta we are indebted for the first voltaic battery, and the distinguished philosopher may truly be said to have laid the foundation of this now *commercially* valuable branch of science.

First Experiment.

If a plate of clean bright zinc is placed in a vessel containing some dilute sulphuric acid, energetic action occurs from the oxidation of the metal, and its union as an oxide with the acid, and the escape of a multitude of bubbles of hydrogen gas. After the action has proceeded some time, the zinc may be removed, and if a little quicksilver is now rubbed over the surface with a woollen rag tied on the end of a stick, it unites with the metal, and the surface of the zinc assumes a brilliant silvery appearance, and is said to be amalgamated. In that condition it is no longer acted upon by dilute sulphuric acid, and for the sake of economy this is the only form in which zinc should be employed in the construction of voltaic batteries or single circles. If a clean plate of copper, with a wire attached, is now placed in the dilute acid opposite to and not touching the amalgamated zinc plate, which may also be furnished with a conducting wire, no bubbles of hydrogen escape until the wires from the two metals are brought in contact, and then, singular to relate, the hydrogen escapes from the copper plate, whilst the oxygen is rapidly absorbed by the zinc, and a current of electricity will now be found to pass from the zinc through the fluid to the copper, and back again through the wire to the starting-point, and if the wires are disconnected, the chemical action ceases, and no more electricity is produced. (Fig. 179.)

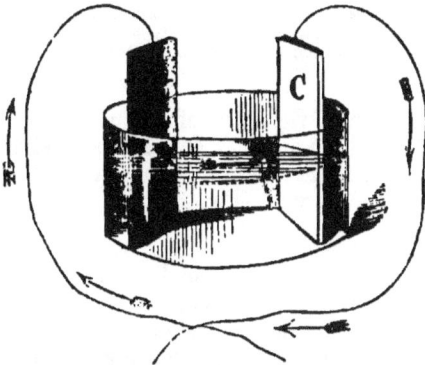

Fig. 179. A single voltaic circle, consisting of a zinc and copper plate (marked z and c) in dilute acid. The arrows show the direction of the current.

The passage of the current of electricity is not discoverable by the electroscope, because it is adapted only to indicate electricity of high tension or intensity, such as that produced from the electrical machine, which will pass rapidly through a certain thickness of air, and cause pith balls to stand out and repel each other; such effects are not producible by a single voltaic circle, or even an ordinary voltaic battery, although one comprising some hundreds of alternations would produce an effect on a delicate electrometer; hence voltaic electricity is said to be of low intensity, and this property makes it much more useful to mankind, because it has no desire to leave a metallic path prepared for it, and does not seize the first opportunity, like the electricity from the electrical machine, to run away to the earth through the best and shortest conductor offered for it. If electricity had only been producible by friction, we should never have heard of electrotyping, and the other useful applications of electrical force of low intensity.

Second Experiment.

To ascertain the passage of a current of voltaic electricity, the instrument called the galvanometer needle is provided, which consists of a coil of copper wire surrounding a magnetic needle, so as to

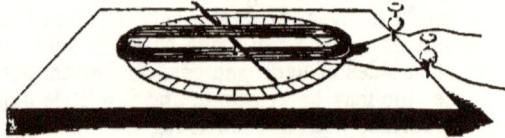

Fig. 180. A galvanometer needle, consisting of a coil of covered copper wire, the ends of which terminate at the binding screws. The magnetic needle is suspended on a point in the centre, and the coil is surrounded with a graduated circle.

leave the latter freedom of motion from right to left, or *vice versâ*. When this coil is made part of the voltaic circle it becomes magnetic, and reacting on the magnetized needle, deflects it to one side or the other, according to the direction of the current. (Fig. 180.)

Third Experiment.

If a number of simple voltaic circles, such as the one described in the first experiment, are connected together, they form a voltaic battery, in which of course the quantity of electricity is greatly increased. Batteries of all kinds, from the original Volta's pile, consisting of round zinc and copper plates soldered together with interposed cloth moistened with dilute sulphuric acid, or his *couronne des tasses*, consisting of zinc and silver wires soldered together in pairs, and placed in glass cups containing dilute acid, to the improved batteries of Cruikshank, Wilkinson, Babington, Wollaston, and the still more perfect arrangements of Daniell, Mullins, Shillibeer, and Grove, have been from time to time recommended for their own peculiar features.

Amongst these several inventions, none will be found more useful than the *constant* battery of Daniell for electrotyping, silvering, gilding, and other purposes, and Grove's battery for all the more brilliant results, such as the deflagration of the metals or the production of the electric light. The construction of the Daniell and Grove batteries will therefore be described. The former consists of a cylindrical vessel made of copper, in which is suspended or placed (as it is open at the top) a membranous, brown-paper, canvas, or porous earthenware tube, containing an amalgamated rod of zinc. To charge this arrangement, a strong solution of sulphate of copper, with some sulphuric acid, is poured into the copper vessel, which is provided usually with a sort of colander at the top to hold crystals of sulphate of copper, and in the porous tube containing the zinc rod is poured dilute sulphuric acid. A number of these cylinders of copper, twenty inches high and three inches and a half in diameter, arranged in wooden frames to the number of

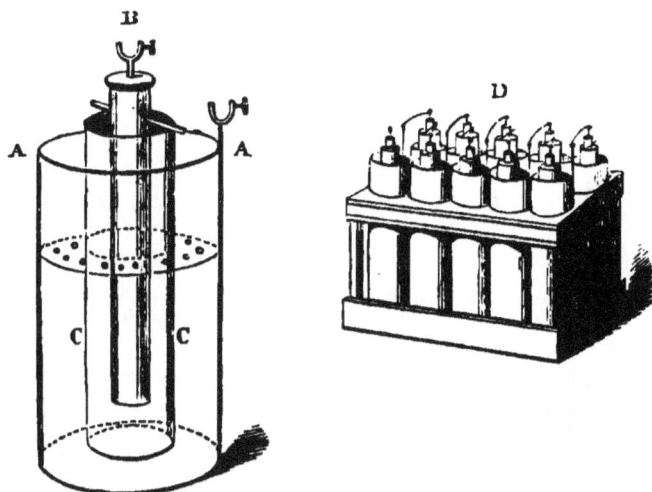

Fig. 181. A A. Copper cylindrical vessel with colander to hold the crystals of sulphate of copper. B. The amalgamated zinc rod inside the porous cell C C. D. A series of single cells forming a Daniell's battery.

twenty, afford a quantity of electricity sufficient to demonstrate all the usual phenomena. (Fig. 181.)

Professor Grove's battery consists of a flat glazed earthenware vessel containing a flat porous cell. An amalgamated zinc plate is placed outside the porous cell, and a platinum plate inside the latter. The arrangement is put in action by pouring dilute sulphuric acid round the zinc and strong nitric acid inside the porous cell. A set of Grove's nitric acid battery, as manufactured by Messrs. Elliott, Brothers, of 30, Strand, with fifty pairs of sheet platinum, five inches by two inches and a quarter, and double amalgamated zinc plates, flat porous cells, and separate earthenware troughs for each pair, and stout mahogany stand, arranged in ten series of five pairs, will evolve with a proper voltameter one hundred cubic inches of the mixed gases per minute from the decomposition of water, and will exhibit a most brilliant electric light, when arranged as a single series of fifty pairs of plates. Even thirty pairs exhibit the most splendid effects, whilst forty may be regarded as the happy medium, giving all the results that can be desired. (Fig. 182.)

The advantage of employing amalgamated zinc is very prominently illustrated whilst using any powerful arrangements of either Daniell's or Grove's batteries, as they will remain for hours quiescent, like a giant asleep, until the terminal wires of the series are brought in contact

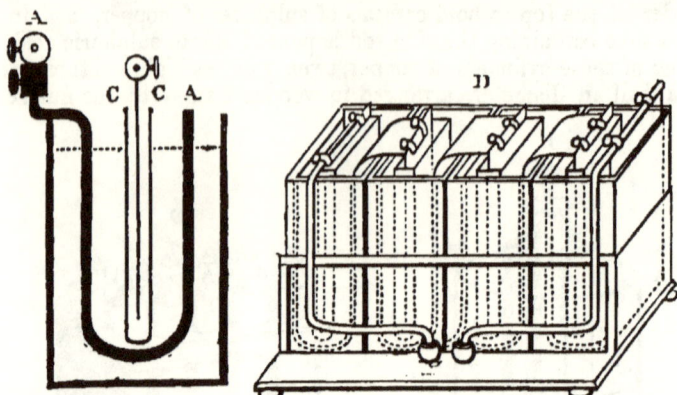

Fig. 182. A A. Amalgamated zinc plate in flat earthenware trough. Attached to a binding screw is the platinum plate in porous cell, c c. D. A series of single cells forming a Grove's battery.

either through the intervention of some fluid under decomposition or by means of charcoal points. The author had the pleasure of witnessing at King's College some of the effects of an enormous battery, prepared by the late Professor Daniell, and consisting of seventy of his cells.

A continuous arch of flame was produced between two charcoal points, when distant from each other three quarters of an inch, and the light and heat were so intense that the professor's face became scorched and inflamed, as if it had been exposed to a summer heat. The rays collected by a lens quickly fired paper held in the focus.*

Fourth Experiment.

It is by "chemical action" the electricity is produced, and as action and reaction are always equal, but contrary, we are not surprised to find that the electricity from the voltaic battery will in its turn decompose chemically many compound bodies, of which water is one of the most interesting examples. It was in the year 1800, and immediately after Volta's announcement to Sir Joseph Banks of his discovery of the pile, that Messrs. Nicholson and Carlisle constructed the first pile in England, consisting of thirty-six half-crowns, with as many discs of zinc and pasteboard soaked in salt water. These gentlemen, whilst experimenting with the pile, observed that bubbles of gas escaped from the platinum wires immersed in water and connected with the extremities of the Volta's pile, and covering the wires with a glass tube full of water, on the 2nd of May, 1800, they completed the splendid discovery of the fact that the Volta's current had the power to decompose water and other chemical compounds.

* By the light from the same battery photogenic drawings were taken, and the heating power was so great as to fuse with the utmost readiness a bar of platinum one-eighth of an inch square; and all the more infusible metals, such as rhodium, iridium, titanium, &c., were melted like wax when placed in small cavities in hard graphite and exposed to the current of electricity.

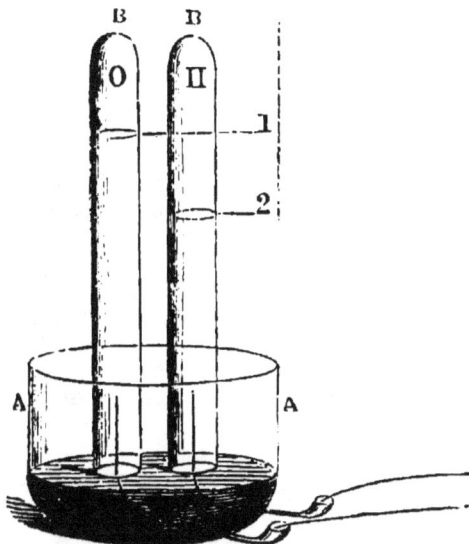

Fig. 133. A A. A finger glass with two holes drilled to pass the wires through, which are imbedded in cement up to the platinum plates. B B. Glass tubes, closed at one end and open at the other, which are placed over the platinum plates to receive the liberated oxygen and hydrogen. The scale at the side shows the respective volumes of two of H to one of O.

In 1801, Davy had succeeded to a vacant post in the Royal Institution, and on Oct. 6th, 1807, made his transcendent discovery of potassium with the aid of the voltaic battery, and from that and other experiments inferred that the whole crust of the globe was composed of the oxides of metals. To exhibit the decomposition of water, two platinum plates with proper connecting wires, passing to small metallic cups full of mercury, are cemented inside a glass vessel, which is then filled with dilute sulphuric acid. Just above the platinum plates and over them, stand two glass tubes also containing the same fluid in contact with the battery. Two measures of hydrogen are found in one tube, and one of oxygen in the other. (Fig. 183.)

To measure the quantity power of the voltaic battery, an important instrument invented by Faraday is used. It consists of separate platinum plates cemented in a wooden stand, and over which a capped air-jar with a bent pipe is also cemented. This apparatus contains dilute sulphuric acid of the same strength as that used in the battery under examination, and by taking the time, the quantity of the mixed oxygen and hydrogen gases producible by a battery per minute is accurately determined, the gases of course being collected in a graduated jar. (Fig. 184.)

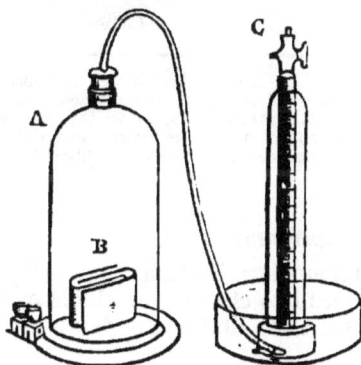

Fig. 184. A. Gas jar with cap and bent tube passing to the graduated tube C; the jar is cemented in the same stand which carries the connecting cups, wires, and platinum plates, which are bent round each other to improve the action of the voltameter.

Fifth Experiment.

By grouping the simple circles forming a voltaic battery in various numerical relations, the *quantity* and *intensity* effects are modified.

Thus, if a series of thirty pairs of Grove's battery are all connected together in consecutive order, the smallest *quantity* and the largest *intensity* effect is produced.

If changed to two groups of fifteen each, the quantity is doubled—that is to say, it will produce double the quantity of the mixed gases from the voltameter with half the intensity.

If arranged in three groups of ten each, it is trebled with a proportional loss of intensity, until the grouping reaches six series of five each, when a maximum supply of the mixed gases is obtained from the voltameter.

In arranging the groups, all the zinc ends of each series are connected, and all the platinum ends are likewise joined by proper wires.

Sixth Experiment.

A plate-glass trough, containing a few grains of iodide of potassium dissolved in water with some starch, is quickly decomposed into its elements by placing in two platinum plates and connecting them with the wires of the voltaic battery. If the glass trough is divided in the centre with a bit of cardboard, the purple colour of the iodine and starch is shown very beautifully on one side, but not on the other, as iodine is liberated at one pole and the alkali at the other. (Fig. 185.)

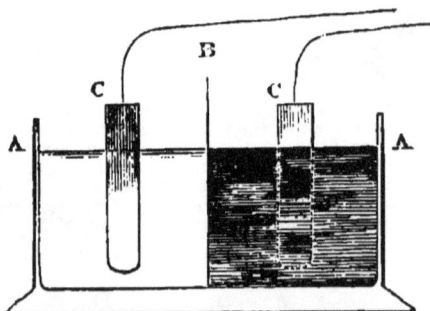

Fig. 185. A A. A glass trough containing the salt dissolved in water, and divided temporarily with a bit of cardboard, B. c c are the two platinum plates connected with the battery, and the shaded side is supposed to represent the liberation of the iodine.

Seventh Experiment.

Some solution of common salt coloured with sulphate of indigo and placed in the trough is decomposed into chlorine, which bleaches one side of the indigo solution, and the alkali liberated on the other does not affect it.

Eighth Experiment.

Some nitrate of potash dissolved in water and coloured with litmus placed in the glass trough, changes red on one side of the cardboard by the liberation of acid, and is not affected on the other.

In these experiments the oxygen, iodine, chlorine, and nitric acid are liberated at the electro-positive pole, and are hence termed electro-negative bodies, whilst hydrogen and the alkalies are set free at the electro-negative pole, and are therefore called electro-positive bodies.

Faraday has modified these terms, and calls the two classes "*anions*" and "*cathions*," and the two poles "anodes" and "cathodes." Anode, from ἀνὰ, up, and ὁδός, a way: the way which the sun rises. Anions, from ἀνὰ, up, εἶμι, to go: that which goes up; a substance which passes to the anode during the passage of a current of electricity. Cathode, from κατὰ, down, and ὁδός, a way: the way which the sun sets. Cathion, from κατὰ, down, and εἶμι, to go: that which goes down; a substance which passes to the cathode during the passage of electricity from the anode to the cathode.

Ninth Experiment.

In the process of the electrotype is presented a valuable application of the chemical power of the voltaic circle or battery, and it may be conducted either as a single cell operation or by distinct batteries. In the former case the most simple arrangement will suffice; the only articles necessary are—a large mug or tumbler; some brown paper and a ruler; a bit of amalgamated zinc, four inches long and half an inch wide; a short length of copper wire; some black lead, blue vitriol, and oil of vitriol.

The mould from which the electrotype is to be taken can be made of common sealing wax, plaster of Paris, white wax, gutta percha, or fusible alloy. Supposing the first to be selected—viz., a common seal, it is first thoroughly black-leaded,* then one end of the copper wire is bent round the top of the amalgamated zinc, and the other is gently warmed and melted into the side of the seal, leaving a small portion uncovered by the wax, which is then well black-leaded. A few ounces of blue vitriol are dissolved in boiling water, and when cold the solution is poured into the tumbler, and the porous cell to contain the mixture of eight parts water to one of sulphuric acid is made by rolling the brown paper three or four times round the ruler and closing the end, and fixing the side with a little sealing wax. The porous cell of brown paper is now filled with the dilute acid, and placed in the tumbler containing the solution of blue vitriol, the amalgamated zinc being arranged in the paper cell, and the attached seal in the copper solution; in about twelve hours a good deposit of copper is produced, and a perfect cast in metal of the seal obtained. (Fig. 186.)

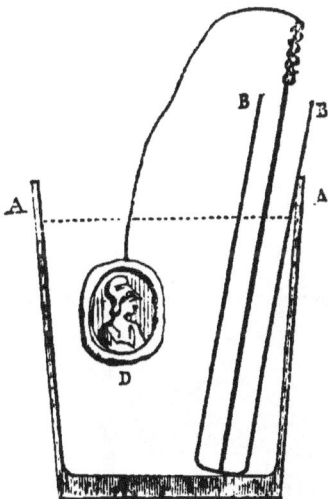

Fig. 186.   A A. The tumbler containing the solution of sulphate of copper.   B B. The brown paper cell containing the dilute sulphuric acid, inside which is the amalgamated zinc with wire attached to the seal D.

* The application of plumbago, or black lead, for electrotype purposes, was first made by the late lamented Mr. Robert Murray.

Messrs. Elliott provide every kind of convenient vessel for the purpose, and in the picture below it will be noticed that the single cell apparatus, though not so economical as the simple tumbler arrangement already described, is perhaps more convenient for electrotyping. (Fig. 187.)

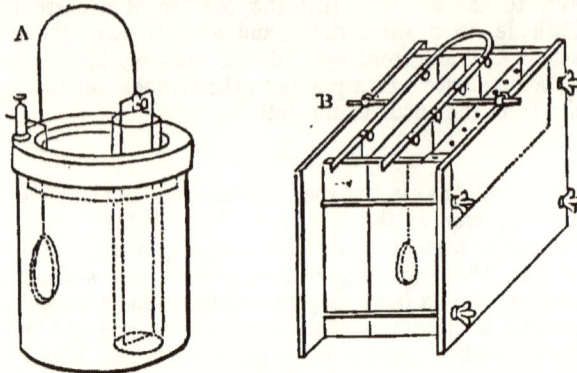

Fig. 187. A. Single cell apparatus with proper vessel, porous tube, and binding screws. B. A large trough divided by a diaphragm of biscuit-ware or very thin porous wood.

Tenth Experiment.

A single cell apparatus is only adapted to produce small electrotypes, but when larger ones are required, a separate battery of three or four

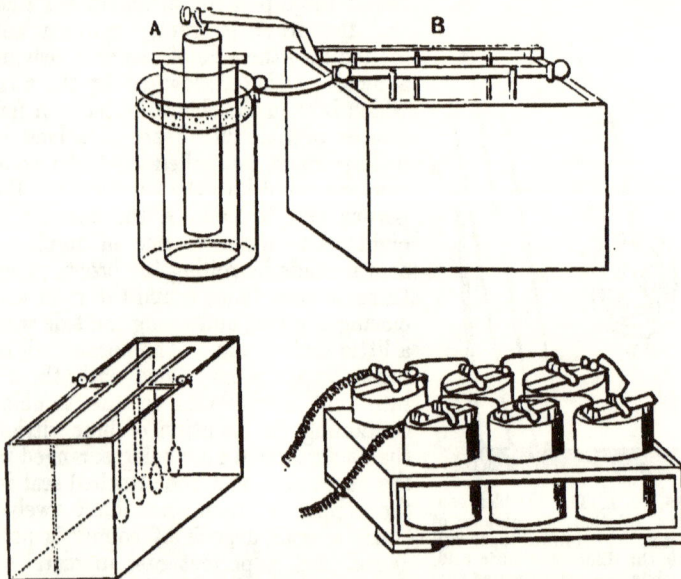

Fig. 188. A. A single cell, Daniell's, attached to B, the trough containing the mould and the plate of copper. Below is a Smee's battery ready to be attached to a larger trough for the purpose of electrotyping a great number of moulds at the same time.

Daniell's or Smee's cells is required; and it is usual to place the mould to be copied in a separate wooden trough, attaching it to the cathode wire, whilst a copper plate is connected with the anode, so that as the solution of sulphate of copper undergoes decomposition by the passage of the electricity, it is kept almost in a normal state, in consequence of the oxygen of the water and the acid passing to the copper plate, which they attack and dissolve as fast as the oxide of copper and hydrogen are liberated at the cathode, where the latter deoxidizes the oxide of copper, and by a secondary action deposits metallic copper; the object being to dissolve fresh metal as the copper is deposited on the mould. (Fig. 188.)

Eleventh Experiment.

To silver electrotypes or other brass and copper articles, the first attention must be paid to the cleanness of them; and when an electrotype is just removed from the copper solution, and washed in clean water, it is at once ready to receive the coating of silver; otherwise, if it has been handled, or is slightly greasy, it should be first boiled in a solution of common washing soda, and then the oxide removed by passing it rapidly in and out of some "Dipping Acid," which is prepared by mixing together equal parts of oil of vitriol and nitric acid; when removed from the "Dipping Acid," it must be well washed in water, and may remain under the surface of the water until the silvering solution is ready. A silver solution may be prepared by dissolving a sixpence in some nitric acid contained in a flask; it is then poured into a solution of common salt, which precipitates the chloride of silver, and leaves the copper in solution—the latter is poured off when the chloride has subsided, and after being well washed in some boiling water, is dissolved in a solution of cyanide of potassium. If a clean electrotype is plunged into this solution, it is immediately covered with a very thin coating of silver, which of course would soon wear off, and in order to increase the thickness of the silver deposit, a single cell arrangement may be constructed of a large gallipot containing a wide porous cell and a circle of amalgamated zinc around it; the arrangement is set in action by pouring a solution of salt (or, still better, sal ammoniac) into and around the porous vessel, and the silvering solution into the latter; a connecting wire passes from the zinc, and the article being attached to it, is now plunged into the porous cell, when a current of electricity slowly passes and deposits the silver on the copper article. (Fig. 189.)

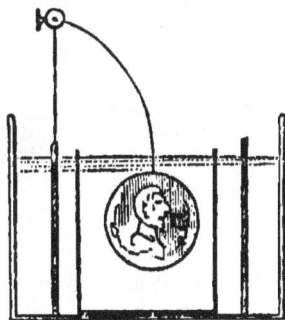

Fig. 189. The gallipot containing the solution of sal ammoniac, with the circular amalgamated zinc with wire and binding screw to which the medal is attached, and contained in the porous vessel holding the silvering solution and medal.

Twelfth Experiment.

Separate batteries and large troughs containing a solution of cyanide of silver in cyanide of potassium are used on a grand scale in the electro-plating establishment of Messrs. Elkington of Birmingham, where the finest specimens of the art are to be obtained; a plate of silver being attached to the anode to supply the loss of silver in these troughs.

Thirteenth Experiment.

The art of gilding by the agency of electricity is quite as simple as the processes already described, although greater care is necessary to avoid any loss of the precious metal. A small bit of gold is dissolved in a mixture of three parts muriatic acid and one of nitric acid, which forms the chloride of gold. This is then digested with an excess of calcined magnesia, and the gold is precipitated as an oxide of the metal; the latter is collected and washed, and then boiled in strong nitric acid to remove the magnesia clinging to it, and being again thoroughly washed with water, is dissolved in a solution of cyanide of potassium, forming a solution of cyanide of gold and potassium, which may be placed in the porous cell of the single cell arrangement already described in the Eleventh Experiment.

Fourteenth Experiment.

The safest and surest mode of making a gilding solution is to dissolve some cyanide of potassium in water in a gallipot, and having placed a porous vessel therein containing the same solution, put a plate of copper into the porous cell, and some thin foil of pure gold into the gallipot; connect the gold with the anode of a single cell of Daniell, and the copper in the porous cell with the cathode, and in a few hours sufficient gold will be dissolved for the purpose of gilding.

It is usually recommended to warm the gilding solution till it reaches a temperature of about 150° Fahr., and a very moderate battery power is employed in Electro Gilding. Indeed the same arrangement, shown in the Eleventh Experiment, Fig. 189, will also answer for the gilding solution. After being gilt, the articles may be rubbed with a little tripoli, or burnished (with taste) by the handle of a key.

Fifteenth Experiment.

Passing on to the more brilliant results obtainable from a powerful voltaic battery (of at least thirty pairs of Grove), the beautiful incandescence of platinum wire may first be noticed. If a wire of this metal is stretched between the brass standards of two ring stands, the length must be proportioned to the power of the battery; the adjustment can be made very conveniently by twisting the platinum wire on one ring stand, and then leaving the other end loose, the second ring stand may be brought nearer and nearer to the first, until the desired intensity of light from the incandescent wire is obtained. (Fig. 190.) If the wire is contained in a glass tube the cooling effect of currents of air is prevented, and a much greater length of wire can be made hot.

Sixteenth Experiment.

With the same arrangement, a chain composed of alternate links of silver and platinum wire presents a very pretty effect, every alternate link of platinum being incandescent, whilst the silver, from its excellent conducting power, remains comparatively cool.

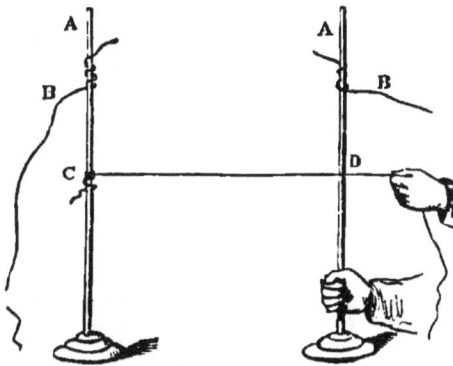

Fig. 190. A A. Two ring stands with the battery wires B B (which should be a convenient length) attached. c. Platinum wire, fixed end. D. The other end held in one hand and shortened as the stand is moved by the other hand.

Seventeenth Experiment.

Fireworks or gunpowder, arranged in proper cases, are fired at a great distance from the voltaic battery by heating a thin iron or platinum wire contained within them by the passage of the electricity; and submarine and other explosions of gunpowder by the same agency have become a common engineering operation. (Fig. 191.)

During the operation of blasting the hard marl rocks in the River Severn by Mr. Edwards, C.E., a number of holes were made side by side in the bed of the river, and cartridges formed of strong duck or canvas, tapered at the bottom, were filled with charges of powder from two to four pounds, according to the depth of the marl;

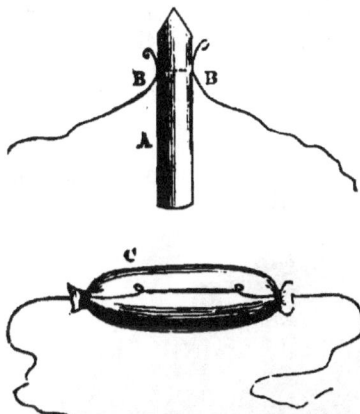

Fig. 191. A. A Gerb firework with two holes punctured, through which the bit of iron wire passes, and is wound round the battery wires tied to the outside of the case. c. A gut bladder containing the thin wire and powder for a miniature submarine explosion.

thus, two pounds for four feet, three pounds for four feet six inches, and four pounds for five feet. Into the bag were conveyed the wires of the voltaic battery, or Bickford's fuse, and being then coated with pitch and tallow, and finally greased all over and dusted with whitening, they rarely failed, and were all fired simultaneously under water. The pitch and tallow first, and afterwards the simple tallow, effectually excluded the water from the gunpowder contained in the canvas bag.

Eighteenth Experiment.

The burning of various metals by the battery is displayed with great effect by De la Rue's discharger, as also the incandescence of the charcoal points producing the *electric light.* The illuminating power derived from a forty-cell Grove's battery of the ordinary size is about equal to the light of 500 candles.

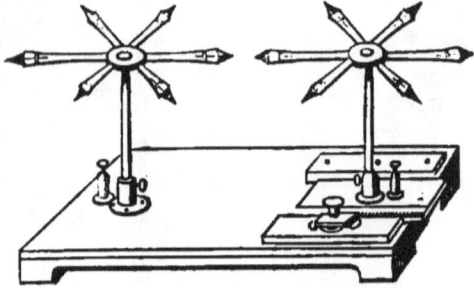

Fig. 192. De la Rue discharger, containing a series of six pairs of different substances, such as charcoal, iron, lead, zinc, copper, antimony, in six pair of crayon holders, and turning on a centre, so as to be charged at pleasure.

Fizeau and Foucault have made a careful comparison of the light obtained from 92 carbon couples as arranged in a Bunsen's battery, and of the oxy-hydrogen, or Drummond Light, as compared with that of the sun, and they state that "On a clear August day, with the sun two hours high, the electric light (assuming the sun as unity) bore to it the ratio of one to two and a half—*i.e.*, the sun was two and a half times more powerful, while the Drummond Light was only $\frac{1}{146}$th that of the sun." Bunsen found the light from 48 carbons equal to 572 candles. In Bunsen's battery carbon is substituted for the platinum in Grove's arrangement; and simultaneously with Bunsen, Cooper (in England) had applied charcoal for the same purpose.

At night the giant ship (Polyphemus like) is to have an electric light at the masthead whilst steaming across the Atlantic.

Fig. 193. *Great Eastern*, with electric light.

CHAPTER XV.

MAGNETISM AND ELECTRO-MAGNETISM.

IF a small helix, or coil of covered wire, is arranged with an unmagnetized steel needle within it, so that the discharge of a large Leyden
jar may take place through the coil, the needle will be found strongly magnetic after the discharge of the electricity. (Fig. 194.) Many years before this was known, it had been noticed that when a ship was struck by lightning, the compasses were generally reversed; and in a special case, where a house was struck, the electricity entered a box of knives, fusing some, tearing the handles off others,

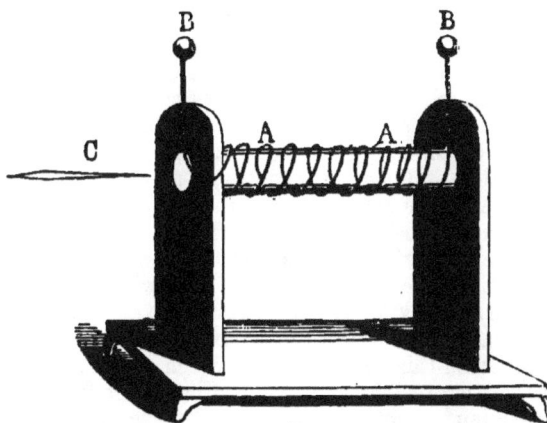

Fig. 194. A A. A glass tube supported on two uprights of wood, with coil of copper wire passing round it, terminating in the balls B B. C. Needle to place inside glass tube.

but leaving them strongly magnetic. Electricians tried to repeat the
effect by sending the discharge of powerful Leyden batteries through
bars of steel without any important result; and it was not until
Oersted, in the year 1819, made his important discovery that the copper
wire conveying the electricity possessed peculiar magnetic power, that
the principle began to be understood, and then the electricians succeeded in imitating the effects of lightning on steel, as already described in the beginning of this chapter. (Fig. 194.)

When the electricity has passed away from the Leyden jar through the coil of

Fig. 195.

copper wire, it no longer possesses any power to affect a piece of steel or iron, but if the wires of the voltaic battery are now connected with the coil of copper wire, which should be covered with cotton or silk, and many yards in length, then a bar of steel or soft iron is not only rendered magnetic, but remains permanently so, as long as the current of electricity continues to pass along the coil of wire, so that if some nails or iron filings are brought to the bar of iron, one end of which projects from the coil, they cling to it with great force, and a great number of nails may be hung on in this manner, but they immediately fall off when the contact is broken with the battery. (Fig. 195.)

Electricity thus becomes a source of magnetism, and the discoverer, Oersted, found that only needles or bars of steel or iron were thus affected, and not those of brass, shell-lac, sulphur, and other substances; he termed the conducting wire "a conjunctive wire," and described the effect of the electric current or *"electric conflict,"* as he called it, as resembling a Helix (from ἑλίσσω, to turn round; a screw or spiral), and that it is not confined to the conducting wire, but radiates an influence at some distance. This latter statement is exactly in accordance with our present notions, and hence the coil conveying the current is said to *induce* magnetism in the iron or steel, just as the phenomena of induction are produced with frictional electricity. The effect of Oersted's discovery, says Silliman, was truly *electric;* the scientific world was ripe for it, and the truth he thus struck out was instantly seized upon by Arago, Ampère, Davy, Faraday, and a crowd of philosophers in all countries. The activity with which this new field of research has been cultivated, has never relaxed even to this hour, while it has borne fruit in a multitude of theoretical and practical truths, and above all, in the electro-magnetic telegraph, truly called, and especially in connexion with the Atlantic telegraph wire, *"the great international nerve of sensation."*

Magnetism is not only the result of a current of electricity through any good conductor, but there are certain oxides of iron, called magnetic iron ores, which have the property of attracting iron filings, and are mostly found in primitive rocks, being abundant at Roslagen, in Sweden, and called the loadstone, from its always pointing, when freely suspended, to the Polar, North, or Load Star. If a tolerably large specimen of this mineral is examined, there will be found usually two points where the iron filings are attracted in larger quantities than in other parts of the same specimen. These attractive points are called poles, and the loadstone being properly mounted with soft iron bars, termed cheeks, bound round it (in old-fashioned loadstones) with silver plate and duly ornamented with

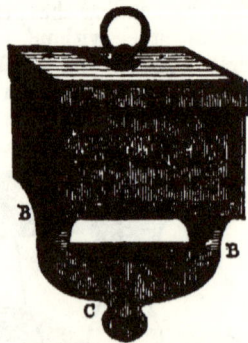

Fig. 196. A loadstone mounted in brass or silver, with the iron cheeks B B attached. C. The bit of soft iron called the armature.

engraving, has its magnetic power greatly increased, and is then said to be endowed with magnetic polarity; and to prevent the loss of power, a soft piece of iron, called the armature, is placed across and attracted to the poles of the loadstone. (Fig. 196.)

Second Experiment.

If a needle of tempered steel (fitted with a little brass cup in the centre to work upon a point) is rubbed with the loadstone in one direction only, it is rendered permanently magnetic, and will now be found to take a certain fixed position, pointing always in a direction due north and south. The end which points towards the north is called the north pole, and the other extremity the south pole, and it is usual to mark the north pole with an indent or scratch to distinguish it at all times.

Third Experiment.

If another bar of steel is magnetized, and the north pole duly marked, and then brought towards the same pole of the suspended magnet, instant repulsion takes place; the magnet, of course, grasped in the hand is not free to move, but the small magnet immediately shows the same fact noticed with electricity, viz., "*that similar magnetisms repel.*" Two n th poles repel each

Fig. 197. A magnetic needle, the north pole N being attracted to the south pole of the bar magnet s, and repelled from the north end.

other, but when the bar of steel is reversed, the opposite effect occurs, and the suspended magnet is attracted, showing that *dissimilar magnetisms attract*, and a north will attract a south pole. (Fig. 197.)

Fourth Experiment.

By contact, the magnetic power is transferred from the magnet to a piece of unmagnetized steel, and it is stated that the highest magnetizing effect is that produced by the simple method of Jacobi. A horseshoe magnet has its poles brought in contact with the intended poles of another bar of steel, likewise bent in the form of a horseshoe, and by

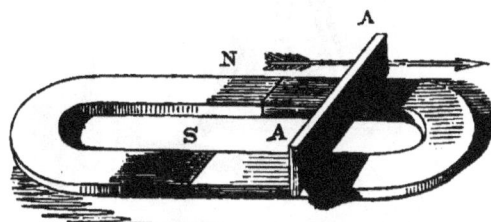

Fig. 198. The horse-shoe magnet, and another one unmagnetized, placed end to end; the one shaded and lettered N and s is the magnet. A A. The piece of soft iron moved in the direction of the arrow.

drawing the feeder over the unmagnetized horse-shoe in the direction of the arrow in the cut, and when it reaches the curve, bringing it back again to the same place, say at least twelve times, and after turning the whole over without separating the poles, and repeating the same operation on the other side likewise twelve times, the steel is then powerfully magnetized; and it is said that a horse-shoe of one pound weight may be thus charged so as to sustain twenty-six and a half pounds, and that by the old method of magnetizing it would only have sustained about twenty-two pounds. (Fig. 198.)

Fifth Experiment.

If the horse-shoe magnet is placed on a sheet of paper, and some iron filings are dusted between the poles, a very beautiful series of curves are formed, called the magnetic curves, which indicate the constant passage of the magnetic power from pole to pole.

Sixth Experiment.

The magnetic force exerted by a horse-shoe-shaped piece of soft iron, surrounded with many strands of covered copper wire in short lengths, is extremely powerful (Fig. 199), and enormous weights have been supported by an electro-magnet when connected with a voltaic battery. Supposing a man were dressed in complete armour, he might be held by an electro-magnet, without the power of disengaging himself, thus realizing the fairy story of the bold knight who was caught by a rock of loadstone, and, in full armour, detained by the unfriendly magician.

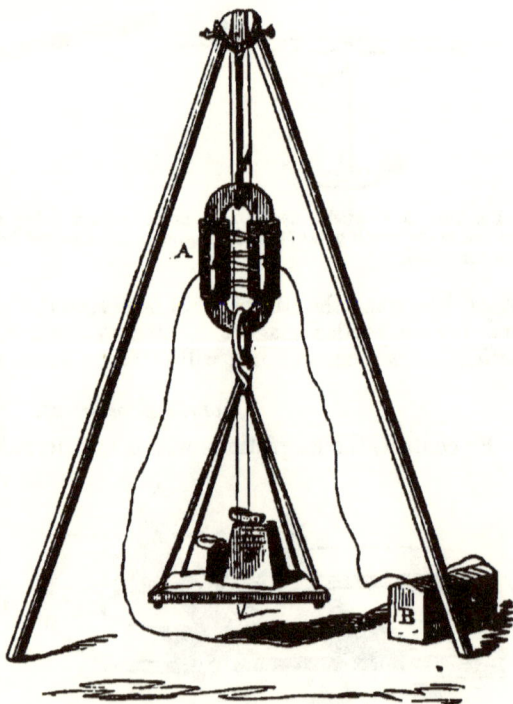

Fig. 199. A. Powerful electro-magnet supporting a great weight. B. The battery.

P

Seventh Experiment.

When a piece of soft iron is held sufficiently near one of the poles of a powerful magnet, it becomes by *induction* endowed with magnetic poles, and will support another bit of soft iron, such as a nail, brought in contact with it. When the magnet is removed, the inductive action ceases, and the soft iron loses its magnetic power. This experiment affords another example of the connexion between the phenomena of electricity and magnetism. It is in consequence of the inductive action of the magnetism of the earth that all masses of iron, especially when they are perpendicular, are found to be endowed with magnetic polarity; hence the reaction of the iron in ships upon the compasses, which have to be corrected and adjusted before a voyage, or else serious errors in steering the vessel would occur, and there is no doubt that many shipwrecks are due to this cause. No other metals beside iron, steel, nickel, cobalt, and possibly manganese, can receive or retain magnetism after contact with a magnet.

The remarkable effect of magnetism upon all matter, so ably investigated by Faraday and others, will be explained in another part of this book—viz., in the article on Dia-Magnetism.

Fig. 200. Magician and his loadstone-rock.—Vide *Fairy Tale.*

CHAPTER XVI.

ELECTRO-MAGNETIC MACHINES.

THE experiments already described in illustration of some of the phenomena of electro-magnetism are of such a simple nature that they may be comprehended without difficulty; but it is not such an easy task to appreciate the curious fact of an invisible power producing motion. It has already been explained that a copper or other metallic wire conveying a current of electricity becomes for the time endowed with a magnetic power, and if held above, or below, or close to, a suspended magnetized steel needle, affects it in a very marked degree, causing it to move to the right or left, according to the *direction* of the electric current; and in order to form some notion of the condition of a metallic wire whilst the electricity is passing through it, the annexed diagrams may be referred to. (Figs. 201, 202.)

Fig. 201. Portion of a square copper conductor, in which A B represents the direction of the electricity, and the small arrows, C C C C, the magnetic current or whirl at right angles to the electrical current, and exercising a tangential action.

Dr. Roget says: "The magnetic force which emanates from the electrical conducting wire is entirely different in its mode of operation from all other forces in nature with which we are acquainted. It does not act in a direction parallel to that of the current which is passing along the wire, nor in any plane passing through that direction. It is evidently exerted in a plane perpendicular to the wire, but still it has no tendency to move the poles of the magnet in a right or radial line, either directly towards, or directly from, the wire, as in every other case of attractive or repulsive agency. The peculiarity of its action is that it produces motion in a circular direction *all round* the wire—that is, in a direction at right angles to the radius, or in the direction of the tangent to a circle described round the wire in a plane perpendicular to it; hence the electro-magnetic force exerts a tangential action, or that which Dr. Wollaston called a vertiginous or whirling motion.

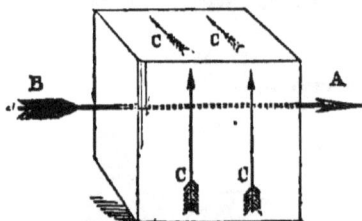

Fig. 202. A round conducting wire, in which the electrical current is flowing in the direction of the large dart A B, and the small arrows indicate the direction of the magnetic force.

P 2

Dr. Faraday concluded that there is no real attraction or repulsion between the wire and either pole of a magnet, the action which imitates these effects being of a compound nature; and he also inferred that the wire ought to revolve round a magnetic pole of a bar magnet, and a magnetic pole round a wire, if proper means could be devised for giving effect to these tendencies, and for isolating the operations of a single pole. For the first idea of electro-magnetic rotation the world is indebted to Dr. Wollaston; but Dr. Faraday, with his usual ingenuity, was the first who carried out the theory practically. The rotation of a wire (conveying a current of voltaic electricity) round one of the poles of a magnet is well displayed with the simple contrivance devised by him. (Fig. 203.)

By a careful observation of the complex action of an electrified wire upon a magnetic needle, Dr. Faraday was enabled to analyse the phenomena with his usual penetration and exhaustive ability, and he found, as Daniell relates,—

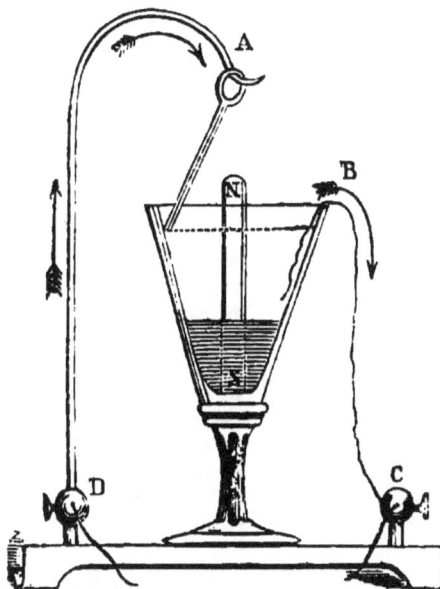

Fig. 203. N. A small bar magnet cemented into a wineglass, the north pole being at N. A is a moveable wire looped over the hook, which is the positive (+) pole of the battery; the free extremity rotates round the pole of the magnet when the current of electricity passes. The dotted line represents the level of the mercury which the glass contains. The electricity passes in at A, and out at the wire B, as shown by the arrows. C is connected with the negative, and D with the positive, pole of the battery.

"That if the electrified wire is placed in a perpendicular position, and made to approach towards one pole of the needle, the pole will not be simply attracted or repelled, but will make an effort to pass off on one side in a direction dependent upon the attractive or repulsive power of the pole; but if the wire be continually made to approach the centre of motion by either the one or the other side of the needle, the tendency to move in the former direction will first diminish, then become null, and ultimately the motion will be reversed, and the needle will principally endeavour to pass in the opposite direction. The opposite extremity of the needle will present similar phenomena in the opposite direction; hence Dr. Faraday drew the conclusion that the direction of the forces was *tangential* to the circumference of the wire, that the pole of the needle is drawn by one force, not in the direction of a radius to its centre, but in that of a line touching its circumference, and that it is repelled by the other force in the opposite direction. In this manner the northern force acted all round the wire in one direction, and the southern in the opposite one. Each pole of the needle, in short, appeared to have a tendency to revolve round the wire in a direction opposite to the other, and, consequently, the wire round the poles. Each pole has the power of acting upon the wire by itself, and not as connected with the opposite pole, and *the apparent attractions* and *repulsions* are merely *exhibitions* of the *revolving motions* in different parts of their circles."

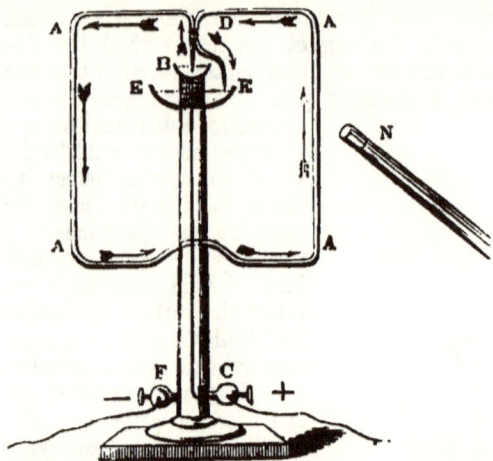

The same fact illustrated at Fig. 203, is also demonstrated in a still more striking manner by means of wire bent into a rectangular form, and so arranged that whilst the current of electricity passes, it is free to move in a circle; and when the poles of a magnet are brought towards the electrified wire, it may be attracted or repelled at pleasure, and in fact becomes a magnetic indicator, and places itself (if carefully suspended) at right angles to the magnetic meridian. (Fig. 204.)

Fig. 204. ▲ ▲ ▲ ▲. The rectangular wire covered with silk and varnished, one end of which being pointed, rests on the little cup B, connected with a covered wire passing down the centre of the brass support to the binding screw C let into ivory. D. The other extremity of the rectangular wire; this being covered and varnished, is not in metallic contact with the end B, but is likewise pointed, and dips into the mercury contained in the large cup E E. The upper and lower cups do not touch, and are separated by ivory, marked by the shaded portion, and the cup E E is in metallic communication with the brass pillar, and is connected with the negative pole of the battery at F, whilst C is connected with the positive pole of the battery, and the electricity circulates round the wire in the direction of the arrows. When a bar magnet, N, is brought towards the wire, the latter is immediately set in motion, and by alternately presenting the opposite poles of the magnet, the rectangular wire rotates freely round the cup E.

These curious movements of a magnetized needle, and rotations of wires and magnets, brought about by the agency of an active current of electricity, have induced Sir David Brewster to advance his admirable theory, which supposes the affection of the mariner's compass needle, and all other suspended pieces of steel, to be due to the agency of *electrical currents* continually *circulating* around the globe; and Mr. Barlow contrived the following experiment in illustration of Brewster's theory. A wooden globe, sixteen inches in diameter, was made hollow, for the purpose of reducing its weight, and while still in the lathe, grooves one-eighth of an inch deep and broad were cut to represent an equator, and parallels of latitude at every four and a half degrees each way from the equator to the poles. A groove of double depth was also cut like a meridian from pole to pole, but only half round. The grooves were cut to receive the copper wire covered with silk, and the laying on was commenced by taking the middle of a length of ninety feet of wire one-sixteenth of an inch in diameter, which was applied to the equatorial groove so as to meet in the transverse meridian; it was then made to pass round this parallel, returned again along the meridian to the next parallel, and then passed round this again, and so on, till the wire was thus led in continuation from pole to pole. The length of wire still remaining at each pole was returned from each pole along the meridian groove to the equator, and at this point, each wire being fastened down with small staples, the wires from the remaining five feet were bound together near their common extremity, when they opened to form separate connexions for the poles of a voltaic battery. When the battery was connected, and magnetic needles placed in different positions, they behaved precisely as they would do on the surface of the earth, the induction set up by the elec·trified wire being a perfect imitation of that which exists on the globe.

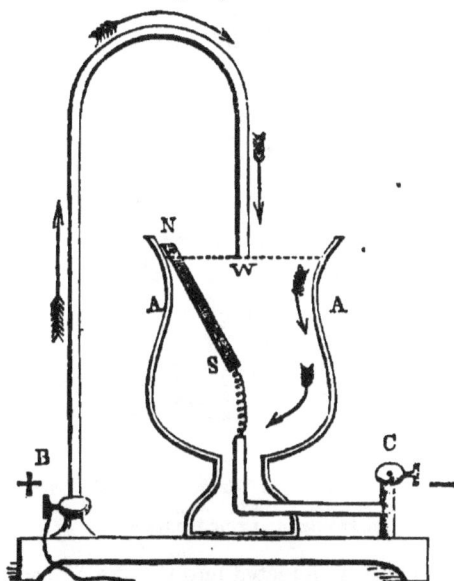

The opposite effect to that already described—viz., the rotation of one pole of a magnet round the electrified wire, was also arranged by Faraday in the following manner. (Fig. 205.)

In the examination of the magnetic phenomena obtained from wires transmitting a current of electricity, it should be borne in mind that any conducting medium which forms part of a closed circuit—i.e., any conductor, such as charcoal, saline fluids, acidulated water, which form a link in the endless chain required for the path of the electricity,—will cause a magnetic needle placed near it to deviate from its natural position.

Fig. 205. N s. A little magnet floating in mercury contained in the glass A A; the north pole is allowed to float above the surface of the quicksilver, and the south pole is attached to the wire passing through the bottom of the glass vessel. The electricity passes in at n, and taking the course indicated by the arrows travels through the glass of quicksilver to the other pole of the battery at c. Directly contact is made with the battery, the little magnet rotates round the electrified wire, w. The dotted line shows the level of the mercury in glass.

These positions of the electrified wire and the magnetic needle are of course almost unlimited, and in order to assist the memory with respect to the fixed laws that govern these relative movements, Monsieur Ampère has suggested a most useful mechanical aid, and he says :—" Let the observer regard himself as the conductor, and suppose a positive electric current to pass from his head towards his feet, in a direction parallel to a magnet ; then its north pole in front of him will move to his right side, and its south pole to his left.

"The plane in which the magnet moves is always parallel to the plane in which the observer supposes himself to be placed.' If the plane of his chest is horizontal, the plane of the magnet's motion will be horizontal, but if he lie on either side of the horizontally-suspended magnet, his face being towards it, the plane of his chest will be vertical, and the magnet will tend to move in a vertical plane."

This very lucid comparison will be seen to apply perfectly to the direction of the rotations in Figs. 203 and 205.

The whole of this apparatus is made in the most elegant and finished manner by Messrs. Elliott, of 30, Strand; and by a modification of the latter arrangement (Fig. 206), the opposite rotations of the opposite poles of the magnets round the electrified wire, are shown in the most instructive manner. The apparatus (Fig. 206) was devised by the late Mr. Francis Watkins, and consists of two flat bar magnets doubly bent in the middle, and having agate cups fixed at the under part of the bend (by which they are supported) upon upright pointed wires, the latter being fixed upright on the wooden base of the apparatus, and the magnets turn round them as upon an axis.

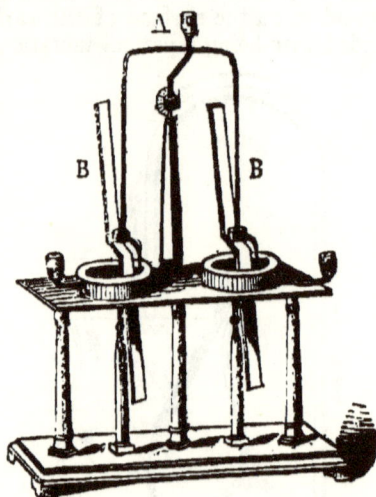

Fig. 206. A. Wire conveying the current of electricity. B B. The magnets balanced on points rotating round the wires.

Two circular boxwood cisterns, to contain quicksilver, are supported upon the stage or shelf above the base. A bent pointed wire is directed into the cup of each magnet, the ends of which dip into the mercury contained in the boxwood circular troughs on the stage. By using a battery to each magnet, and taking care that the currents of electricity flow precisely alike, they will then rotate in opposite directions.

Directly after the ingenious experiments of Faraday became known, a great number of electro-magnetic engine models were constructed, and many thought that the time was fast approaching when steam would be superseded by electricity; and really, to see the pretty electro-magnetic models work with such amazing rapidity, it might be supposed that if they were constructed on a larger scale, a great amount of hard work could be obtained from them. This idea, however, has been proved to be a fallacy, for reasons that will be presently explained. The figure on p. 216 displays two of these engines, one of which represents the rotation of electro-magnets within four *fixed steel magnets*, and the other the rotation of steel magnets by the *fixed electro-magnets*. The latter (No. 2) moves with such great velocity, that unless the strength of the battery is carefully adjusted, the connexions are soon destroyed. (Fig. 207.)

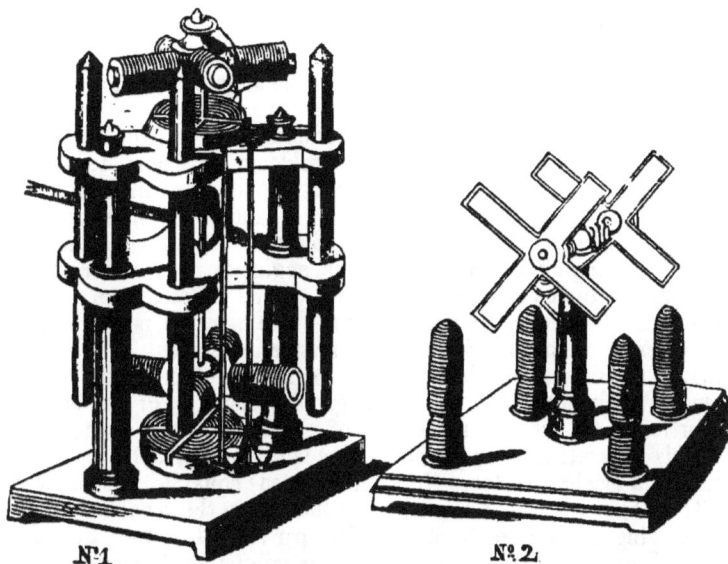

Fig. 207.—No. 1 consists of vertical permanent steel magnets and horizontal soft-iron electro-magnets which rotate.
No. 2 consists of two fixed soft-iron electro-magnets, and four bent permanent steel magnets, which rotate, in both cases of course, only when connected with the battery.

Considering the prodigious power or *pull* of a soft-iron electro-magnet, and its capability of supporting considerable weight, the most reasonable expectations of success might be entertained with machines acting by the direct pull. It was, however, discovered that they soon became inefficient, from the circumstance that the repeated blows received by the iron so altered its character, that it eventually assumed the quality of steel, and had a tendency to retain a certain amount of permanent magnetism, and thus to interfere with the principle of making and unmaking a magnet. It was this fact that induced Professor Jacobi, of St. Petersburg, after a large expenditure of money, to abandon arrangements of this kind, and to employ such as would at once produce a rotatory motion. The engine thus arranged was tried upon a tolerably large scale on the Neva, and by it a boat containing ten or twelve people was propelled at the rate of three miles an hour.

Various engines have been constructed by Watkins, Botta, Jacobi, Armstrong, Page, Hjorth; the engine made by the latter (Hjorth) excited much attention in 1851–52, and consisted of an electro-magnetic piston drawn within or repelled from an electro-magnetic cylinder; and by this motion it was thought that a much greater length of stroke could be secured than by the revolving wheels or discs, but the loss of power (not only in this engine, but in others) through space is very great, and the lifting power of any magnet is greatly reduced and altered at the smallest possible distance from its poles. This loss of power is therefore a great obstacle in the way of the useful application of electro-magnetic force, and can be appreciated even with the little models, all of which may be stopped with the slightest friction, although they may be moving at the time with great velocity.

In the second place, supposing the reduced force exerted by the two magnets, a few lines apart, was considered available for driving machinery, the moment the magnets begin to move in front of one another there is again a great loss of power, and as the speed increases, there is curiously a corresponding diminution of available mechanical power, a falling-off in the *duty* of the engine as the rotations become more rapid. In the third place, the cost of the voltaic battery, as compared with the consumption of coal in the steam-engine, is very startling, and extremely unfavourable to electro-magnetic engines.

Mr. J. P. Joule found that the economical duty of an electro-magnetic engine at a given velocity and for a given resistance of the battery is proportioned to the mean intensity of the several pairs of the battery. With his apparatus, every pound of zinc consumed in a Grove's battery produced a mechanical force (friction included) equal to raise a weight of 331,400 pounds to the height of one foot, when the revolving magnets were moving at the velocity of eight feet per second. Now, the *duty* of the best Cornish steam-engine is about one million five hundred thousand pounds raised to the height of one foot by the combustion of each pound of coal, or nearly five times the extreme *duty* that could be obtained from an electro-magnetic engine by the consumption of one pound of zinc. This comparison is therefore so very unfavourable, that the idea of a successful application of electricity as an *economic* source of power, is almost, if not entirely abandoned.

By instituting a comparison between the different means of producing power, it has been shown that for every shilling expended there might be raised by

Pounds.

Manual power . . .	600,000	one foot high in a day.	
Horse	3,600,000	,,	,,
Steam	56,000,000	,,	,,
Electro-magnetism .	900,000	,,	,,

A powerful magnet has been compared to a steam-engine with an enormous piston but with an exceedingly short stroke. Although motive power cannot be produced from electricity and applied successfully to commercial purposes, like the steam-engine, yet the achievements of the electric telegraph as an application of a small motive power must not be lost sight of, whilst the fall of the ball at Deal and other places, by which the chronometers of the mercantile navy are regulated, as also the means of regulating the time at the General Post Office and various railway stations, are all useful applications of the power which fails to compete in other ways with steam.

CHAPTER XVII.

THE engineering and philosophical details of this important instrument
have grown to such formidable dimensions, that any attempt (short of
devoting the whole of these pages to the subject) to give a full account
of the history and application of the instrument, the failures and successes
of novel inventions, and the continued onward progress of this mode
of communication, must be regarded as simply impossible, and therefore a very brief account of the *principle* only will be attempted in these
pages.

For the complete history of the discovery and introduction of the
principle of the Electric Telegraph the reader is referred to the Society
of Arts Journal (Nos. 348-9, vol. viii.), where it is stated that it is *half
a century*, dating from August, 1859, since the first galvanic telegraph
was made. "It was the Russian Baron Schilling's electro-magnetic
telegraph which, without its being known to be his, was brought to
London, and caused the establishment of the first practically useful
telegraph lines, not only in Great Britain, but in the world." Dr.
Hamel says: "The small sprout nursed on the Neva, which had been
exhibited on the Rhine, and thence brought to the Thames, grew up
here to a mighty tree, the fruit-laden branches of which, along with
those from trees grown up since, extend more and more over the lands
and seas of the Eastern hemisphere, whilst kindred trees planted in the
Western hemisphere have covered that part of the world with their
branches, some of which will, ere long, be interwoven with those in our
hemisphere."

The first telegraph line in England was constructed by Mr. Cooke
from Paddington along the Great Western Railroad to West Drayton
in 1838-39; and it must be remembered that it was in February, 1837,
that Mr. Cooke first consulted Professor Charles Wheatstone, having
previously visited Dr. Faraday and Dr. Roget, and on the 19th
November, 1837, a partnership contract was concluded between Messrs.
Cooke and Wheatstone.

To the distinguished philosopher, Professor Wheatstone, the merit of
the ingenious construction of the vertical-needle telegraph is due;
whilst Mr. Cooke's name will always be associated with the practical
establishment of the first telegraph lines in England. The first line in
the United States, from Washington to Baltimore, was completed in
1844, being arranged and worked by Professor Morse.

In British India, in April and May, 1839, the first long line of
telegraph, twenty-one miles in length, and embracing 7000 feet of
river surface, was constructed by Dr. (now Sir William) O'Shaughnessy

The construction of the electric telegraph may be considered under three heads:

1st. The Battery, *the motive power.*
2nd. The Wires, *the carriers of the force.*
3rd. The Instruments to be worked—*the bell* and the *needle telegraph.*

THE BATTERY.

The construction and rationale of the batteries generally in use have been explained in another part of this work; those used for telegraphic purposes consist of one or more couples, of which zinc is one, the second being copper, silver, platinum, or carbon. Each couple is termed an *element,* and a series of such couples a *battery.*

The batteries employed chiefly on the English lines consist of a plate of cast-zinc four inches square and $\frac{3}{16}$ths of an inch thick, attached by a copper strap one inch broad to a thin copper plate four inches square. The zinc is well amalgamated with mercury. Twelve of these couples are arranged in a trough of wood, porcelain, or gutta-percha, divided by partitions into twelve water-tight cells, $1\frac{1}{4}$ inch wide. The zinc and copper preserve the same order and direction throughout, and when arranged, the trough is filled with the finest white sand, and then moistened with water previously mixed with five per cent. by measure of pure sulphuric acid. This mode of applying the acid is the clever practical improvement of Mr. Cooke, and prevents any inconvenience from the spilling of the acid, and at the same time renders the battery quite portable. The voltaic arrangement thus prepared is found to remain in action for several weeks, or even months, with the occasional addition of small quantities of acid, and answers well for working needle telegraphs in fine and dry weather. In fogs and rains, at distances exceeding 200 miles at most, their action is not so perfect, and a vast number of couples must be employed, 144 to 288 being frequently in use. In France, Prussia, and America, sand batteries do not appear to answer, and Daniell's arrangement is preferred. Sixty couples suffice in France for some of the long lines—viz., from Paris to Bordeaux, 284 miles; Paris to Brussels, $231\frac{1}{4}$ miles; and in fact, the advantages of the Daniell's battery have become so apparent, that they are now being used on English lines. In Prussia, Bunsen's carbon battery is much used; in India, a modification of Grove's battery is preferred, the zinc being acted upon by a solution of common salt in water. Two of these *elements* were found sufficient to work a line of forty miles totally uninsulated, and including the sub-aqueous crossing of the Hooghly River, 6200 feet wide.

The continual energy of the battery, whatever may be its construction, depends on the circulation of the electricity, the object being to pass the force from the positive end of the series through the wires, back again to the negative extremity of the voltaic series.

The wire (the carrier of the force) must be continuous throughout, unless, of course, water or earth forms a part of the endless conducting chain.

THE CONDUCTING WIRES.

These roads for the electricity may be of any convenient metal, and the one preferred and used is iron, which is well calculated from its great tenacity (being the most tenacious metal known) and cheapness to convey the electricity, although it is not such a good conductor as copper, and offers about six times more resistance to the flow of the current than the latter metal. The wire does not appear to be made of iron, because it is galvanized or passed through melted zinc, which coats the surface and defends it from destructive rust, at the same time does not destroy its valuable property of tenacity or power of resisting a strain. About one ton of wire is required for every five miles, and to support this weight, stout posts of fir or larch are erected about fifty yards apart, and from ten to twenty-five feet high. At every quarter mile, on many lines, are straining - posts with ratchet wheel winders, for tightening the wires. On some of the lines the wires are attached to the posts by side brackets carrying the insulators invented by Mr. C. V. Walker, which are composed of brown salt-glazed stoneware of the hourglass shape, as shown in the drawing. (Fig. 208.)

There are some objections to the hourglass insulators, and they have been modified by Mr. Edwin

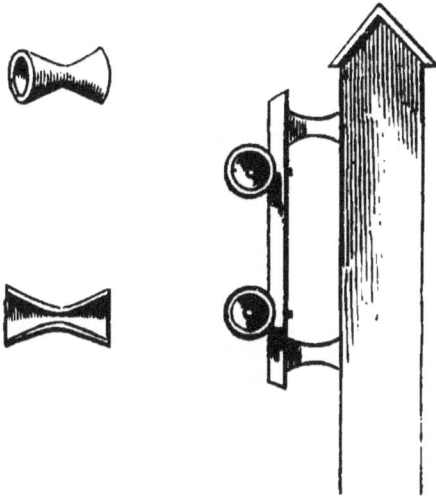

Fig. 208. Walker's insulator.

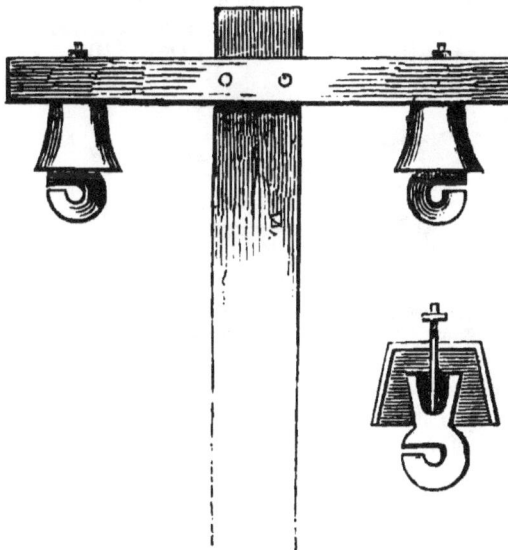

Fig. 209. Clark's insulator.

Clark, who employs a very strong stone-ware hook open at the side, so that the wire can be placed on the hook without threading, and the hooks can be replaced in case of breaking, without cutting the telegraph wire, which is securely fastened to each insulator by turns of thinner wire. An inverted cap of zinc is used to keep the insulator dry. (Fig. 209.)

In India the conductor is rather a rod than a wire, and weighs about half a ton per mile; it is erected in the most substantial manner, and many miles of the rod are supported on granite columns, other portions on posts of the iron-wood of Arracan, or of teak.

The number of wires required by the electric telegraph often puzzles the railway traveller, and people ask why so many wires are used on some lines and so few on others? The answer is very simple: they are for convenience. Two wires only are required for the double needle telegraph, and one for the single needle instrument. But as so many instruments are required at the terminal stations, an increased number of wires, like rails for locomotives, must be provided; thus, on the Eastern Counties, seven wires are visible, and are thus employed. The two upper wires pass direct from London to Norwich; the next pair connect London, Broxbourne, Cambridge, Brandon, Chesterfield, Ely; the third pair all the small stations between London and Brandon; and the seventh wire is entirely devoted to the bell.

If the earth was not a conductor of electricity, and employed in the telegraphic circuit, four wires would be required for the double needle telegraph, and two for the single instrument. To understand this, let us suppose a battery circuit extending from Paddington to the instrument at Slough, and the wire returning from Slough to Paddington, it is evident that one wire would take the electricity to Slough, and the other return it to London, as in the diagram below. (Fig. 210.)

LONDON   SLOUGH

Fig. 210. A. The battery. B. The instrument. The arrows show the passage of the electricity to the single needle telegraph instrument by one wire, and the return current by the other.

If the whole of the return wire is cut away except a few feet at each end, which are connected by plates of copper with the damp earth, the current not only passes as before, but actually has increased in intensity, and will cause a much more energetic movement of the needle in the telegraph instrument. (Fig. 211.) These plates are called "Earth Plates;" and Steinheil, in 1837, was the first who proved that the earth might perform the function of a wire.

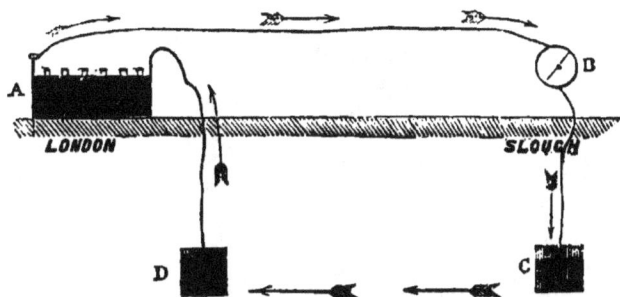

Fig. 211. A. The battery. B. The instrument. C. Earth plate at Slough. D. Earth plate at London. The arrows show the direction of the electric current.

It must be obvious that a message may be received at any station without a battery, but in order to be able to return an answer, every station must have its own battery.

Ingeniously-constructed lightning-conductors are attached to the posts which carry the wires, so that in case of a storm, the natural electricity is conveyed to the earth, whilst the voltaic electricity artificially produced pursues its own course without deviation. Protectors are also required for the instruments at the stations, and the plan devised by Mr. Highton is thus described by the inventor:—

" A portion of the wire circuit—say for six or eight inches—is enveloped in blotting-paper or silk, and a mass of metallic filings, in connexion with the earth, is made to surround it. This arrangement is placed on each side of the telegraph instrument at a station. When a flash of lightning happens to be intercepted by the wires of the telegraph, the myriads of infinitesimally fine points of metal in the filings surrounding the wire at the station, on having connexion with the earth, at once draw off nearly the whole charge of lightning, and carry it safely to the earth."

THE INSTRUMENTS TO BE WORKED—THE BELL AND THE TELEGRAPH.

The bell or alarum resembles in construction that of an ordinary clock, and is in fact a piece of clockwork wound up and ready to ring a bell, when the *detent* or preventive is removed. The detent is connected with a piece of soft iron placed before an electro-magnet, and directly the current passes, the electro-magnet attracts the soft piece of iron attached to a perpendicular lever which the bell-crank lever rests upon; the detent is removed, and the bell rings, and again stops when the current of electricity ceases to pass.

One of the most simple alarum clocks is a common American clock, wound up daily. A small electro-magnet surrounded with thick wire is placed below a moveable piece of tinned iron, so that when this is attracted, the fly of the clock is released, and its bell tolls unceasingly while the magnet is excited. This arrangement is employed by Sir W. O'Shaughnessy in the Indian telegraph system. (Fig. 212.)

It will readily be comprehended from this description that the alarum is sounded by ordinary mechanism, and that the duty of the current of the electricity is simply comprised in the act of removing the lever and liberating machinery, which may be large or small; and if it were thought necessary, the bells of the great clock-tower of the Houses of Parliament, which chime the quarters, or even "Big Ben" himself (when his constitution is restored), could be rung by a person at York or Edinburgh, supposing wires, batteries, and a powerful electromagnet with a detent mechanism for the bells, were properly arranged and connected with the clockwork.

Fig. 212. A. The soft iron tinned, which is attracted to the electro-magnet B, and liberates the detent.

In certain cases, Mr. Charles V. Walker states that a single and distinct wire is used for the bell only, with his special mechanism, called the *ringing key*. If the bell was always on the same wire as the needle-coil, the bell would not only call the attention of, but seriously annoy the clerk (unless, of course, he happened to be a very deaf person) by its ringing whilst he was reading the signals of the needle. The nuisance is prevented by what is termed *joining over* or making the *short circuit*—in fact, by providing for the current a shorter and much more capacious road to the needle coil than by going through that of the bell-magnet, which is made with very fine wire; and the control of the short circuit is put in the hands of the clerk.

COOKE AND WHEATSTONE'S DOUBLE NEEDLE TELEGRAPH.

The principle of this instrument, as already explained, is involved in the elementary experiment of Oersted—viz., the deflection of a magnetic needle from the inside of a coil of wire conveying a current of electricity, and as it is difficult to give a good description and drawing of the interior of the instrument that can really be understood, it may be sufficient to state that the handles give the operator the power of reversing the current of electricity, so that the needles are deflected with the utmost certainty to one side or the other, either separately or simultaneously. (Fig. 213.)

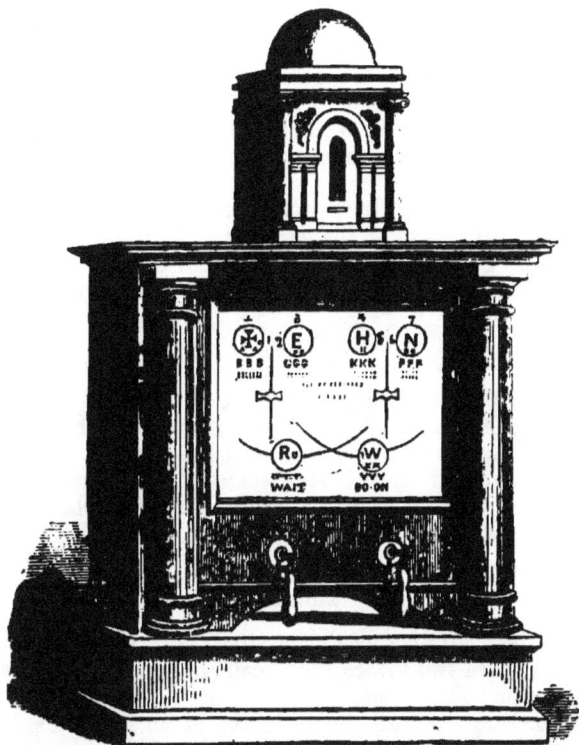

Fig. 213. The letters of the alphabet, figures, and a variety of conventional signals, are indicated by the single and combined movements of the needles on the dial. The left-hand needle moving once to the left indicates the +, which is given at the end of a word. Twice in the same way, A; thrice, B; first right, then left, C; the reverse, D. Once direct to the right, E; twice, F; thrice, G. In the same order with the other needle for H, I, K, L, M, N, O, P. The signals below the centre of the dial are indicated by the parallel movements of both needles simultaneously. Both needles moving once to the left indicate R; twice, S; thrice, T. First right, then left with both, U; the reverse, V. Both moving once to the right, W; twice, X; thrice, Y. The figures are indicated in the same way as the letters nearest to which they are respectively placed. To change from letters to figures the operator gives H, followed by the +, which the recipient returns to signify that he understands. If, after the above signs (H and +) were given, C R H L were received, 1845 would be understood. A change from figures to letters is notified by giving I, followed by the +, which the recipient also returns. Each word is acknowledged. If the recipient understand, he gives B; if not, the +, in which case the word is repeated. Attention to a communication by this instrument is called by the ringing of a bell (of any size), which is effected through the agency of an electric current. The upper case contains the bell.

Sir W. O'Shaughnessy, in his excellent work on the electric telegraph in British India, gives a description of a telegraphic instrument of remarkable simplicity, which is successfully employed in India, and is highly spoken of by Mr. E. V. Walker and other gentlemen practically acquainted with the working of telegraphs. It consists of a coil of fine wire on a card or ivory frame, a magnetic needle with a light index of paper pasted across it; two stops of thin sheet lead to limit the vibrations of the index; a supporting board eight inches square, and a square of glass in a frame of wood, or a common glass tumbler placed over it as a shade, to prevent the index being moved by currents of air. It is stated that the office boys, with the assistance of a native Indian carpenter, make up these telegraphs at a price not exceeding two shillings each.

In England of course they would be more expensive; but the simplicity and perfection of the arrangement are so much to be commended that we give the details for the benefit of those boys who might wish to establish a telegraph on a small scale for amusement.

THE FRAME.

This is a piece of mahogany eight inches square and one inch thick, with a hollow groove cut in its centre two inches and a half long, half an inch wide, and a quarter of an inch deep; a ledge of the same wood one inch wide and half an inch deep surrounds the frame, leaving the inner surface seven inches square; this is stained black with ink to make the motions of the index more conspicuous.

THE COIL.

This consists of fifty feet of the finest silk-covered copper wire wound on a frame of card two inches long, half an inch broad, three-eighths deep in the open part.

An edge or flange of card, three-eighths of an inch wide, is attached to it at each side to keep the wire in its place. The frame may be of thin wood or ivory, and the winding of the wire commences at the lower left corner, and it is coiled from left to right, as the hands of a watch would move in the same plane. (Fig. 214.)

Fig. 214. The coil.

Two inches of each end of the coil wire are now stripped of their silk covering by being rubbed with sand-paper. The coil is mounted in the frame by inserting its lower edge or flange in the groove, so that the lower part or floor of the inside of the coil is level with that of the

Q

frame, as shown below, and it is now ready to receive the magnetized needle. (Fig. 215.)

Fig. 215. The coil fitted into frame.

THE NEEDLE.

This is one inch long, one-twelfth of an inch wide, of the thinnest steel, and fitted with a little brass cap turned to a true cone to receive the point on which it is balanced. These needles are of hard tempered steel, and are magnetized by a single contact with the poles of an electro-magnet or other ordinary powerful magnet.

The magnet is now to be balanced on a steel point one-eighth of an inch high; these are nipped off with cutting pliers from common sewing needles, and soldered into a slip of thin copper three inches long, half an inch wide. (Fig. 216.)

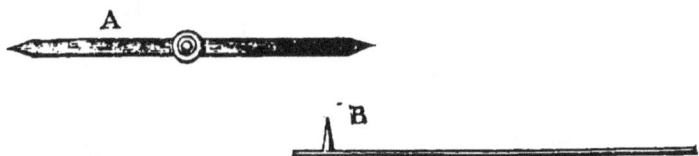

Fig. 216. A. The needle. B. The point on the slip of copper.

As the north end of the needle will be found to dip, it is advisable to counteract this by touching the south end with a little shell-lac varnish, which dries rapidly, and soon restores the needle to a perfect equilibrium.

The needle is completed for use by fixing to it an index of paper (cut from glazed letter paper) two inches long, tapering from one-eighth of an inch to a point, and fastened at right angles on to the needle with lac varnish, so as to be truly balanced, and pointing the sharp end to the east, when the needle placed on the point settles due north and south, its north pole being opposite the observer's right hand, the observer facing west. (Fig. 217.)

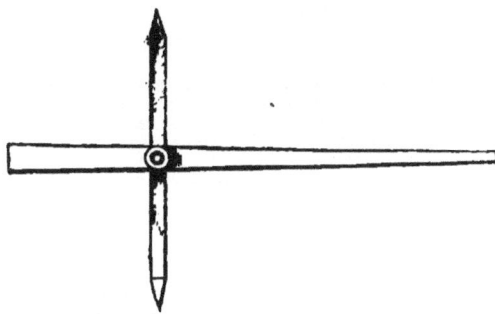

Fig. 217. The needle with the paper index.

The coil frame is placed north and south, and the needle is now introduced by sliding the end of the slip of copper into the opening in the frame.

To limit the vibrations of the paper index a *stop* is placed at each side. The stops are made of a strip of thin sheet-lead or copper, a quarter of an inch broad, one inch and a half long, and turned up at a right angle, so that one inch rests on the board and half an inch is vertical. For ordinary practice these stops are placed each at half an inch from the index.

The telegraph is placed in a box, which may have a piece of looking-glass in the lid, so that the readings can be taken with the needle in the vertical instead of the horizontal position, if required. (Fig. 218.)

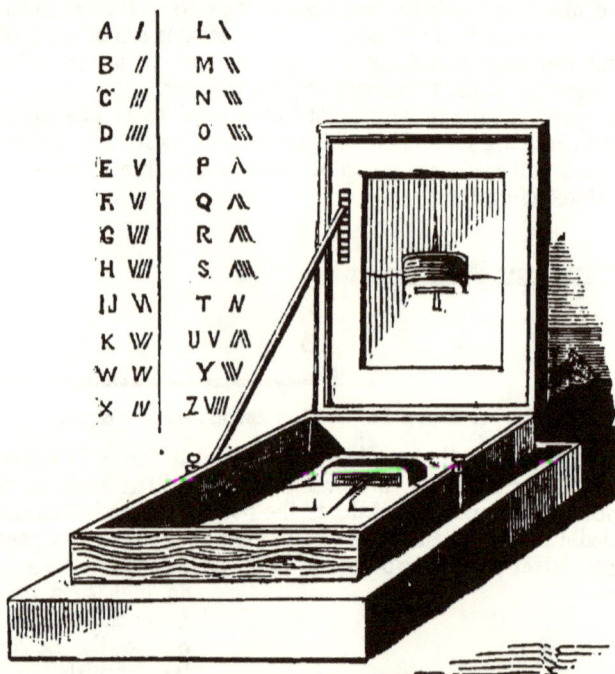

A	/	L	\
B	//	M	\\
C	///	N	\\\
D	////	O	\\\\
E	V	P	Λ
F	W	Q	Λ
G	VII	R	Λ\
H	VIII	S	Λ\\
IJ	\\	T	N
K	\\/	UV	/Λ
W	W	Y	\\\
X	\\	Z	V///

Fig. 218. Box containing the telegraph, with the looking-glass in the lid. A small steel magnet is placed on or near the frame, if required, the south pole of this magnet being opposite to the north pole of the needle in the telegraph coil. The bar is four inches long, half an inch broad, three-sixteenths of an inch thick, and it is only used to counteract any local deviation which may arise in using the instrument with miles of wire. It would not be required under ordinary circumstances. The alphabet used is shown to the left.

The ends of the fine wire of the telegraph coil are joined on to the wires from the *reversing* instrument, and this is connected with a voltaic series of one or more elements, so that by the employment of the reverser the needle is caused to move right or left at pleasure. The white paper index on the black ground can be followed with the greatest certainty, and Sir W. O'Shaughnessy states that with this instrument a telegraph clerk may read at the rate of twenty words per minute with a double needle wire, being equal to forty words per minute.

THE REVERSER

consists of a block of wood, two inches and a half square, in which four hollows, half an inch deep, are cut, and these hollows are joined diagonally by copper wires let into the substance of the wood, and most carefully insulated from each other by melted cement, but exposing a clean metallic surface in each cell, which is filled with mercury. (Fig. 219.)

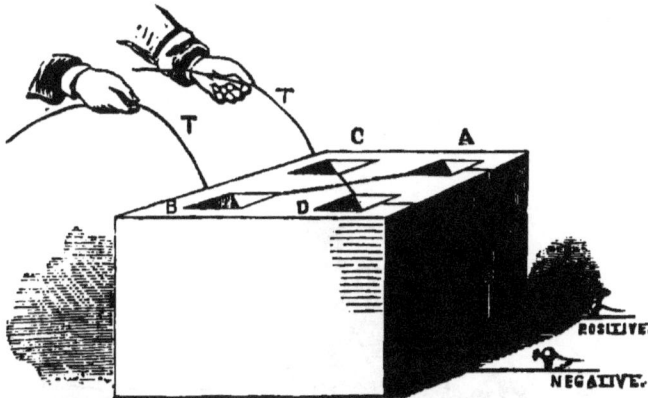

Fig. 219. Block of wood with four holes; the positive terminal is connected with the holes A and B, the negative with C and D; the hollows are filled with mercury. T T are the wires from the telegraph box, and it is obvious that by dipping them alternately into C B and A D the current is reversed, and the needle deflected right or left at pleasure.

In practice a more elaborate reverser is employed, but to demonstrate the principle the simple block above described is quite sufficient.

With the telegraph placed at the top of a house, or in a distant cottage, and a single cell of Grove's battery, or at most two, for any short distances, with the reverser, messages may be passed with great rapidity from the bottom of the house to the top, or from a mansion to the lodge, it being understood that a battery, reverser, and telegraph, are required at both places where messages are received and *answered;* but if no answers are required, the battery and reverser are placed at one end of the wire in the house, and the telegraph at the other extremity in the cottage, and earth plates may be arranged to return the current, or another wire used for that purpose. *

Whilst lauding to the utmost the invention of the electric telegraph, we must remember "there is nothing new under the sun," and that after all Nature claims the *principle* of telegraphing, and with the silent gesture, the speaking eye, interpreted and answered by others, she proclaims herself to be the originator of communication by signs.   Whilst the language of flowers, and the mournful requirements of the deaf and dumb in the use of the finger alphabet, show how readily man has adopted the important principle, till he has brought it to the highest state of perfection in the electric telegraph.

When the telegraph was first adopted on the Great Western Railway, the most ridiculous ideas were formed of its capabilities, and many persons firmly believed that the wires were used for the purpose of dragging letters and different articles from station to station. " Wife," said a man, looking at the telegraph wires, " I don't see, for my part, how they send letters on them wires, without tearin' 'em all to bits." " Oh, you stupid!" exclaimed his intellectual spouse; " why, they don't send the paper: they just send the writin' in a *fluid* state."

Fig. 220. One of the ideas of telegraphic communication.

CHAPTER XVIII.

RUHMKORFF'S, HEARDER'S, AND BENTLEY'S COIL APPARATUS.

IN the course of the popular articles on frictional and voltaic electricity, it has already been mentioned that whilst the *intensity effects*—such as the capability of the spark to pass through a certain thickness of air, or the production of the peculiar physiological effect of the shock—belong especially to the phenomena of frictional electricity, they are not apparent with the *quantity effects*, such as may be produced by an ordinary voltaic battery, unless the latter consists of an immense number of elements, such as the famous water battery of the late respected Mr. Crosse, which consisted of two thousand five hundred pairs of copper and zinc cylinders, well insulated on glass stands, and protected from dust and light. If, however, the feeble intensity current of voltaic electricity, from four or five elements, is permitted to pass into a coil of a peculiar construction, fitted with a condenser, and manufactured either by Ruhmkorff of Paris, or Mr. Hearder of Plymouth, then the most remarkable effects are producible, which have created quite a new and distinct series of phenomena, and further established in the most satisfactory manner the connexion between the electricities derived from *friction* and *chemical action.*

The construction of these coils does not differ very materially, and great merit is due to Messrs. Ruhmkorff, Hearder, and Bentley, who have separately and independently worked out the construction of the most formidable machines of this class. In a letter to the author Mr. Bentley says:—

"I commence the formation of my coil by using as an axis an iron tube ten inches long and half an inch diameter; around this is placed a considerable number of insulated iron wires the same length as the tube, and sufficiently numerous to form a bundle one inch and three quarters diameter. This core is wrapped carefully in eight or nine layers of waxed silk, the necessity of which will be obvious presently.

"My primary helix, which is formed of thirty yards of No. 14 cotton-covered copper wire, is wound carefully on this core, and consists of two layers, each layer being carefully insulated one from the other by waxed silk, for I find that if a wet string or fine platinum wire be con-nected with the two ends of the primary wires of an induction coil in action, there is scarcely an indication of an induced current to be obtained from the secondary wire. That this is not owing to any decrease of magnetic power is proved by testing the iron core before and after the experiment, but is simply owing to the central magnet or coil exerting the whole of its inductive powers upon the nearest closed circuit; it therefore follows that if the two layers of primary wire are connected by the cotton covering becoming moist, the whole of the induced current will take this path instead of traversing the secondary wire.

"Before describing my secondary wire I must again call attention to the important fact that the magnetism of the iron exerts its inductive power upon the nearest conducting medium; and I have constructed an instrument to demonstrate this fact. It consists simply of an ordinary coil, giving the third of an inch spark, but having the four inner layers of secondary wire brought out separately. Now, I find that when I keep the ends of this wire separate I obtain nearly the third of an inch spark, but when I connect them metallically I can obtain no intensity spark whatever from the seventeen coils which surround them.

"It follows from this that before winding the secondary wire the striking distance of a single layer must be ascertained, and I find that with my coil I can get a spark one-tenth of an inch long from one coil of wire, and sufficiently intense to penetrate with facility six layers of waxed silk.

"Waxed silk is therefore unsuited for the insulation of large coils, and I find, after numerous experiments, that there is no substance so fitted for the purpose as gutta-percha tissue, and I use five layers of this substance to each layer of wire.

"The secondary helix then consists of three thousand yards of No. 35 silk-covered copper wire, and is insulated in the manner described above; but as I do not use cheeks to my coil it assumes the form of a cylinder having rounded ends.

"For the protection of this instrument I place it in a mahogany box of the proper size, and it is supported and retained in its position by an iron rod, which is thrust through the hollow axis of the core and the two ends of the box, leaving half an inch of the iron projecting to work the contact breaker, which is fixed to one end of the box, while the two ends of the secondary wire are brought out of the other through gutta percha tubes.

"The condenser is contained in a separate box, and is formed of one hundred and twenty sheets of tinfoil between double that number of sheets of varnished paper, the alternate sides of the foil being brought out and connected to appropriate binding screws.

"This condenser forms a convenient stand for the coil, and can be used for many interesting experiments."

The shock which the condenser gives to the system depends in a great measure on the size of the coatings. The primary wire alone does not produce any physiological results, or at least very feeble ones. Mr. Hearder's coil is wound on a bobbin six inches in length, and four inches and a half thick, and includes three thousand yards of covered wire (No. 35). The iron core consists of a bundle of small wires capped with solid ends, and the sparks obtained from it were five-eighths of an inch in air when the primary coil was excited by four pairs of Grove's series; and when connected with the Leyden jar, the most vigorous and brilliant results were produced. The condenser is made of cartridge paper, coated in the proper manner with tinfoil. The secondary coil is quite independent of the primary one, which is laid on in different lengths, so that the coil can be adjusted to any battery power, whether for quantity or intensity.

For the successful exhibition of the capabilities of the machine, it is required to perform the experiments in a darkened room. (Fig. 221.)

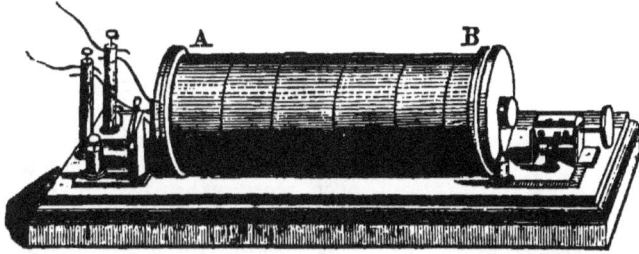

Fig. 221. Ruhmkorff's apparatus. A B. The coil, containing more than a mile of insulated wire. The stand it rests upon, and with which it is in communication, contains the *condenser*.

In using this apparatus, eight pairs of Grove's battery will be quite sufficient to produce the effects, and the greatest care must be taken to avoid the shock, which is most severe and painful, and might do a great deal of harm to a weakly, sensitive, and nervous person. To avoid any accidents of this kind, the convenient arrangement at one end shown in Fig. 222 must be carefully attended to, and when manipulating with any part of the apparatus, if the battery is attached, the contact should first be broken by bringing the ivory (the non-conducting) part of the cylinder A (Fig. 222) in communication with the conductors, B B, where the wires from the battery are attached.

Fig. 222. One end of Ruhmkorff's coil. B B. Connexion to receive the battery wires. A is the cylinder, one half of which is ivory and the other metal. In this position no shock can be received, because the electricity is cut off by the ivory from the coil.

First Experiment.

It is at the other extremity of the coil that the experiments are performed; for instance, if an exhausted globe is connected with the pillars B B (Fig. 223), and the connexion made with the battery, a beautiful faint blue light is apparent on one of the knobs and wires, and by reversing the current the light appears on the other knob and wire.

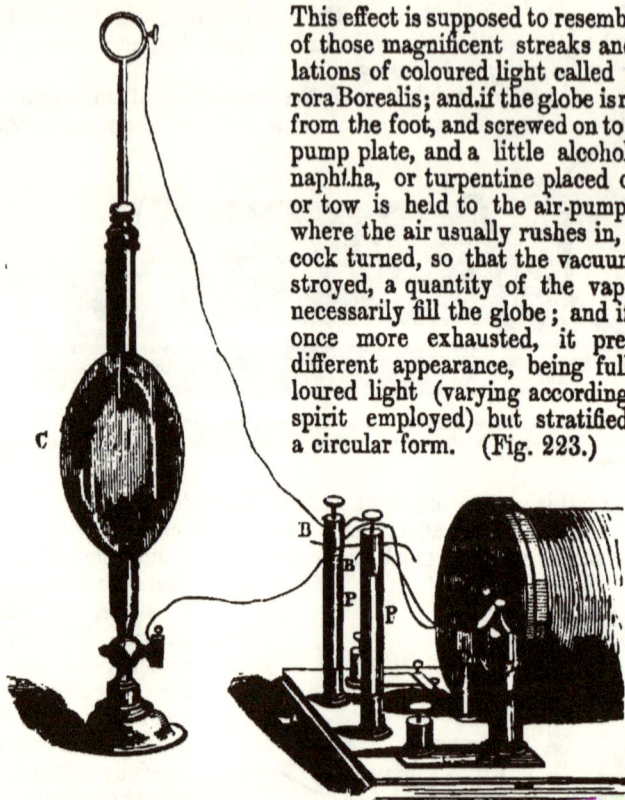

This effect is supposed to resemble some of those magnificent streaks and undulations of coloured light called the Aurora Borealis; and.if the globe is removed from the foot, and screwed on to the air-pump plate, and a little alcohol, ether, naphtha, or turpentine placed on wool or tow is held to the air-pump screw, where the air usually rushes in, and the cock turned, so that the vacuum is destroyed, a quantity of the vapour will necessarily fill the globe; and if this is once more exhausted, it presents a different appearance, being full of coloured light (varying according to the spirit employed) but stratified and of a circular form. (Fig. 223.)

Fig. 223. End of coil where the experiments are performed. в в. Connecting screws and wires passing to the exhausted globe, c. The screws are supported on insulating glass pillars, p p.

Second Experiment.

The appearance of these bands of light is modified by the nature of the glass tubes employed, and the subject has been carefully investigated by Mr. Gassiott. At the last meeting of the British Association at Aberdeen, Dr. Robinson made various experiments, arranged by Mr. Ladd, for the purpose of showing the connexion between these miniature effects of bands of light in tubes containing various gases, and the phenomena of the Aurora Borealis. The title of the discourse, which was specially delivered in the Music Hall by the learned Doctor, was "On Electrical Discharges in Highly-rarefied Media," and it was illustrated by experiments prepared by Mr. Gassiott and Mr. Ladd.

The kind of tubes employed may be understood from the next figure. They are made in Germany, and by approaching a powerful magnet to

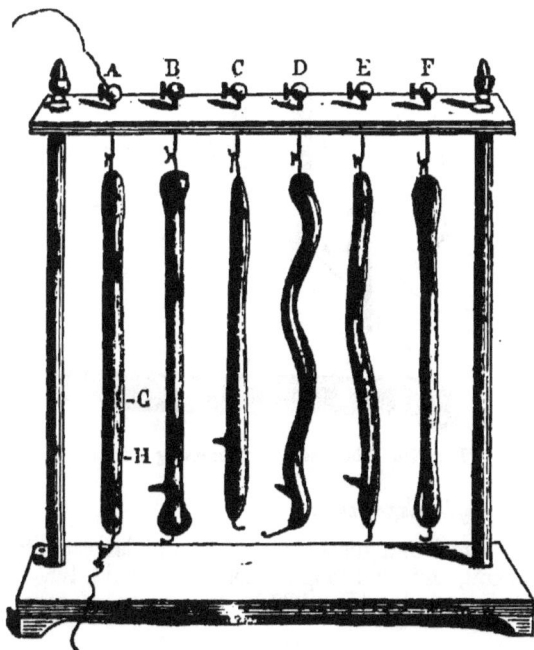

Fig. 224. A, B, C, D, E, F. Various tubes of different kinds of glass, and containing gases and vapours. Each tube has a platinum wire inserted at both ends, with which the contact is made with the coil. The tube A contains mercury, which has been boiled in it, and the air expelled. By moving the conducting wire to G or H, the light which otherwise passes through the whole of the tubes stops at these points.

the outside of any of the glass tubes whilst the bands of light are being produced, the most remarkable modifications of them are obtained. Mr. Ladd has mounted one of these tubes in a rotatory arrangement similar to that described at page 186. When connected with the coil and battery, it furnishes one of the most lovely "electric fire-wheels" that can possibly be described. (Fig. 224.) Mr. Grove placed a piece of carefully-dried phosphorus in a little metallic cup, and covered it with a jar having a cap and wire. On removing the air from the receiver, and passing the current of electricity through it from the Ruhmkorff coil, he obtained a light completely stratified, and blended transversely with straight but vibrating dark bands.

Third Experiment.

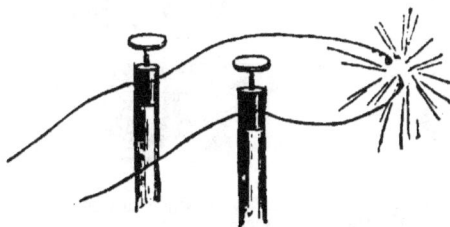

Fig. 225. Melting of the iron wire.

When two very thin iron wires are arranged in the upright pillars (Fig. 223), and held sufficiently close to each other, as in Fig. 225, light passes from one to the other. The wire from which the light passes remains *cold*, the other becomes so *hot* that it melts into a little globule of liquid iron, and if paper is held between the wires it rapidly takes fire. (Fig. 225.)

Fourth Experiment.

Remove the break.
Attach two wires to
× × (Fig. 226). Hold
them so as at pleasure
to complete and inter-
rupt the galvanic circle.
Two other wires are at-
tached at P P, their
ends being about three-
quarters of an inch
asunder. When the cur-
rent is closed or broken
at A A, a spark passes
between B B. (Fig. 226.)

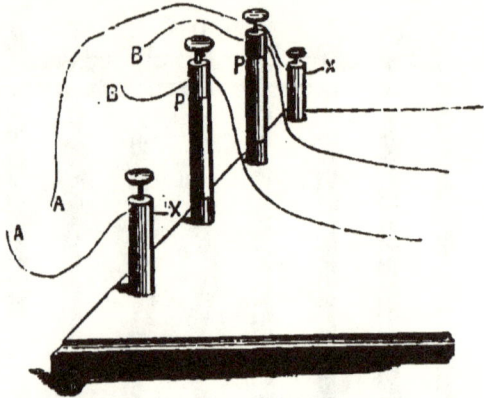

Fig. 226. The making and breaking of the circuit.

Fifth Experiment.

A Leyden jar may be charged and discharged with singular rapidity
when connected with the coil, and the snapping noise is so rapid, that it
produces a continuous sharp sound. (Fig. 227.) If a piece of paper

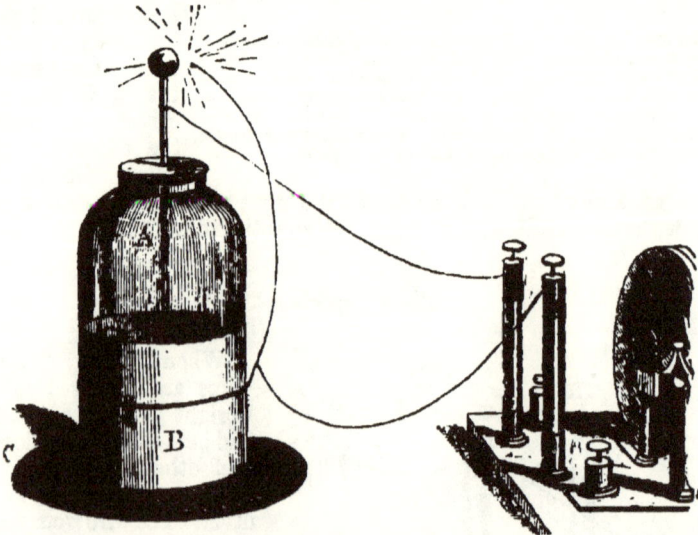

Fig. 227. A B. Leyden jar coated with tinfoil, and standing on any non-conductor,
such as gutta percha or the resinous or glass plate, c.

is held between the ball of the Leyden jar and the wire, it is instantly
perforated, but not set on fire.

Sixth Experiment.

When the Leyden jar is coated with spangles of tinfoil, a spark appears at each break, and the whole jar is lit up with hundreds of brilliant sparks each time it is charged and discharged, and as this occurs with amazing rapidity, the light is almost continuous. (No. 1. Fig. 228.) The larger the Leyden jar, the shorter the spark, and *vice versâ.* By the employment of a nicely-made screw and inch-scale, the distance between the discharging points connected with a Leyden jar can be accurately determined; and Mr. Hearder states that supposing a Leyden jar has one square foot of charging surface, it will give a spark of one inch in length, but if a smaller jar is used, with only half a square

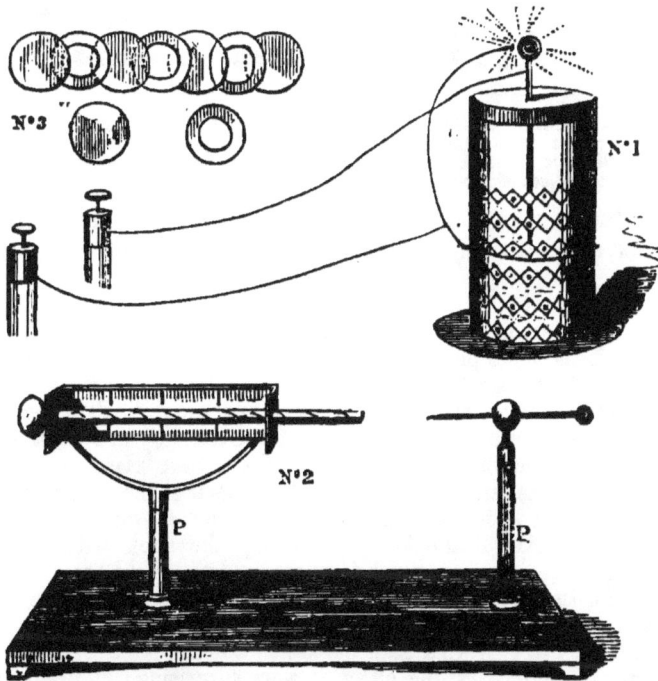

Fig. 228.—No. 1. Spangled Leyden jar.   No. 2. Hearder's apparatus for measuring the length of spark for Leyden jar and coil.   P F. Glass pillars.   No. 3. Two best forms of spangles to paste on a Leyden jar.

foot of charging surface, the spark would be about one inch and a quarter in length. (Fig. 228.)

Seventh Experiment.

The direction and rapidity of the current appear to influence greatly the heating and fire-giving power of the coil, and the following experiment, devised by Mr. Hearder, furnishes a curious illustration of this fact.

When the current passes in the direction of the arrows (Fig. 229), the platinum wire remains perfectly cool whilst the gunpowder is fired; and the contrary takes place if the current is reversed—viz., the gun-

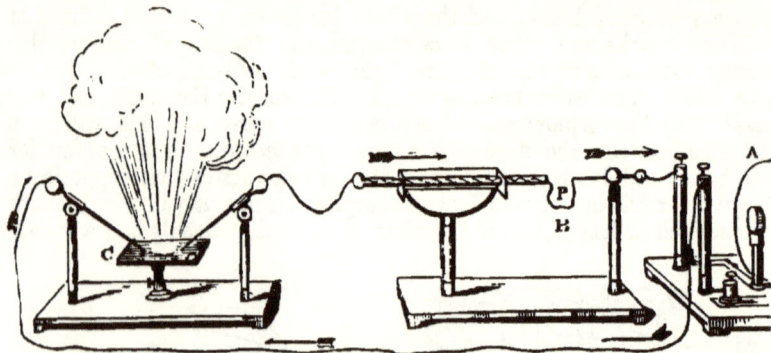

Fig. 229. A. The coil. B. Hearder's discharger, with thin platinum wire, P, hanging between the points. C. Another discharger, and powder going off between the points from the little table. The pillars of the dischargers are glass. The arrows show the direction of the current of electricity.

powder does not blow up, but the platinum wire is heated. In the second experiment, a Leyden jar is included in the circuit. (Fig. 229.)

Eighth Experiment.

Amongst so many beautiful experiments, it is somewhat difficult to say which is the most pleasing, but for softness and exquisite colouring, with the continuous vibrating motion of the flowing current of electricity, nothing can surpass "the cascade experiment." [This beautiful experiment is usually termed "Gassiott's Cascade," and is thus described by that gentleman. Two-thirds of a beaker glass, four inches deep by two inches, are coated with tinfoil, leaving one inch and a half of the upper part uncoated. On the plate of an air-pump is placed a glass plate, and over it the beaker, covering the whole with an open-mouthed glass receiver, on which is placed a brass plate having a thick wire passing through a collar of leather; the portion of the wire within the receiver is covered with a glass tube; one end of the secondary coil is attached to this wire, and the other to the plate of the pump. As the vacuum improves the effect is very surprising; at first a faint clear blue light appears to proceed from the lower part of the beaker to the plate; this gradually becomes brighter, until by slow degrees it rises, increasing in brilliancy until it arrives at that part which is opposite, or on a line with the inner coating, the whole being intensely illuminated; a discharge then commences, as if the electric fluid were itself a material body running over.] This result is obtained by coating the inside of a handsome glass goblet with tinfoil, and placing it under a jar fitted with a collar of leather and ball, and arranged in the usual manner on the air-pump. Directly a vacuum is obtained, the ball is moved down to the inside of the goblet, and the wires from the coil being attached, a continuous series of streams of electric light seem to overflow the goblet all round the edge, and it stands then the very embodiment of the brimming cup of *fire*, and emblematical of the dangers of the wine-cup.   (Fig. 230.)

Fig. 230. Gassiott's Cascade.

Ninth Experiment.

If a piece of wood five inches long and half an inch square is placed on the table of the discharger, and one wire brought on to the top edge

Fig. 231. Burning the piece of wood moistened with the strongest nitric acid.

and the other approached to within three inches of it, and touching the wood, and the space between them moistened with the strongest nitric acid, a curious effect is visible from the creeping along of the fire, which gradually carbonizes and burns the wood. (Fig. 231.)

Tenth Experiment.

A glass plate wetted with gum, and then sprinkled with various filings of iron, zinc, lead, copper, &c., produces a very pretty effect of deflagration as one of the conducting wires is moved over its surface, the other of course being in contact with the plate. The gum quickly dries by putting the plate in a moderately-heated oven.

Eleventh Experiment.

When the continuous discharges from the Leyden jar are made to pass through the centre of a large lump of crystal of alum, blue vitriol, or ferroprussiate of potash, &c., the whole of the crystal is beautifully

Fig. 232. A. The Leyden jar. B. Large lump of alum, with a hole bored through it in a line with C D. The discharging wires are brought within three-eighths of an inch of each other, and the whole crystal is lighted up with the brilliant electric sparks.

lighted up during the passage of the electricity from one wire of the discharger to the other. (Fig. 231.)

Twelfth Experiment.

When a piece of paper slightly damped is placed between the wires of the discharger, the spark is increased to a much greater length, on account of the conducting power of the water contained in the pores of the paper; and taking all things into consideration, the author considers he has witnessed the grandest effects from the coil invented and constructed by Mr. Hearder, the talented lecturer and electrician of the West of England.

Thirteenth Experiment.

Electro-magnetic coil machines have been employed for a very considerable time in alleviating certain of "the ills which flesh is heir to," by the administration of shocks. These may be so regulated as to be hardly perceptible, or may be so powerful that the pain becomes absolutely intolerable.

These coils are now made self-acting, and consist of two coils of covered and insulated wire wound round a bundle of soft-iron wires, with the necessary connecting screws for the voltaic battery. The contact with the battery is made and broken with great rapidity by a simple form of break, consisting of a tinned disc of iron held by a spring over the axis of the bundle of iron wires; and the continual noise of the break, which is alternately attracted down to the bundle and brought back by the spring, when the coil is in contact with the battery, demonstrates (without the pain of taking the shock) when the instrument is in full working order.

The coil machine is not only useful in a medical point of view, but when properly arranged offers a good reception to a run-away bell-ringer, and is an excellent preventive against illicit attempts at cheap rides by small boys.

Fig. 233. Boy, *evidently shocked*, behind doctor's carriage provided with a small coil machine.

CHAPTER XIX.

MAGNETO-ELECTRICITY.

Fig. 234. Clarke's magneto-electrical machine.

THE correlation of the physical forces, heat, light, electricity, magnetism, and motion, is one of the most interesting subjects for study that can be suggested to the lover of science. The examination of the precise meaning of the term correlation, so ably considered by Professor Grove, indicates a necessary mutual or reciprocal dependence of one force on the other. Thus, electricity will produce heat, and *vice versâ*; motion, such as friction, produces electricity, and the latter, by its attraction and repulsion, establishes itself as a source of motion. Electricity produces light, also magnetism, and contrariwise light is said to possess

B

the power of magnetizing steel, whilst magnetism again produces light and electricity. Such are the intimate connexions that exist between these imponderable agents, and we may trace cause and effect and its reversal amongst these forces, until the mind is lost in the examination of the bewildering mazes, and is content to return to the beaten track and work out experimentally the practical truths. We have had occasion to notice in another part of this playbook the fact that a current of electricity causes the evolution of magnetism in its passage through various conducting media, and the truth has been specially illustrated by the various experiments in the chapter devoted to electro-magnetism. In commencing this portion of electrical science, we have no new terms to coin for the title of the discourse, as we merely reverse the other when we examine the nature and peculiarities of

MAGNETO-ELECTRICITY.

The source of the power must necessarily be a bar or horseshoe-shaped piece of steel permanently endowed with magnetism. If the former is thrust into a cylinder of wood or pasteboard, around which coils of covered copper wire have been carefully wound, so that the extremities communicate with a galvanometer, an immediate deflection of the needle occurs, which, however, quickly returns to its first position, but is again deflected in the opposite direction on the withdrawal of the steel magnet from the coil of copper wire. (Fig. 235.)

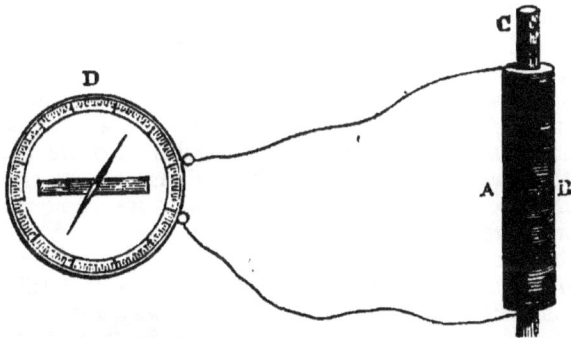

Fig. 235.   A B. Coil of copper wire.   c. Permanent bar magnet placed inside the coil, when the galvanometer needle, D, is deflected.

The rapid entrance and exit of the steel magnet in the helix of copper wire would be insufficient to produce any quantity of electricity, and the ingenuity of man has been taxed to arrange a method by which a magnet may be suddenly formed and destroyed inside a coil of insulated copper wire. The difficulty, however, has been surmounted by several ingenious contrivances, based on the principles first discovered by Faraday; and the one especially to be noticed is the revolution of a coil of copper wire enclosing a piece of soft iron, called the *armature*, before the poles of a powerful magnet. The first machine was invented by M. Hypolyte Pixii, of Paris, and in 1833, Mr. Saxton improved upon this machine, and three years afterwards, Mr. E. M. Clarke de. scribed a very ingenious modification of the electro-magnetic machine, which is depicted at page 241 of this chapter. In this picture, the letter A is the permanent fixed horseshoe magnets, which are very appropriately termed the *battery* magnets, because they take the position that would otherwise be occupied by a voltaic battery, and they are indeed the prime source of the electrical power that is evoked. D is the intensity *armature* which screws into a brass mandril seated between the poles of the magnets A, motion being communicated to it by the multiplying wheel, E. This armature or *inductor* has two coils of fine insulated copper wire of 1500 yards in length, coiled on its cylinders, the commencement of each coil being soldered to the bar D, from which projects a brass stem, also soldered into D, carrying the break-piece H, which is made fast in any position by a small binding-screw in a hollow brass cylinder to which the other terminations of the coils, F F, are soldered, these being insulated by a piece of hard wood attached to the brass stem. O is an iron wire spring pressing against one end of the hollow brass cylinder; P is a square brass pillar; Q is a metal spring that rubs gently on the break piece H; T is a copper wire for connecting the brass pieces with the wood L between them, and out of which P and O pass; R R are two handles of brass with metallic wires, the end of one being inserted into either of the brass pieces connected with P and O, and the other into the brass stem that carries the break H, delivers a most severe shock directly the wheel is set in motion.

In Saxton's electro-magnetic machine, the permanent steel magnets are placed horizontally instead of perpendicularly, and are composed of six or more horseshoe-shaped pieces of steel. The armatures, or inductors, or electro-magnets (for they consist of pieces of soft round iron with wire wound round them), are two in number, and are adapted to exhibit either *quantity* or *intensity* effects. The quantity armature is constructed of stout iron, and covered with thick insulating wire. The intensity armature is made of slighter iron, and covered with from one thousand to two thousand yards of fine copper wire coated with silk. The *quantity* armature is intended for the exhibition of results similar to those which are procurable from a voltaic battery, such as the magnetic spark, inducing magnetism in soft iron, heating platinum wire. The intensity armature is employed for the chemical decomposition of water and other bodies, and likewise for the administration of those terrible blows to the nervous system which cause strong men of the mildest deportment to become painfully excited, and to make those ejaculations which are so peculiar to the genus John Bull.

EXPERIMENTS WITH THE MAGNETO-ELECTRIC MACHINE.

First Experiment.

The decomposition of water by the passage of electricity from one platinum plate to another, has already been illustrated at page 198. The

same fact may likewise be displayed by the following arrangement of
the machine.   (Fig. 236.)

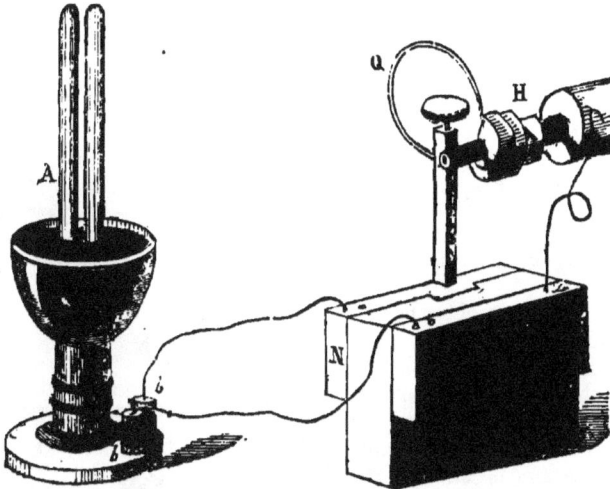

Fig. 236.   A. Apparatus for decomposing water and collecting the gases separately.
B B. Wires proceeding from the machine at M, N.   Q, works on the single break, H.

Second Experiment.

The electric light obtained by the passage of the electricity from the
battery through the charcoal points, is also an effect that can be produced by magneto-electric machines, the wires leading from the points
A B being insulated by glass handles, and placed in the holes M N.
(Fig. 237.)

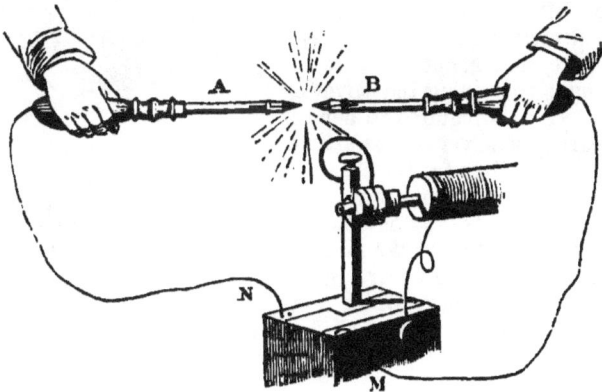

Fig. 237.   The electric light obtained from the magneto machine.

Third Experiment.

The scintillation of iron wire is one of the most pleasing experiments with this apparatus, and is performed by pressing gently one end of a piece of thin iron wire (attached by means of a binding-screw to the upright bar A) against the armature, D. (Fig. 238.)

Fourth Experiment.

The combustion of ether or other inflammable spirit may also be demonstrated with the aid of this powerful apparatus, and the arrangement, in common with the others employed by Mr. Clarke, is shown in Fig. 239.

With the assistance of the magneto-electric machine, telegraphic communication may be conducted without the assistance of a battery. It has also been applied to the art of electro-plating by Mr. J. P. Woolrich, of Birmingham; and whilst visiting that place, the author had the opportunity of witnessing the arrangement employed.

It consists of a very powerful magneto-electric machine turned by a steam-engine, and connected with the large troughs containing the silvering solution. If it is required to deposit a thin coating of silver on the article, a short period suffices for the action of the machine, whilst a thick deposit of the precious metal is only obtained by the constant operation of the magnets for several hours. At Mr. Woolrich's factory, the goods which were

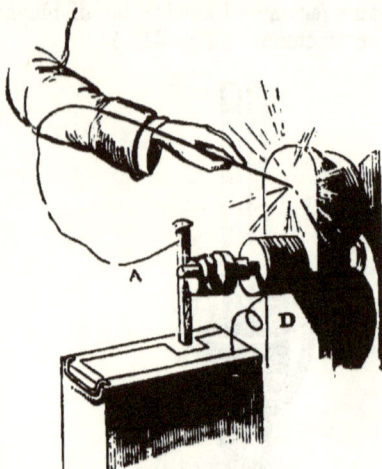

Fig. 238. Deflagration of iron wire.

Fig. 239. The break is removed, and the double blades, B, fixed in its place. The brass cup, A, containing mercury is so adjusted that the points will leave the surface of the mercury when the armature is vertical. Ether or alcohol poured on the surface is quickly inflamed by the electric spark.

being coated with silver were all kept in motion, moving slowly backwards and forwards in the trough by means of an eccentric connected with the same steam-engine that worked the electro-magnetic machine.   (Fig. 240.)

Fig. 240.   Silvering and plating by the magneto machine, turned by a steam-engine.

The magneto-electric telegraph patented by Mr. Henley in 1848, offers another example of the application of the electric current induced in electro-magnetic coils, when they rotate in close proximity to the poles of a powerful steel magnet.   This telegraph is now in constant use by the English and Irish Magnetic Telegraph Company, through a distance of more than 2100 miles.   The whole length of wires in use amounts to the astonishing quantity of 13,900 miles, of which 6350 miles are hidden underground, and 7500 conducted above.

This telegraph is considered to be one of the simplest and most economical yet brought into practical working.

CHAPTER XX.

DIA-MAGNETISM.

AT the end of the chapter devoted to the subject of light, will be found an experiment devised and carried out by Dr. Faraday, in which it is shown that if a bar of a peculiar glass (called after the inventor, *Faraday's heavy glass*, or silicated borate of lead) is subjected to the inductive action of a very powerful electro-magnet, that it has the power of changing the direction of a ray of polarized light transmitted through it. This effect is not confined to the poles of an electro-magnet, but is also perceptible (though in a diminished degree) with ordinary magnets.

The result of this important experiment was communicated to the Royal Society by Dr. Faraday on the 27th November, 1845, the enunciation of the fact by this learned philosopher being, "that when '*the line of magnetic force*' is made to pass through certain transparent bodies parallel to a ray of polarized light traversing the same body, the ray of polarized light experiences a rotation." Now, "*the line of magnetic force*" means that continual flow of the magnetic current which passes from pole to pole, and is indicated by iron filings sprinkled on paper placed above the poles of a magnet, and usually termed *magnetic curves*, or the curved lines of magnetic force. (Fig. 241.)

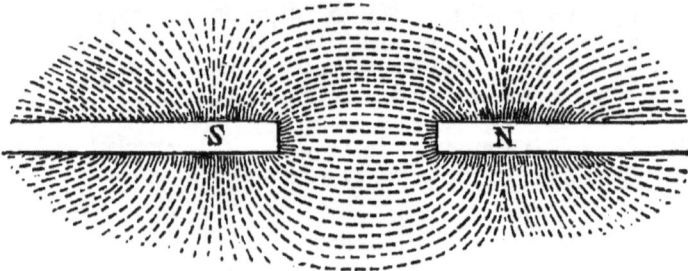

Fig. 241. The curved lines of magnetic force.

The heavy glass already alluded to, upon which the magnet exerts a certain influence, is called

THE DIA-MAGNETIC;

and by this term is meant a body through which the lines of magnetic force are passing without affecting it like iron or steel. At page 212 is a picture representing (at Figs. 201 and 202) the direction of the electricity and that of the magnetic current or whirl at right angles to it. If, then, Fig. 202 be considered as a piece of glass, the arrow A B will show "the line of magnetic force," the point B being the north pole, and the shaft A the south pole of the magnet, and the arrows traced round will represent direction. This simple drawing expresses the whole of the law of the action of the magnet on the glass, and if kept in view, will give every position and consequence of direction resulting from it.

The phenomenon of the affection of the beam of polarized light is immediately connected with the magnetic force, and this is supposed to be proved by the *brightness* of the polarized ray being developed *gradually*, as the iron coiled with wire requires about two seconds to acquire its greatest power after being connected with the battery.

In another experiment of Faraday's, where a beam of polarized light was sent through a long glass tube containing water, and introduced as a core *inside* a powerful electro-magnetic coil, the image of a candle viewed with a proper eye-piece, appeared or disappeared as the battery connexion was made or broken with the coil; but this result is not considered by many philosophers to be conclusive of the action of magnetism on light, but rather as an alteration of the *refracting* power of the medium through which the light passes. These experiments were the precursors of the other effects of magnetism upon different kinds of matter which Faraday discovered, and he commenced his examination with a small bar of heavy glass suspended by a filament of silk between the poles of an electro-magnet, and when the twisting or effects of torsion had ceased, the battery was connected. Directly the current passed, Faraday's keen eye detected a movement of the glass, and on repeating the experiment, he discovered that the movement was not accidental, but always took place in a certain fixed direction—viz., a direction at right angles to a line drawn across and touching the two poles of a horseshoe-shaped magnet—i.e., supposing the feeder or bit of soft iron usually placed in contact with the poles of the horseshoe-magnet to represent the "*axial line*," any line drawn across it at right angles would be called the *equatorial line*, whilst the general space included between the poles of the magnet is called "the *magnetic field*." The movement of the heavy glass was therefore *equatorial*, and it pointed east and west instead of north and south, like iron and steel.

By the use of the apparatus (Fig. 242) Faraday proved that every

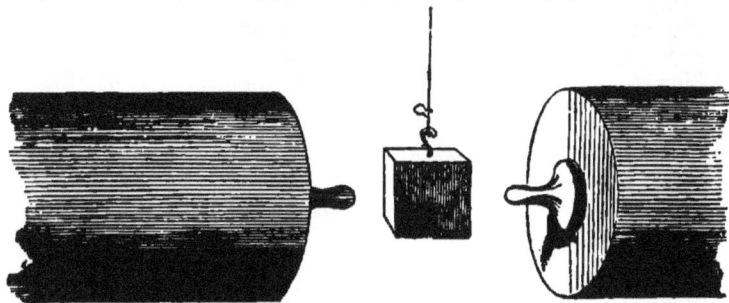

Fig. 242. A cube of copper suspended between the poles of a powerful electro-magnet.

substance, whether solid, fluid, or gaseous, was subject to magnetic influences, assuming either the axial or equatorial position. The apparatus consists of a prolongation of the poles of a powerful electromagnet, between which *the* cube of copper, weighing from a quarter to half a pound, suspended by a thread, may be set spinning or rotating. If the electro-magnet is connected with the battery, the cube stops immediately, and whilst still in the same position or in the *magnetic field*, with the magnet in full action, it is impossible to set it spinning or twisting round again. (Fig. 242.)

A large number of other substances, solid, liquid, and gaseous, were submitted to the action of the magnet, the liquids and gases being hermetically sealed in glass tubes, and some of the results are detailed in the following list :

Bodies that point axially, or are paramagnetic, like a suspended needle.

Iron.	*Red-lead.*
Nickel.	Sulphate of zinc.
Cobalt.	Shell-lac.
Manganese.	Silkworm-gut.
Chromium.	Asbestos.
Cerium.	Vermilion.
Titanium.	Tourmaline.
Palladium.	Charcoal.
Platinum.	All salts of iron, when the latter is
Osmium.	basic.
Paper.	Oxide of titanium.
Sealing-wax.	Oxide of chromium.
Fluor spar.	Chromic acid.
Peroxide of lead.	Salts of manganese.
Plumbago.	Salts of chromium.
China ink.	Oxygen, which stands alone as a
Berlin Porcelain.	paramagnetic gas.

Bodies that point equatorially, or are diamagnetic, like Faraday's heavy glass.

Bismuth.	Uranium.
Antimony.	Rhodium.
Zinc.	Iridium.
Tin.	Tungsten.
Cadmium.	Rock crystal.
Sodium.	The mineral acids.
Mercury.	Alum.
Lead.	Glass.
Silver.	*Litharge.*
Copper.	Common salt.
Gold.	Nitre.
Arsenic.	Phosphorus.

Sulphur.	Apple.
Resin.	Bread.
Spermaceti.	Leather.
Iceland spar.	Fresh blood.
Tartaric acid.	Dried blood.
Citric acid.	Caoutchouc.
Water.	Jet.
Alcohol.	Turpentine.
Ether.	Olive oil.
Sugar.	Hydrogen.
Starch.	Carbonic acid.
Gum-arabic.	Carbonic oxide.
Wood.	Nitrous oxide (moderately).
Ivory.	Nitric oxide (very slightly).
Dried mutton.	Olefiant gas.
Fresh beef.	Coal gas.
Dried beef.	

Nitrogen is neither paramagnetic nor diamagnetic, and is equivalent to a vacuum. Magnetically considered, it is like space itself, which may be considered as zero.

The term *magnetic* Faraday proposes should be a general one, like that of *electricity*, and include *all* the phenomena and effects produced by the power, and he proposes that bodies magnetic in the sense of iron should be called *paramagnetic*, so that the division would stand thus :

Magnetic . . . . $\begin{cases} \text{Paramagnetic,} \\ \text{Diamagnetic ;} \end{cases}$

and it is this division which has been observed in the preceding tables.

All space above and within the limits of our atmosphere may be regarded as traversed by lines of force, and amongst others are the lines of magnetic force which affect bodies, as shown in the table of paramagnetic and diamagnetic bodies, which have the same relation to each other as positive and negative, or north and south, in electricity and magnetism.

The lines of magnetic force are assumed to traverse void space without change; but when they come in contact with matter of any kind they are either concentrated upon it or scattered according to the nature of the matter.

The power which urges bodies to the axial or equatorial lines is not a central force, but a force differing in character in the axial or radial directions. If a liquid paramagnetic body were introduced into the field of force, it would dilate axially, and form a prolate spheroid like a lemon, while a liquid diamagnetic body would dilate equatorially, and form an oblate spheroid like an orange. Plücker has demonstrated that if magnetic solutions are placed in watch glasses across the poles of the electro-magnet, they are heaped up in a very curious manner. The poles of the electro-magnet are pieces of soft iron, which may be drawn away or approached at pleasure, and according as the poles are nearer or further asunder, the magnetic liquids, such as solution of iron, are heaped up in one or two directions, as shown at B and C in Fig. 243.

Fig. 243.   Glass dish holding magnetic solution of iron, and placed in the magnetic field.

"The diamagnetic power, doubtless," says Faraday, " has its appointed office, and one which relates to the whole mass of the globe.   For though the amount of the power appears to be feeble, yet, when it is considered that the crust of the earth is composed of substances of which by far the greater portion belongs to the diamagnetic class, it must not be too hastily assumed that their effect is entirely overruled by the action of the magnetic matters, whilst the great mass of waters and the atmosphere must exert their diamagnetic action uncontrolled."

Plücker has also announced—what at the time he believed to be true—the highly interesting and important fact that the optic axis of Iceland or calcareous spar is repelled by the magnet and placed equatorially—a fact which Plücker thought true of many other crystals when the magnetic axis is parallel to the longer crystallographic axis. A piece of kyanite, which is a mineral composed of sand, clay, often lime, iron, water, and is used in India, being cut and polished as a gem, and sold frequently as an inferior kind of sapphire, will, it is said, even under the influence of the earth's magnetism, arrange itself like a magnetic needle.

Plücker believed that he had discovered an existing relation between the forms of the ultimate particles of matter and the magnetic forces, and he imagined that the results he obtained would lead gradually to the determination of crystalline form by the magnet.   The experiments of Tyndal and Knoblauch lead, however, to a very opposite series of conclusions, and by ingeniously powdering the crystals with water, and making them into a paste, which was afterwards dried and suspended as a model in "the magnetic field;" also by taking a slice of apple about as thick as a penny-piece, with some bits of iron wire through it, in a direction perpendicular to its flat surface, they were found to set equatorially not by repulsion but by the attraction of the iron wires ; or instead of the iron by placing bismuth wires, the apple now settled axially, not by attraction but by the repulsion of the bismuth. Ipecacuanha lozenges, Carlisle biscuits also, suspended in the magnetic field, exhibited a most striking directive action. The materials in these two cases were *diamagnetic;* but owing to the pressure exerted in their formation their largest horizontal dimensions set from pole to pole, the line of compression being equatorial ; and it is a universal law "*that in diamagnetic bodies the line along which the density of the mass has been induced by compression sets equatorial, and in magnetic bodies axial.*" Hence they assume, from these and many other conclusive experiments, that crystallized bodies, such as Iceland spar, take their position in the magnetic field without reference to the existence of an "optic axis."

At the conclusion of a brilliant lecture at the Royal Institution by Dr. Tyndal "On the influence of material aggregation upon the manifestations of force," in which Plücker's experiments respecting the repulsion of the optic axis were gracefully discussed and his theory refuted, the learned doctor said : "This evening's discourse is in some measure connected with this locality; and thinking thus, I am led to inquire wherein the true value of a scientific discovery consists? Not in its immediate results alone, but in the prospect which it opens to intellectual activity—in the hopes which it excites—in the vigour which it awakens. The discovery which led to the results brought before us to-night was of this character. *That* magnet* was the physical birthplace of these results; and if they possess any value they are to be regarded as the returning crumbs of that bread which in 1846 was cast so liberally upon the waters. I rejoice, ladies and gentlemen, in the opportunity here afforded me of offering my tribute to the *greatest workman* of the age, and of laying some of the blossoms of that prolific tree which he planted at the feet of the great discoverer of diamagnetism."†

It was first observed by Father Bancalari, of Genoa, that when the flame of a candle is placed between the poles of a magnet it is strongly repelled. The flames of combustible gases from various sources are differently affected, both by the nature of the combustible and by the nearness of the poles. Faraday repeated Bancalari's experiments, and by a certain arrangement of the poles of this magnet he obtained a powerful effect in the *magnetic field*, and having the axial line of the magnetic force horizontal, he found that when the flame of a wax taper was held near the axial line (but on one side or the other), and about one-third of the flame rising above the level of the upper surface of the

---

* Alluding to a splendid magnet made by Logeman, which was sent to the Exhibition in Hyde-park in 1851. It could sustain a weight of 430 pounds, and was purchased by the Royal Institution for Dr. Faraday.
† Dr. Faraday.

poles, as soon as the magnetic force was exerted the flame receded from the axial line, moving equatorially until it took an inclined position, as if a gentle wind was causing its deflection from the upright position.

When the flame was placed so as to rise truly across the magnetic axis, the effect of the magnetism was very curious, and is shown at A, Fig. 244.

On raising the flame a little more the effect of the magnetic force was to intensify the results already mentioned, and the flame actually became of a *fish-tailed shape*, as at C, Fig. 244; and when the flame was raised until about two-thirds of it were above the level of the axial line, and the poles approached very close, the flame no longer rose between the poles, but spread out right and left on each side of the axial line, producing a double flame with two long tongues, as at B, Fig. 244.

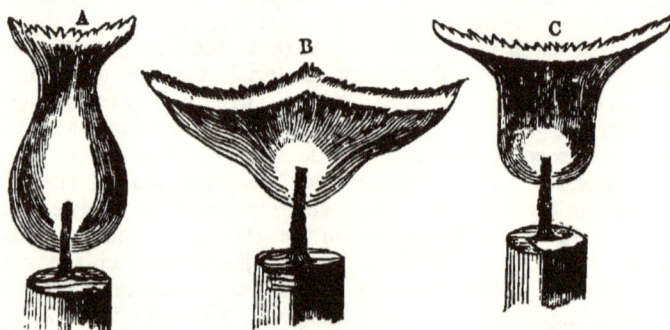

Fig. 244. Effect of magnetism on candle-flame between the poles of the magnet.

It was these experiments that led to the important discovery of the paramagnetic property of oxygen, and proved in a decided manner that gaseous bodies when heated became more highly diamagnetic. Oxygen, which (tried in the air) is powerfully magnetic, becomes diamagnetic when heated. A coil of platinum wire heated by a voltaic current, and placed beneath the poles of Faraday's apparatus, occasioned a strong upward current of air; but directly the magnetic action commences the ascending current divides, and a descending current flows down *between* the upward currents.

The discovery, says Silliman, of the highly paramagnetic character of oxygen gas, and of the neutral character of nitrogen, the two constituents of air, is justly esteemed a fact of great importance in studying the phenomena of terrestrial magnetism. We thus see that one-fifth of the air by volume consists of an element of eminent magnetic capacity, after the manner of iron, and liable to great physical changes of density, temperature, &c., and entirely independent of the solid earth. In this medium hang the magnetic needles used as tests, and as this magnetic medium is daily heated and cooled by the sun's rays, its power of transmitting the lines of magnetic force is then affected, influencing undoubtedly the diurnal changes of the magnetic needle.

For a complete digest of Faraday's discoveries in diamagnetism the reader is referred to the second edition of Dr. Noad's comprehensive and learned work entitled " A Manual of Electricity."

Coming always from the highest walks of philosophy to lower and " *common things*," one cannot help being reminded of the old-fashioned method of *drawing up* a sluggish fire, and the natural query is suggested whether the poker is to be considered as a weak magnet, and does influence and draw towards the fire a greater supply of magnetic oxygen gas? (Fig. 245.)

Fig. 245.

The interior of the optical box at the Polytechnic—looking towards the screen. The assistants are supposed to be showing the dissolving views.

p. 255.

Fig. 246. "The moon shines bright :—In such a night as this."—*The Merchant of Venice.*

CHAPTER XXI.

LIGHT, OPTICS, AND OPTICAL INSTRUMENTS.

"To gild refined gold, to paint the lily,
To throw a perfume on the violet,
To smooth the ice, or add another hue
Unto the rainbow, or with taper light
To seek the beanteous eye of heaven to garnish,
Is wasteful and ridiculous excess."

PERFECTION admits of no addition, and it is just this feeling that might check the most eloquent speaker or brilliant writer who attempted to offer in appropriate language, the praises due to that first great creation of the Almighty, when the Spirit of God moved upon the face of the waters and said, "Let there be light." If any poet might be permitted to laud and glorify this transcendant gift, it should be the inspired Milton; who having enjoyed the blessing of light, and witnessed the varied and beautiful phenomena that accompany it, could, when afflicted by blindness, speak rapturously of its creation, in those sublime strains beginning with—

"'Let there be light,' said God, and forthwith light
Ethereal, first of things, quintessence pure,
Sprung from the deep: and from her native east
To journey through the airy gloom began,
Sphered in a radiant cloud, for yet the sun
Was not; she in a cloudy tabernacle

Sojourn'd the while. God saw the light was good,
And light from darkness by the hemisphere
Divided: light the day, and darkness night,
He named."

There cannot be a more glorious theme for the poet, than the vast utility of light, or a more sublime spectacle, than the varied and beautiful phenomena that accompany it. Ever since the divine command went forth, has the sun continued to shine, and to remain, "till time shall be no more," the great source of light to the world, to be the means of disclosing to the eye of man all the beautiful and varied hues of the organic and inorganic world. By the help of light we enjoy the prismatic colours of the rainbow, the lovely and ever changing and ever varied tints of the forest trees, the flowers, the birds, and the insects ; the different forms of the clouds, the lovely blue sky, the refreshing green fields; or even the graceful adornment of "the fair," their beautiful dresses of exquisite patterns and colours. Light works insensibly, and at all seasons, in promoting marvellous chemical changes, and is now fairly engaged and used for man's industrial purposes, in the pleasing art of photography; just as heat, electricity, and magnetism, (all imponderable and invisible agents,) are employed usefully in other ways.

The sources from whence light is derived are six in number. The first is the sun, overwhelming us with its size, and destroying life, sometimes, with his intense heat and light, when the piercing rays are not obstructed by the friendly clouds and vapours, which temper and mitigate their intensity, and prevent the too frequent recurrence of that quick and dire enemy to man, the *coup de soleil*.

The body of the sun is supposed to be a habitable globe like our own, and the heat and light are possibly thrown out from one of the atmospheric strata surrounding it. There are probably three of these strata, the one believed to envelope the body of the sun, and to be directly in contact with it, is called the *cloudy stratum ;* next to, and above this, is the luminous stratum, and this is supposed to be the source of heat and light; the third and last envelope is of a transparent gaseous nature. These ideas have originated from astronomers who have carefully watched the sun and discovered the presence of certain black spots called *Maculæ*, which vary in diameter from a few hundreds of miles to 40 or 50,000 miles and upwards. There is also a greyish shade surrounding the black spots called the *Penumbra*, and likewise other spots of a more luminous character termed *Faculæ* ; indeed the whole disc of the sun has a mottled appearance, and is stippled over with minute shady dots. The cause of this is explained by supposing that these various spots represent openings or breaks in the atmospheric strata, through which the black body of the sun is apparent or other portions of the three strata, just as if a black ball was covered with red, then with yellow, and finally with blue silk : on cutting through the blue the yellow is apparent ; by snipping out pieces of the blue and yellow, the red becomes visible ; and by slicing away a portion of the three silk coverings the black ball at last comes into view. On a similar principle it is supposed that the variety of spots and eruptions on the sun's face or disc may be explained. The evolution of light is not, however, confined to the sun, and it emanates freely from terrestrial matter by mechanical action, either by friction, or in some cases by mere percussion. Thus the axles of railway carriages soon become red hot by friction if the oil holes are stopped up; indeed hot axles are very frequent in railway travelling, and when this happens, a strong smell of burning oil is apparent, and flames come out of the axle box. The knifegrinder offers a familiar example of the production of light by the attrition of iron or steel against his dry grindstone.

The same result on a much grander scale is produced by the apparatus invented by the late Jacob Perkins; the combustion of steel ensues under the action, viz., the friction of a soft iron disc revolving with great velocity against a file or other convenient piece of hardened steel. (Fig. 247)

The stand has a disc of soft iron fixed upon an axis, which revolves on two anti-friction wheels of brass. The disc, by means of a belt worked over a wheel immediately below it, is made to perform 5000 revolutions per minute. If the hardest file is pressed against the edge of the revolving disc, the velocity of the latter produces sufficient heat by the great friction to melt that portion of the file which is brought in contact with it, whilst some particles of the file are torn away with violence, and being

Fig. 247. Instrument for the combustion of steel.

8

projected into the air, burn with that beautiful effect so peculiar to
steel. If the experiment is performed in a darkened room, the periphery of the revolving disc will be observed to have attained a
luminous red heat. Thirty years ago every house was provided with a
"tinder-box" and matches to "strike a light." Since the advent of
prometheans and lucifers, the flint and steel, the tinder, and the matches
dipped in sulphur, have all disappeared, and now the box might be
deposited in any antiquarian museum under the portrait of Guy Fawkes,
and labelled, "an instrument for procuring a light, extensively used in
the early part of the nineteenth century." (Fig. 248.)

Fig. 248.   c. The steel.   b. The flint.   e. The tinder.   d. The matches of the
old-fashioned tinder-box, a.

The rubbing of a piece of wood (hardened by fire, and cut to a point)
against another and softer kind, has been used from time immemorial
by savage nations to evoke heat and light; the wood is revolved in the
fashion of a drill with unerring dexterity by the hands of the savage,
and being surrounded with light chips, and gently aided by the breath,
the latent fire is by great and incessant labour at last procured. How
favourably the modern lucifers compare with these laborious efforts of
barbarous tribes! a child may now procure a light with a chemically
prepared metal, and great merit is due to that person who first devised
a method of mixing together phosphorus and chlorate of potash and so

adjusted these dangerous materials that they are as safe as the "old tinder-box," and have now become one of our domestic necessaries. Ignition, or the increase of heat in a solid body, is another source of light, and is well illustrated in the production of illuminating power from the combustion of tallow, oil, wax, camphine or coal gas. The term *ignition* is derived from the Latin (*ignis*, fire), and is quite distinct, and has a totally different meaning from that of *combustion*. If a glass jar is filled with carbonic acid gas, and a little tray placed in it containing some gun cotton, it will be found impossible to fire the latter with a lighted taper, *i.e.* by combustion (*comburo*, to burn), because the gas extinguishes flame which is dependent on a supply of oxygen; whereas if a copper or other metallic wire is made red hot or ignited, the carbonic acid has no effect upon the heat, and the red hot wire being passed through the gas, the gun cotton is immediately fired.

Flame consists of three parts—viz., of an outer film, which comes directly in contact with the air, and has little or no luminosity; also of a second film, where carbon is deposited, and, first by *ignition*, and finally by combustion, produces the light; and thirdly, of an interior space containing unburnt gas, which is, as it were, waiting its turn to reach the external air, and to be consumed in the ordinary manner. (Fig. 249.)

Chemical action and electricity have been so frequently mentioned in this work as a source of heat and light, that it will be unnecessary to do more than to mention them here, whilst phosphorescence (the sixth source of light) in dead and living matter, a spontaneous production of light, is well known and exemplified in the "glow-worm," the "fire-fly," the luminosity of the water of the ocean, or the decomposing remains of certain fish, and even of human bodies. Phosphorescence is still more curiously exemplified by holding a sheet of white paper, a calcined oyster-shell, or even the hand, in the sun's rays, and then retiring quickly to a darkened room, when they appear to be luminous, and visible even after the light has ceased to fall upon them.

For the purpose of examining the temporary phosphorescence of various bodies, M. Becquerel has invented a most ingenious instrument, called the "phosphorescope." It

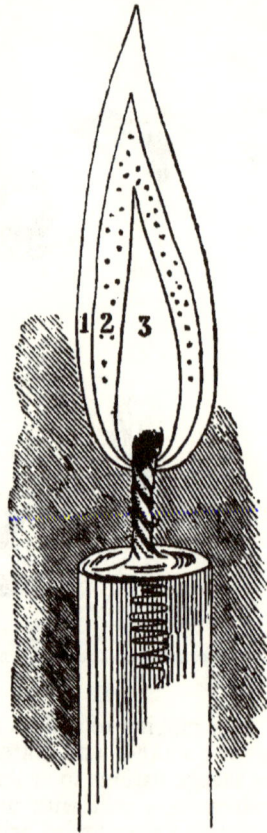

Fig. 249. A candle flame. 1. Outer flame. 2. Inner flame, which is badly supplied with oxygen, and where the carbon is deposited and *ignited*. 3. The interior, containing unburnt gas.

consists of a cylinder of wood one inch in diameter and seven inches long, placed in the angle of a black box with the electric lamp inside, so that three-fourths of the cylinder are visible outside, and the remaining fourth exposed to the interior electric light.

By means of proper wheels the cylinder, covered with any substance (such as Becquerel's phosphori), is made to revolve 300 times in a second, and by using this or a lesser velocity, the various phosphori are first exposed to a powerful light and then brought in view of the spectator outside the box.

It is understood that light is produced by an emanation of rays from a luminous body. If a stone is thrown from the hand, an arrow shot from a bow, or a ball from a cannon, we perfectly understand how either of them may be propelled a certain distance, and why they may travel through space; but when we hear that light travels from the sun, which is ninety-five millions of miles away from the earth, in about seven minutes and a half, it is interesting to know what is the kind of force that propels the light through that vast distance, and also what is supposed to be the nature of the light itself.

There are two theories by which the nature of light, and its propagation through space, are explained; they are named after the celebrated men who proposed them, as also from the theoretical mechanism of their respective modes of propulsion: thus we have the Newtonian or *corpuscular* theory of light, and the Huyghenian or *undulatory* theory; the first named after Sir Isaac Newton, and the second after Huyghens, another most learned mathematician. Many years before Newton made his grand discovery of the composition of light in the year 1672, mathematicians were in favour of the *undulatory* theory, and it numbered amongst its supporters not only Huyghens, but Descartes, Hook, Malebranche, and other learned men. Mankind has always been glad to follow renowned leaders, it is so much easier, and is in most cases perhaps the better course, to resign individual opinion when more learned men than ourselves not only adopt but insist upon the truth of their theories; and this was the case with the corpuscular theory, which had been written upon systematically and supported by Empedocles, a philosopher of Agrigentum in Sicily, who lived some 444 years before the Christian era, and is said to have been most learned and eloquent; he maintained that light consisted of particles projected from luminous bodies, and that vision was performed both by the effect of these particles on the eye, and by means of a visual influence emitted by the eye itself. In course of time, and at least 2000 years after this theory was advanced, philosophers had gradually rejected the corpuscular theory, until the great Newton, about the middle of the seventeenth century, advanced as a champion to the rescue, and stamping the hypothesis with his approval, at once led away the whole army of philosophers in its favour, so that till about the beginning of the nineteenth century the whole of the phenomena of light were explained upon this hypothesis.

The corpuscular theory, reduced to the briefest definition, supposes light to be really a material agent, and requires the student to believe that this agent consists of particles so inconceivably minute that they could not be weighed, and of course do not gravitate ; the corpuscles are supposed to be given out bodily (like sparks of burning steel from a gerb firework) from the sun, the fixed stars, and all luminous bodies ; to travel with enormous velocity, and therefore to possess the property of *inertia ;* and to excite the sensation of vision by striking bodily upon the expanded nerve, the retina, the quasi-mind of the eye. Dr. Young remarks, "that according to this projectile theory the force employed in the free emission of light must be about a million million times as great as the force of gravity at the earth's surface, and it must either act with equal intensity on all the particles of light, or must impel some of them through a greater space than others, if its action be more powerful, since the velocity is the same in all cases—for example, if the projectile force is weaker with respect to red light than with respect to violet light, it must continue its action on the red rays to a greater distance than on the violet rays. There is no instance in nature besides of a simple projectile moving with a velocity uniform in all cases, whatever may be its cause ; and it is extremely difficult to imagine that such an immense force of repulsion can reside in all substances capable of becoming luminous, so that the light of decaying wood, or two pebbles rubbed together, may be projected precisely with the same velocity as the light emitted by iron burning in oxygen gas, or by the reservoir of liquid fire on the surface of the sun." Now one of the most striking circumstances respecting the propagation of light, is the *uniformity* of its velocity in the same medium. These and other difficulties in the application of the corpuscular theory aroused the attention of the late Dr. Young, and in the year 1801 he again revived and supported the neglected undulatory theory with such great ability that the attention of many learned mathematicians was directed to the subject, and now it may be said that the corpuscular theory is almost, if not entirely, rejected, whilst the undulatory theory is once more, and deservedly, used to explain the theory of light, and its propagation through space. By this hypothesis it is assumed that the whole universe, including the most minute pores of all matter, whether solid, fluid, or gaseous, are filled with a highly elastic rare medium of a most attenuated nature, called *ether*, possessing the property of *inertia* but not of gravitation. This *ether* is not light, but light is produced in it by the excitation on the part of luminous bodies of a vibratory motion, similar to the undulation of water that produces waves, or the vibration of air affording sound. Water set in motion produces waves. Air set in motion produces waves of sound. Ether, *i.e.* the theoretical ether pervading all matter, likewise set in motion, produces light. The nature of a vibratory medium is indeed better understood by reference to that which we know possesses the ordinary properties of matter—viz., the air ; and by tracing out the analogy between the propagation of sound and light, the difficulties of the undulatory theory very quickly vanish. To illustrate vibration it is only necessary to procure a finger glass, and having supported a little ebony ball attached to a silk thread by a bent brass

wire directly over it, so that the ball may touch either the outside or
the inside of the glass, attention must be directed to the quiescence of
the ball when a violin bow is lightly moved over the edge of the glass
without producing sound, and to the contrary effect obtained by so
moving and pressing the bow that a sharp sound is emitted, when immediately the little ball is thrown off from the edge, the repulsive action
being continued as long as the sound is produced by the vibration of
the glass. (Fig. 250.)

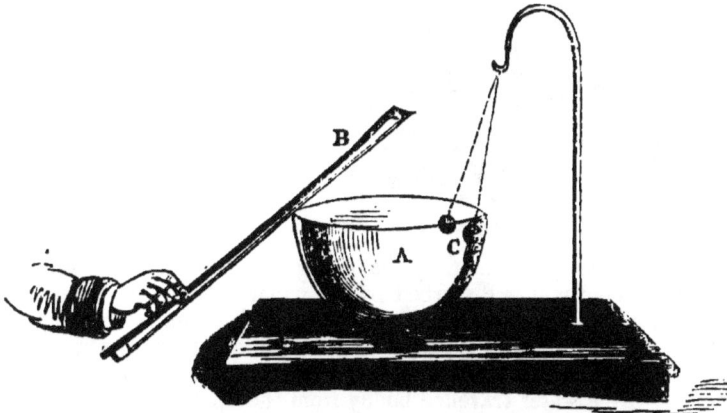

Fig. 250. A. The finger glass. B. The violin bow. C. The ebony ball. The dotted ball
shows how it is repelled during the vibration of the glass.

Here the vibrations are first set up in the glass, and being communicated to the surrounding air, a sound is produced; if the same experiment could be performed in a vacuum, the glass might be vibrated, but
not being surrounded with air, no sound would be produced. This fact
is proved by first ringing a bell with proper mechanism fixed under the
receiver placed on the air-pump plate; the sound of the bell is audible
until the pump is put in motion and the receiver gradually exhausted,
when the ringing noise becomes fainter and fainter, until it is perfectly
inaudible. This experiment is made more instructive by gradually
admitting the air again into the exhausted vessel, and at the same time
ringing the bell, when the sound becomes gradually louder, until it
attains its full power. The sun and other luminous bodies may be
compared to the finger glass, and are supposed to be endowed
naturally with a vibratory motion (a sort of perpetual ague), only
instead of the air being set in motion, the *ether* is supposed to be
thrown into waves, which travel through space, and convey the
impression of light from the luminous object. Another familiar
example of an undulatory medium is shown by throwing a stone
into a pool of water; the former immediately forces down and displaces
a certain number of the particles of the latter, consequently the surrounding molecules of water are heaped up above their level; by the
force of gravitation they again descend and throw up another wave, this
in subsiding raises another, until the force of the original and loftier

wave dies away at the edge of the pool into the faintest ripples. It must however be understood that it is not the particles of water first set in motion that travel and spread out in concentric circles;

Fig. 251. Boy throwing stones into water and producing circular waves.

but the force is propagated by the rising and falling of each separate particle of water as it is disturbed by the momentum of the descending wave before it. When standing at a pier-head, or on a rock against which the sea dashes, it is usual to hear the observer cry out, if the weather is stormy and the waves very high, "Oh! here comes a great wave!" as if the water travelled bodily from the spot where it was first noticed, whereas it is simply the force that travels, and is exerted finally on the water nearest the rock. It is in fact a progressive action, just as the wind sweeps over a wide field of corn, and bends down the ears one after the other, giving them for the time the appearance of waves. The principle of successive action is well shown by placing a number of billiard balls in a row, and touching each other; if the first is struck the motion is communicated through the rest, which remain immovable, whilst the last only flies out of its place. The force travels through all the balls, which simply act as carriers, their motion is limited, and the last only changes its position. Progressive movement is also well dis-

Fig. 252. A B. Series of needles arranged as described. c. The bar magnet, with the north pole N towards the needles. The dotted lines show the direction gradually assumed by all the needles, commencing at D.

played by arranging six or eight magnetized needles on points in a row, with all their north poles in one direction. (Fig. 252.)

On approaching the north pole of a bar magnet to the same pole of one end of the series of needles, it is very curious to see them turn in the opposite direction progressively, one after the other, as the repulsive power of the bar magnet gradually operates upon the similar poles in the magnetic needles. The undulations of the waves of water are also perfectly shown by using the apparatus consisting of the trough with the glass bottom and screen above it, as described at page 10. The transmission of vibrations from one place to another is also admirably displayed in Professor Wheatstone's Telephonic Concert (see page picture), where the musical instruments, as at the Polytechnic, were placed by the author in the basement, and the vibration only conducted by wooden rods to the sounding-boards above, so that the music was laid on like gas or water. These vibrations or undulations in air, water, and the theoretical ether, have therefore been called waves of water, waves of sound, and waves of light, just as if three clocks were made of three different metals, the mechanism would remain the same, though the material, or in this case the medium, be different in each.

Any increase in the number of vibrations of the air produces acute, whilst a decrease attends the grave sounds, and when the waves succeed each other not less than sixteen times in a second, the lowest sound is produced. Light and colours are supposed to be due to a similar cause, and in order to produce the red ray, no less than 477 millions of millions of vibrations must occur in a second of time; the orange, 506; yellow, 535; green, 577; blue, 622; indigo, 658; violet, 699; and white light, which is made up of these colours, numbers 541 millions of millions of undulations in a second.

Although light travels with such amazing rapidity, there is of course a certain time occupied in its passage through space—there is no such thing as instantaneity in nature. A certain period of time, however small, must elapse in the performance of any act whatever, and it has been proved by a careful observation of the time at which the eclipses of the satellites of Jupiter are perceived, that light travels at the rate of 192,500 miles per second, and by the aberration of the fixed stars, 191,515, the mean of these two sets of observations would probably afford the correct rate. Such a velocity is, however, somewhat difficult to appreciate, and therefore, to assist our comprehension of their great magnitude, Sir J. Herschel has given some very interesting comparative calculations, and coming from such an authority we can readily believe them to be correct.

"A cannon-ball moving uniformly at its greatest velocity would require seventeen years to reach the sun. Light performs the same distance in about seven minutes and a half.

"The swiftest bird, at its utmost speed, would require nearly three weeks to make the tour of the earth, supposing it could proceed without stopping to take food or rest. Light performs the same distance in less time than is required for a single stroke of its wing."

Dismissing for the present the theory of undulations, it will be necessary to examine the phenomena of light, regarding it as radiant matter, without reference to either of the contending theories.

Light issues from the sun, passes through millions of miles to the earth, and as it falls upon different substances, a variety of effects are apparent. There is a certain class of bodies which obstruct the passage of the rays of light, and where light is not, a shadow is cast, and the substance producing the shadow is said to be opaque. Wood, stone, the metals, charcoal, are all examples of opacity; whilst glass, talc, and horn allow a certain number of the rays to travel through their particles, and are therefore called transparent. Nature, however, never indulges in sudden extremes, and as no substance is so opaque as not (when reduced in thickness) to allow a certain amount of light to pass through its substance, so, on the other hand, however transparent a body may be, a greater or lesser number of the rays are always stopped, and hence opacity and transparency are regarded as two extremes of a long chain; being connected together by numerous intermediate links, they pass by insensible gradations the one into the other.

If a gold leaf, which is about the one two-hundredth part of an inch in thickness, is fixed on a glass plate and held before a light, a green colour is apparent, the gold appearing like a green, semi-transparent substance. When plates of glass are laid one above the other, and the flame of a candle observed through them, the light decreases enormously as the number of glass-plates are increased. Even in the air a considerable portion of light is intercepted. It has been estimated that of the horizontal sunbeams passing through about two hundred miles of air, one two-thousandth part only reaches us, and that no sensible light can penetrate more than seven hundred feet deep into the sea; consequently, the vast depths discovered in laying the Atlantic telegraph must be in absolute darkness.

Light is thrown out on all sides from a luminous body like the spokes of a cart-wheel, and in the absence of any obstruction, the rays are distributed equally on all sides, diverging like the radii drawn from the centre of a circle. As a natural consequence arising from the divergence of each ray from the other, the intensity of light decreases as the distance from the luminous source increases, and *vice versâ*. Perhaps the best mechanical notion of this law is afforded by an ordinary fan; the point from which the sticks radiate, and where they all meet, may be

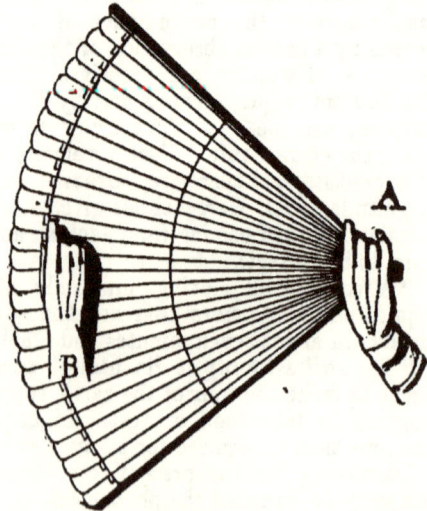

Fig. 253.

termed the light; the sticks are the rays proceeding from it. (Fig. 253.)

The fan is held in one hand, and the first finger of the other can be made to touch all the sticks if placed sufficiently near to A; and supposing the sticks are called rays of light, the intensity must be great at that point, because all the rays fall upon it; but if the finger is removed towards the outer edge—viz., to B, it now only touches some three or four sticks; and pursuing the analogy, a very few rays fall upon that point—hence the light has decreased in intensity, or to speak correctly, "Light decreases inversely as the squares of the distance." This law has already been illustrated at page 13; and as an experiment, the rays from the oxy-hydrogen lantern may be permitted to pass out of a square hole (say two inches square), and should be thrown on to a transparent screen divided into squares by dark lines, so that the light at a certain distance illuminates one of them; then it will be found that at twice the distance, four may be illuminated, at three times nine, and so on. (Fig. 254.)

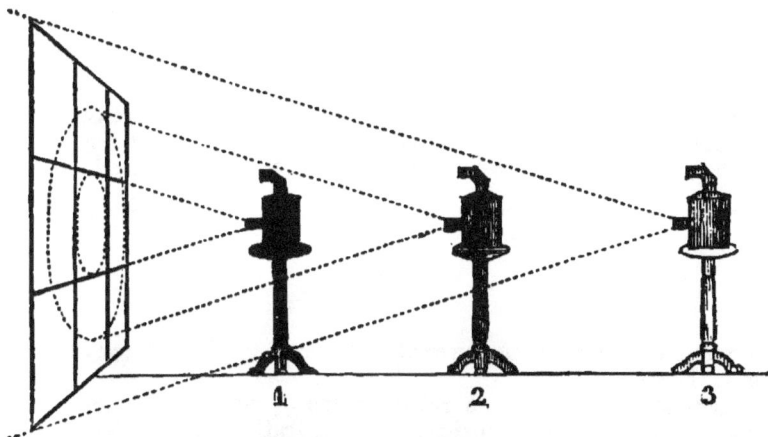

Fig. 254. Lantern at the three distances from the transparent screen, which is divided into nine equal squares.

Upon this law is based the use of photometers, or instruments for measuring light, and supposing it was required to estimate roughly the illuminating power of any lamp, as compared with the light of a wax candle six to the pound, the experiment should be conducted in a dark room, from which every other light but that from the lamp and candle under examination must be excluded.

The lamp, with the chimney only, is now placed say twelve feet from the wall, and a stick or rod is placed upright and about two inches from the latter, so that a shadow is cast on the wall; if the candle is now lighted and allowed to burn up properly, two shadows of the stick will be apparent, the one from the lamp being black and distinct, and the other from the candle extremely faint, until it is approached nearer the wall—say to within three feet—when the two shadows may be now equal in blackness. (Fig. 255.) After this is apparent to one or more

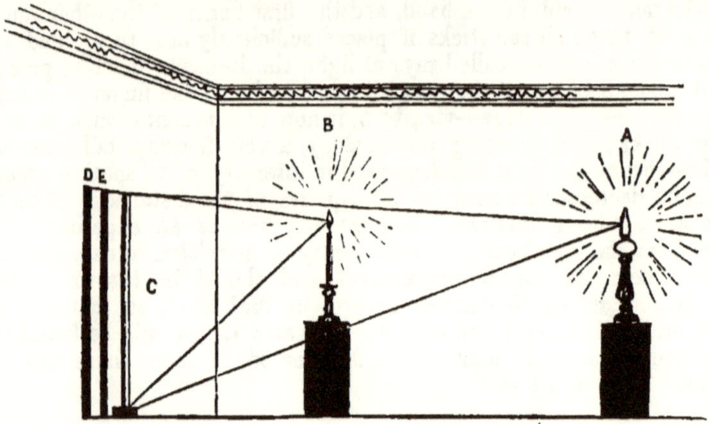

Fig. 255. A. The lamp. B. The candle. C. The rod throwing the two shadows, marked D and E, on a white wall or a sheet of paper.

persons, the distances of the lamp and candle from the wall are carefully measured, and being squared, and the greater divided by the lesser number, the quotient gives the illuminating power. For example:

The lamp was 12 feet from the wall . . . $12 \times 12 = 144.$
The candle was 3 feet   „   . . . $3 \times 3 = 9.$

$$9) \frac{144}{16}$$

Therefore the illuminating power of the lamp is equal to 16 wax candles six to the pound.

There are other and more refined means of working out the same fact, but for a rough approximation to the truth, the plan already described will answer very fairly.

A most amusing effect can be produced on the principle that every light casts its own shadow, called the "dance of death," or the "dance of the witches ;" either of these agreeable subjects are drawn, and the outlines cut out of a sheet of cardboard. If a wet sheet is stretched or hung on one side of a pair of folding doors partly open, and between which the cardboard is tacked up, and the space left at the top and bottom closed with a dark cloth, directly the room before the sheet is darkened and a lighted candle held behind the figure cut out in the cardboard, one shadow or image is thrown upon the sheet, and these shadows may be increased according to the number of candles used, and if they are held by two or three persons, and moved up and down, or sideways, the shadows follow the direction of the candles, and present the appearance of a dance. (Fig. 256.)

Fig. 256.　"Before the curtain."

Fig. 257.　"Behind the curtain."

Another very comic effect of shadow is that called "jumping up to the ceiling," and when carried out on a large scale by the author on an enormous sheet suspended in the centre transept of the Crystal Palace, Sydenham, it had a most laughable effect, and caused the greatest amusement to the children of all ages.   (Fig. 258.)

Fig. 258.   The laughable effect of the shadows at the Crystal Palace.

This very telling result is produced by placing an oxy-hydrogen light some feet behind a large sheet, and of course if any one passes between the two a shadow of the individual is cast upon the sheet, then by walking towards the light the figure diminishes in size, and by jumping over it the shadow appears to go up to the ceiling, and to come down when the jump is made in the opposite direction over the light and towards the sheet. The *rationale* of this experiment is very simple, and is another proof of the distribution of light from a luminous source being in every direction. By jumping over the light the radii projected from the candle over the sheet are crossed, and the shadow rises or falls as the figure passes upwards or downward. (Fig. 259.)

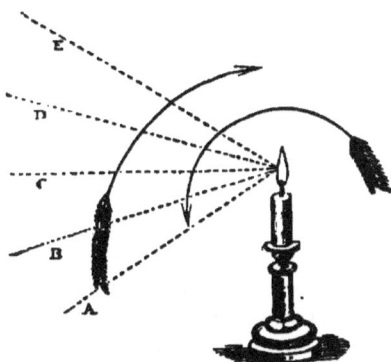

A beam of light is defined to be a collection of rays, and it is a convenient definition, because it prevents confusion to speak only of one ray in attempting to explain how light is disposed of under peculiar circumstances.

The smallest portion of light which it is supposed can be separated is therefore called a ray, and it will pass through any medium of the same density in a perfectly straight line; but if it passes out of that medium into another of a different density, or into any other solid, fluid, or gaseous matter, it may be disposed of in four different ways, being either reflected, refracted, polarized, or absorbed.

Fig. 259. The rays of light marked A B C D E proceeding from a lighted candle or oxy-hydrogen light. The arrow pointing to the right shows how these rays are crossed in jumping up to the ceiling; and the second arrow, pointing to the left, shows the reverse.

The reflection of light is the first property that will be considered, and it will be found that every substance in nature possesses in a greater or lesser degree the power of throwing off the rays of light which fall upon them. Thus if we go into a room perfectly darkened, containing every kind of work produced by nature or art, such as flowers, birds, boxes of insects, rich carpets, hangings, pictures, statuary, jewellery, &c., they cannot excite any pleasure because they are invisible, but directly a lighted lamp is brought into the chamber, then the rays fall upon all the surrounding objects, and being reflected from their surfaces enter the eye, and there produce the phenomena of vision.

This connexion between luminous and non-luminous bodies becomes very apparent when we consider that the sun would appear only as an intense light in a dark background, if the earth was not surrounded with the various strata of air, in which are placed clouds and vapours that collectively reflect and scatter the light, so as to cause it to be endurable to vision. It is when the sky is very clear during July or August that the heat becomes so intense, directly clouds begin to form and float about, the heat is then moderated.

Many years ago, Baron Alexander Funk, visiting some silver mines in Sweden, observed, that in a clear day it was as dark as pitch underground in the eye of the pit at sixty or seventy fathoms deep; whereas, on a cloudy or rainy day he could even see to read at 106 fathoms deep. Inquiring of the miners, he was informed that this is always the case, and reflecting upon it he imagined very properly that it arose from this circumstance—that when the atmosphere is full of clouds, light is reflected from them into the pit in all directions, so that thereby a considerable proportion of the rays are reflected perpendicularly upon the earth; whereas when the atmosphere is clear there are no opaque bodies to reflect the light in this manner, at least, in a sufficient quantity, and rays from the sun itself can never fall perpendicularly in Sweden. The use of reflecting surfaces has now become quite common in all crowded cities, and especially in London, where even the rays of light are too few to be lost, and flat or corrugated mirrors are placed at various angles, either to throw the light from the outside on the white-washed ceiling within, and thus obtain a better diffused light through the apartment, or it is reflected bodily to some back room, or rather dark brick box, where perhaps for half a century candles have been required at an early hour in the afternoon. The brilliant cut in diamonds is such an arrangement of the posterior facets, or cut faces of the jewel, that all light reaching them shall be thrown back and reflected, and thus impart an extraordinary brilliancy to the gem.

The intense glare of snow in the Alpine regions has long been noticed, and the reflected light is so powerful, that philosophers were even disposed to believe that snow possessed a natural or inherent luminosity, and gave out its own light. Mr. Boyle, however, disproved this notion by placing a quantity of snow in a room from which all foreign light was excluded, and neither he nor his companion could observe that any light was emitted, although, on the principle of momentary phosphorescence, it is quite possible to conceive that if the snow was suddenly brought into a darkened room after exposure to the rays of the sun, that it would give out for a few seconds a perceptible light. In trying such an experiment, one person should expose the snow to the sun, and bring it into a perfectly darkened room to a second person, whose eyes would be ready to receive the faintest impression of light, and if any phosphorescence existed, it must be apparent.

The property of reflection is also illustrated on a grand scale in the illumination of our satellite, the moon, and the various planetary bodies which shine by light reflected from the sun, and have no inherent self-luminosity. Aristotle was well aware that it is the reflection of light from the atmosphere which prevents total darkness after the sun sets, and in places where the sun's rays do not actually fall during the daytime. He was also of opinion that rainbows, halos, and mock suns, were all occasioned by the reflection of the sunbeams in different circumstances, by which an imperfect image of the sun was produced, the colour only being exhibited, but not the proper figure.

The image, Aristotle says, is not single, as in a mirror, for each drop of rain is too small to reflect a visible image, but the conjunction of all the images is visible. Aristotle ascribed all these effects to the *reflection* of light, and it will be noticed when we come to the consideration of the refraction of light, that of course his views must be seriously modified.

The reflection of light is affected rather by the condition of the surface than the whole body of a substance, as a piece of coal may be covered with gold or silver leaf and caused to shine, whilst the brightest mirror is dimmed by the thinnest film of moisture.

From whatever surface light is reflected, it always takes place in obedience to two fixed laws.

First. *The incident and reflected rays always lie in the same plane.*

Second. *The angle of incidence is equal to the angle of reflection.*

With a single jointed two-foot rule, both of these laws are easily illustrated. The rule may be held in the hand, and one end being marked with a piece of white paper may be called the incident ray, *i.e.*, the ray that falls upon the surface; and the other is the reflected ray, the one cast off or thrown back. A perpendicular is raised by holding a stick upright at the joint. (Fig. 260.)

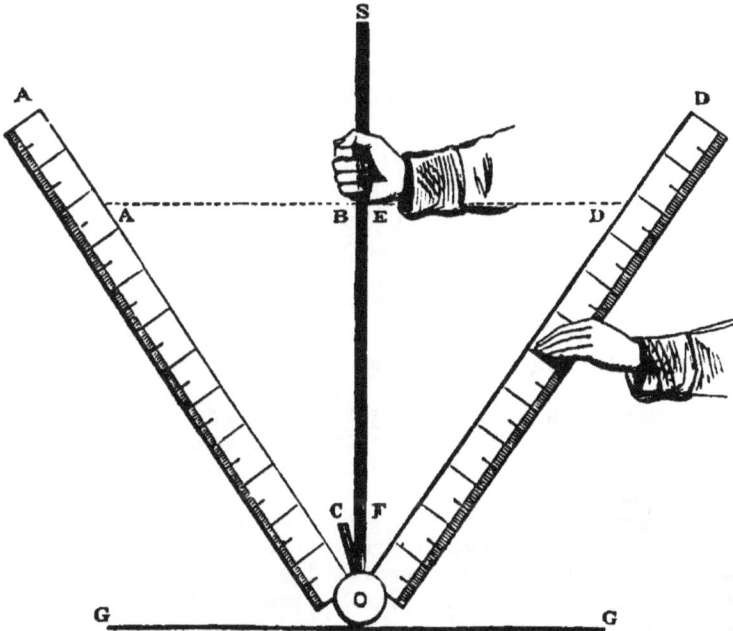

Fig. 260. A D. A two foot rule; the end A may be termed the incident ray, and the end D the reflected ray. S. The stick held perpendicularly. The angle A B C is equal to the angle D B F, and the whole may be moved in any direction or plane, either horizontal or perpendicular. G G. The reflecting surface.

One of the most simple and pleasing delusions produced by the reflection of light, is that afforded by cutting through the outline of a vase, or statuette, or flower, drawn on cardboard, and if certain points are left attached, so that the design may not fall out, all the effect of solidity is given by bending back the edges of the cardboard, so that the light from a candle placed behind it, may be reflected from the back edge of one cardboard on to the design, which is bent back. The light reflected from one surface on to the other, imparts a peculiarly soft and marble-like appearance, and when the design is well drawn and cut, and placed in a good position, the illusion is very perfect, and it appears like a solid form instead of a mere design cut out of cardboard. (Fig. 261.)

Fig. 261. Cardboard design in frame, cut and bent back. The lighted candle is behind.

The leaf at the side of the above picture is intended to give an idea of the mode of cutting out the designs, and in this case the leaf would be cut and bent back, and a small attachment slip of cardboard left to prevent it falling out.

The cardboard design is always bent toward the light, which is placed behind it. As a good illustration of the importance of reflected light and its connexion with luminous bodies, a beam of light from the oxy-hydrogen lantern may be allowed to pass above the surface of a table, when it will be noticed that the latter is lighted up only when the beam is reflected downward by a sheet of white paper.

By reference to the two laws of reflection already explained, it is easy to trace out on paper, with the help of compasses and rule, the effect of plane, concave, and convex surfaces on parallel, diverging, or converging rays of light, and it may perhaps assist the memory if it is remembered that a *plane* surface means one that is flat on both sides, such as a looking-glass: a *convex* surface is represented by the outside of a watch-glass; a *concave* surface, by the inside of a watch-glass; parallel rays are like the straight lines in a copy-book; diverging and converging

T

rays, are like the sticks of a fan spread out as the sticks separate or diverge; the sticks of the fan come together, or converge at the handle. The reflection of rays from a plane surface may be better understood by reference to the annexed diagram. (Fig. 262.)

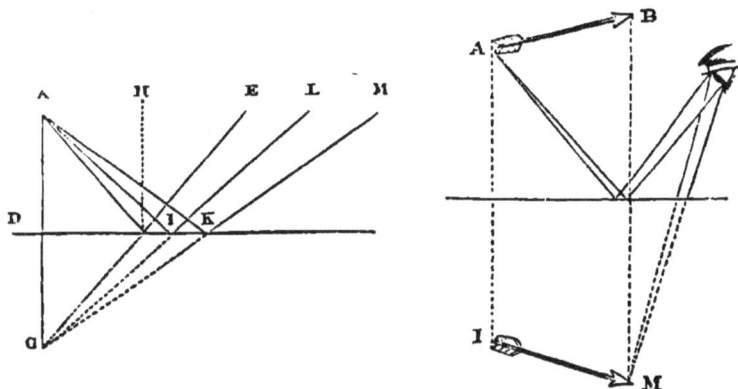

Fig. 262. A I, A K. Two diverging rays incident on the plane surface, D. A D is perpendicular, and is reflected back in the same direction. A I is divergent, and is thrown off at I L. The incident and reflected rays forming equal angles, as proved by the perpendicular, H. Any image reflected in a plane mirror appears as far behind it as the object is before it, and the dotted lines meeting at G show the apparent position of the reflected image behind the glass, as seen at G. The same fact is also shown in the second diagram, where the reflected picture, I M, appears at the same distance behind the surface of the mirror as the object, A B, is before it.

By the proper arrangement of *plane* mirrors, a number of amusing delusions may be produced, one of which is sometimes to be met with in the streets, and is called "the art of looking through a four-inch deal board." The spectator is first requested to look into a tube, through which he sees whatever may be passing the instrument at the time; the operator then places a deal board across the middle of the tube, which is cut away for that purpose, and to the astonishment of the juveniles the view is not impaired, and the spectator still fancies he is looking through a straight tube; this however is not the case, as the deception is entirely carried out by reflection, and is explained in the next cut. (Fig. 263.)

During the siege of Sebastopol numbers of our best artillerymen were continually picked off by the enemy's rifles, as well as by cannon shot, and in order to put a stop to the foolhardiness and incautiousness of the men, a very ingenious contrivance was invented by the Rev. Wm. Taylor, the coadjutor of Mr. Denison in constructing the first "Big Ben" bell. It was called the reflecting spy-glass, and by its simple construction rendered the exposure of the sailors and soldiers, who would look over the parapet or other parts of the works to observe the effect of their shot, perfectly unnecessary; whilst another form was constructed for the purpose of allowing the gunner to "lay" or aim his gun in safety. The instruments were shown to Lord Panmure, who was so convinced of the importance of the invention, that he immediately

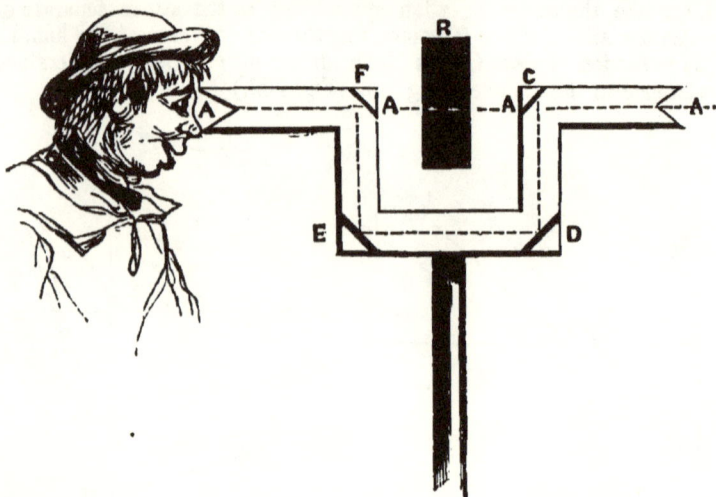

Fig. 263. ▲ ▲ ▲ ▲. The apertures through which the spectator first looks. B. The piece of wood, four inches thick. C, D, E, F, are four pieces of looking-glass, so placed that rays of light entering at one end of the tube are reflected round to the other where the eye of the observer is placed.

commissioned the Rev. Wm. Taylor to have a number of these telescopes constructed; and if the siege had not terminated just at the time the invention was to have been used, no doubt a great saving of the valuable lives of the skilled artillerymen would have been effected in the allied armies. The principle of the reflecting spy-glass may be comprehended by reference to the next cut. (Fig. 264.)

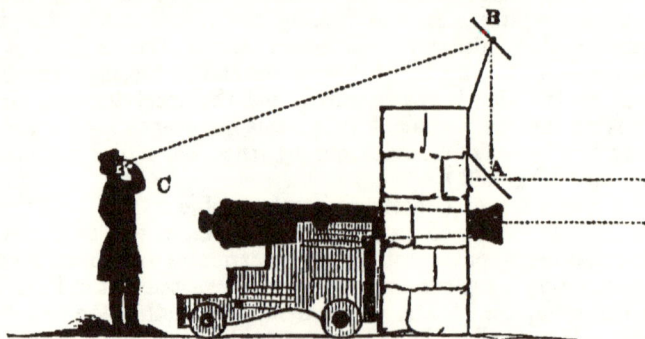

Fig. 264. A picture of enemy's battery is supposed to be on the mirror, A, whence it is reflected to B, and from that to the artilleryman at C.

By placing two mirrors at an angle of 45°, the reflected image of a person gazing into one is thrown into the other, and of course the effect is somewhat startling when a death's head and cross bones, or other

T 2

cheerful subject, is introduced opposite one mirror, whilst some person who is unacquainted with the delusion is looking into the other. Two adjoining rooms might have their looking-glasses arranged in that manner, provided there is a passage running behind them. (Fig. 265.)

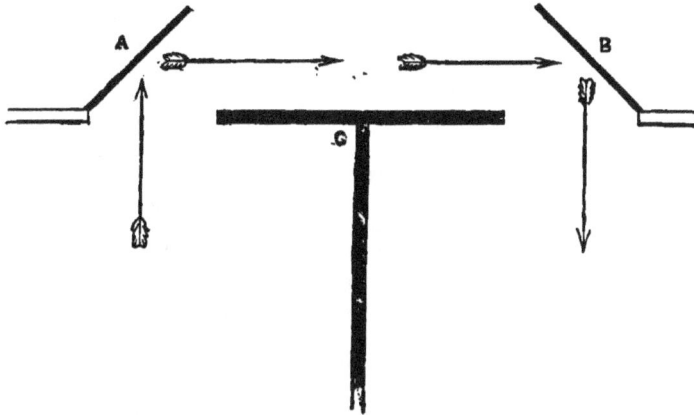

Fig. 265. A. A mirror at an angle of 45 degrees. The arrows show the direction of the reflected image. B. The second mirror, also at an angle of 45 degrees; the face of the person looking in at A is reflected at B. C is the partition between the rooms.

One of the most startling effects that can be displayed to persons ignorant of the common laws of the reflection of light, is called the "magic mirror," and is described by Sir Walter Scott in his graphic story of that name. The apparatus for the purpose must be well planned and fixed in a proper room for that purpose, and if carefully conducted, may surprise even the learned. A long and somewhat narrow room should be hung with black cloth, and at one end may be placed a large mirror, so arranged that it will turn on hinges like a door. The magician's circle may be placed at the other end of the chamber in which the spectators must be rigidly confined, and there is very little doubt that the arrangement about to be described was formerly used by clever astrologers who pretended to look into the future, and to hold communication with the supernatural powers. The credulity of the persons who consulted these "wise men," is not surprising when we consider the ignorance of the public generally of common physical laws, and of the wonders that may be worked without the assistance of the "evil one;" moreover, the initiated took great care to conceal the machinery of their mysteries, never imparting the illusive tricks even to their most faithful dependents except under solemn oaths of secresy, because they derived in many cases considerable profit by their pretended conjurations and juggling tricks, and therefore were interested in keeping the outer world in ignorance. The wizards were always careful to impress those who came to consult them with the awful nature of the incantations they were about to perform, and with such a powerful auxiliary as fear, and a well-darkened room, they diverted the thoughts of the more curious, and prevented them watching the proceedings too closely. Theatrical effects were not disdained, such as suppressed and dismal groans, sham thunder, and the wizard usually heightened his own inspiring personal appearance by wearing of course a long beard and flowing robe trimmed with hieroglyphics, and with the assistance of a ponderous volume full of cabalistic signs, a few skulls and cross bones, an hour-glass, a pair of drawn swords, a black cat, a charcoal fire, and sundry drugs to throw into it, a very tolerable collection of imps, familiars, and demons, might be expected to attend without the modern practice of spirit-rapping. As before stated, the delusion must be carefully conducted, and a confederate is necessary in order to use the phantasmagoria, or magic lantern. The slides of course were painted to suit the fortune to be unfolded—an easy road to riches for the gentlemen, a tale of love, ending in matrimony, for the ladies.

The spectators being placed in the magic circle, are directed to look into the mirror; they may even be ordered singly to fetch a skull off the mantel-shelf beside the mirror, and whilst doing so to look full into the mirror, and then return to the circle. Absolute silence is enjoined, and soft music is now heard; the darkened room is lit up for the moment by a little yellow or green fire thrown on to the charcoal fire, and now looking into the mirror, it no longer reflects surrounding objects, but a picture, at first small and faint, and then gradually becoming large and clearer, is apparent. The picture is made visible by the confederate gently drawing the mirror from its position parallel with the frame to an angle of 45 degrees, and then throwing on from the side a picture from a magic-lantern. The picture is small and indistinct whilst the confederate holds it near the mirror and out of focus, but as he moves backwards and focuses the lenses, the picture gradually increases in size, and the reflecting angles having been well planned beforehand, only those in the circle will be able to see the picture, and great fun may be elicited from the magic mirror by pretending to tell the future fate of a very slim person, and introducing him by a succession of pictures which gradually assume a John Bull rotundity of figure, surrounded by dozens of children; whilst to young ladies who are engaged, a provoking picture of an old maid may be introduced; indeed, there is no end to the innocent fun that may be extracted from the magic mirror, and the whole plan of the delusion may be better understood by reference to the next picture. (Fig. 266.)

Monsieur Salverte very properly remarks that "man is credulous from his cradle to his tomb; but the disposition springs from an honourable principle, the consequences of which precipitate him into many errors and misfortunes. . . . . . The novelty of objects, and the difficulty of referring them to known objects, will not shock the credulity of unsophisticated men. They are some additiona. sensations which he receives without discussion, and their singularity is perhaps a charm which causes him to receive them with greater pleasure. *Man almost always* loves and seeks the marvellous. Is this taste natural?

The magic mirror.

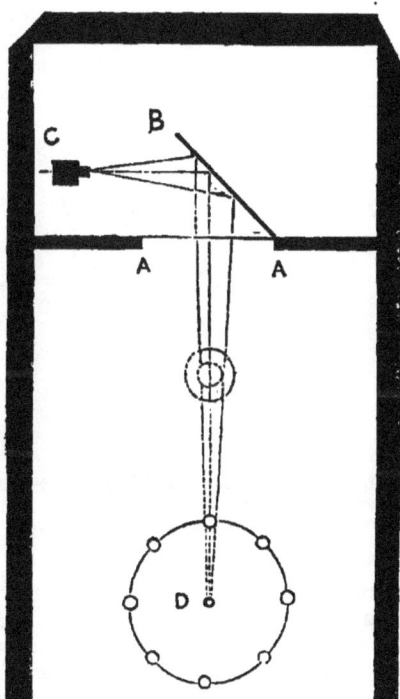

Fig. 266.   Plan of room.   A A. The frame of the looking-glass.   A B. Mirror put back to an angle of 45 degrees.   c. The confederate who manages the lantern and shuts the glass to the frame after each fortune is told.   D. The magic circle, to which the rays are reflected.

Does it spring from the education which during many ages the human race has received from its first instructors? A vast and novel question, but with which I have nothing to do. It is sufficient to observe that as the lover of the wonderful always prefers the most surprising to the most natural account, this last has been too frequently neglected, and is irrevocably lost. Occasionally, however (and we shall cite more than one instance), simple truth has escaped from the power of oblivion. Credulous man may be deceived once, or more frequently; but his credulity is not a sufficient instrument to govern his whole existence. The wonderful excites only a transient admiration. In 1798, the French *savans* remarked with surprise how little the spectacle of balloons affected the indolent Egyptian. . . . . But man is led by his passions, and particularly by *hope* and *fear*."

When parallel rays fall upon a convex mirror, they are scattered and dispersed in all directions, and the image of an object reflected in a convex mirror appears to be very small, being reduced in size because the reflected picture I M is nearer the surface of the mirror than the object A B. No. 1. (Fig. 267.)

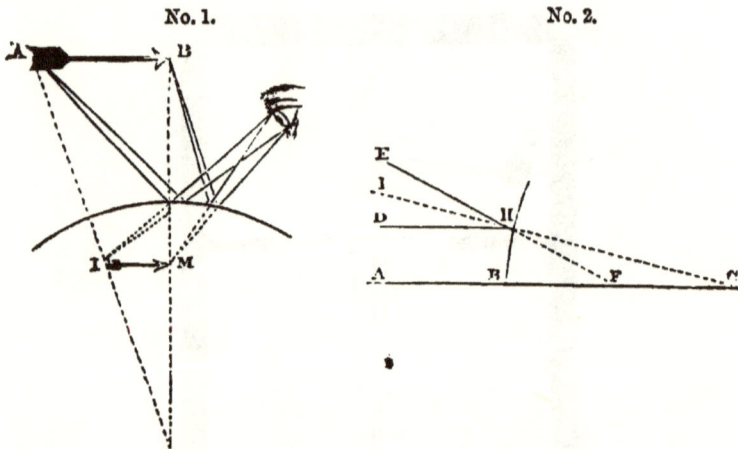

No. 1. No. 2.

Fig. 267. A B, D H. (No. 2) represent two parallel rays incident on the convex surface D H, the one (A B) perpendicularly, the other (D H) obliquely. O is the centre of convexity. H E is the reflected ray of the oblique incident one, D H; whilst C H I is the perpendicular.

Convex mirrors are not employed in any optical deception on a large scale, although some ingenious delusions are producible from cylindrical and conical mirrors, and are thus described by Sir David Brewster:

"Among the ingenious and beautiful deceptions of the seventeenth century, we must enumerate that of the re-formation of distorted pictures by reflection from cylindrical and conical mirrors. In these representations, the original image from which a perfect picture is pro-

duced, is often so completely distorted, that the eye cannot trace in it
the resemblance to any regular figure, and the greatest degree of
wonder is of course excited, whether the original image is concealed or
exposed to view. These distorted pictures may be drawn by strict
geometrical rules, and I have shown a simple method of executing
them. Let M be an accurate cylinder made of tin-plate or of thick

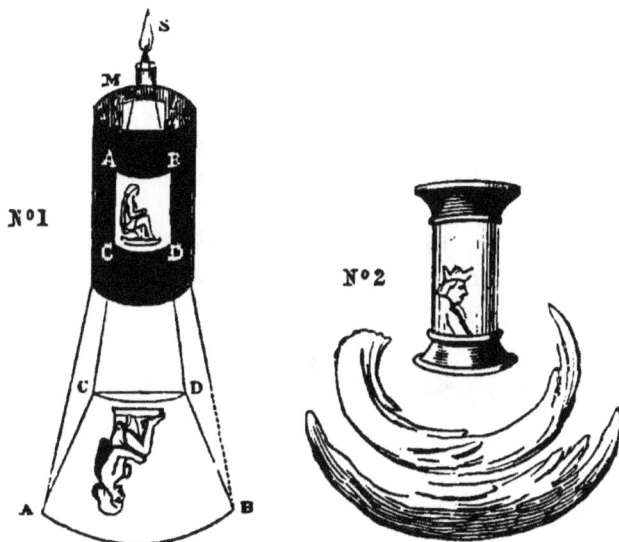

Fig. 268.

pasteboard. Out of the further side of it cut a small aperture, *a b c d*, and
out of the nearer side cut a larger one, A B C D (white letters), the size of
the picture to be distorted; having perforated the outline of the picture
with small holes, place it in the opening A B C D (white letters), so that its
surface may be cylindrical; let a candle or a bright luminous object—the
smaller the better—be placed at s, as far behind the picture A B C D (white
letters) as the eye is afterwards to be placed before it, and the light passing
through the small holes will represent on a horizontal plane a distorted
image of the picture at A B C D, which, when sketched in outline with a
pencil, shaded, and coloured, will be ready for use. If we now substitute
a polished cylindrical mirror of the same size in place of M, then the
distorted picture, when laid horizontally at A B C D, will be restored to
its original state when seen by reflection at A B C D (white letters) in
the polished mirror." The effect of a cylindrical mirror on a distorted picture is shown at No. 2, being copied from an old one seen by Sir D.
Brewster.

By looking at a reflection of the face in a dish-cover or the common
surface of a bright silver spoon or of a silver mug, the latter truly
becomes ugly as the image is seen reflected from its surface, and

assumes the most absurd form as the mouth is opened or shut, and the face advanced or removed from the silver vessel. (Fig. 269.)

Fig. 269.   Distorted image produced by an irregular convex surface.

In the writings of the ancients there are to be found certain indications of the results of illusions produced by simple optical arrangements, and the sudden and momentary apparition (from the gloom of perfect darkness) of splendid palaces, delightful gardens, &c., with which—the concurrent voice of antiquity assures us—the eyes of the beholders were frequently dazzled in the mysteries, such as the evocation and actual appearance of departed spirits, the occasional images of their *umbræ*, and of the gods themselves. From a passage in "Pausanias," (Bœotic xxx.), when, speaking of Orpheus, he says there was anciently at Aornos, a place where the dead were evoked, νεκυομαντείον, we learn that in those remote ages there were places set apart for the evocation of the dead. Homer relates, in the eleventh book of the "Odyssey," the admission of Ulysses alone into a place of this kind, when his interview with his departed friend was interrupted by some fearful voice, and the hero, apprehending the wrath of Proserpine, withdrew; the priests who managed these deceptive exhibitions no doubt adopted this method of getting rid of their visitor, who might become too inquisitive, and discover the secret of the mysteries.

Of all the reflecting surfaces mentioned, none produce more interesting deceptions than the concave mirror, and there is very little doubt that silver mirrors of this form were known to the ancients, and employed in some of their sacred mysteries. Mons. Salverte has industriously collected in his valuable work the most interesting proofs of their use, and quotes the following passage of "Damascius," in which the results obtainable from a concave mirror are clearly apparent. (Fig. 270.)

Fig. 270. The picture of a human face, possibly reflected from a concave mirror concealed below the floor of the temple; the opening being hidden by a raised mass of stone, and the worshippers confined to a certain part of the temple, and not allowed to approach it.

He says:—"In a manifestation which ought not to be revealed . . . . . there appeared on the wall of the temple a mass of light which at first seemed very remote; it transformed itself in coming nearer into a face evidently divine and supernatural, of a severe aspect, but mixed with gentleness, and extremely beautiful. According to the institution of a mysterious religion, the Alexandrians honoured it as Osiris and Adonis."

Parallel rays thrown upon a concave surface are brought to a focus or converge, and when an object is seen by reflection from a concave surface, the representation of it is various, both with regard to its magnitude and situation, according as the distance of the object from the reflecting surface is greater or less. (Fig. 271.) When the object is placed between the *focus* of parallel rays and the centre, the image falls on the *opposite* side of the centre, and is larger than the object, and in an inverted position. The rays which proceed from any remote terrestrial object are nearly parallel at the concave mirror — not strictly so, but come diverging to it in separate pencils, or, as it were, bundles

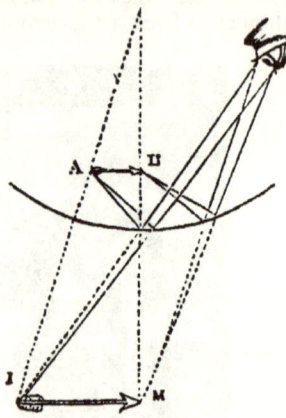

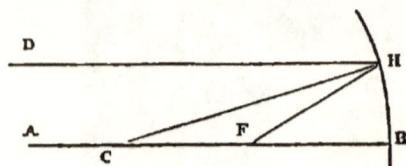

No. 1.   No. 2.

Fig. 271. No. 1. A B, D H represent two parallel rays incident on the concave surface B H, whose centre of concavity is C. B F and H F are the reflected rays meeting each other in F, and A B being perpendicular to the concave surface, is reflected in a straight line. No. 2. A B. The object. I M. The image.

of rays, from each point of the side of the object next the mirror; therefore they will not be converged to a point at the distance of half the radius of the mirror's concavity from its reflecting surface, but in separate points at a little greater distance from the concave mirror. The nearer the object is to the mirror, the further these points will be from it, and an inverted image of the object will be formed in them, which will seem to hang pendant in the air, and will be seen by an eye placed beyond it (with regard to the mirror), in all respects like the object, and as distinct as the object itself. No. 2. (Fig. 271.)

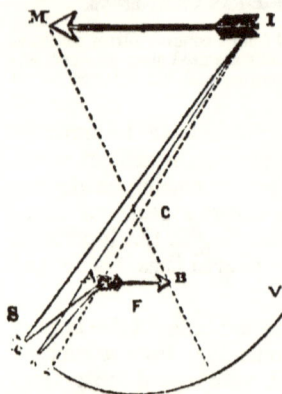

Fig. 272. A B represents the object, S V the reflecting surface, F its focus of parallel rays, and C its centre. Through A and B, the extremities of the object, draw the lines C B and C H, which are perpendicular to the surface, and let A B, A G, be a pencil of rays flowing from A. These rays proceeding from a point beyond the focus of parallel rays, will, after reflection, converge towards some point on the opposite side of the centre, which will fall upon the perpendicular, B C, produced, but at a greater distance from C than the radiant A from which they diverged. For the same reason, rays flowing from B will converge to a point in the perpendicular N C produced, which shall be further from C than the radiant B, from whence it is evident that the image I M is larger than the object A B, that it falls on the *contrary* side of the centre, and that their positions are inverted with respect to each other.

It appears, from a circumstance in the life of Socrates, that the effects of burning-glasses were known to the ancients; and it is pro-

bable that the Romans employed the concave speculum for the purpose of lighting the "sacred fire." This is very likely to be true, considering that the priests who conducted the heathen worship of Osiris and Adonis were acquainted with the use of concave metallic specula, as already described at page 282. The effects that can be produced with the aid of concave mirrors are very impressive, because they are not merely confined to the reflection of inanimate objects, but life and motion can be well displayed by them; thus, if a man place himself directly before a concave mirror, but further from it than its centre of concavity, he will see an inverted image of himself in the air between him and the mirror

Fig. 273. A concave mirror, showing the appearance of the inverted and reflected image in the air.

of a less size than himself; and if he hold out his hand towards the mirror the hand of the image will come out towards his hand and coincide with it, being of an equal bulk when his hand is in the centre of concavity, and he will imagine he may shake hands with his image. (Fig. 273.)

By using a large concave mirror of about three feet in diameter, the author was enabled to show all the results to a large audience that were usually visible to one person only. Whilst experimenting with a concave mirror, by holding out the hand in the manner described, a bystander will see nothing of the image, because none of the reflected rays that form it enter his eyes. This circumstance is well illustrated by placing a concave mirror opposite the fire, and allowing the image of the flames projected from it to fall upon a well-polished mahogany table. If the door of the room opens towards the mirror, and a spectator unacquainted with the properties of concave mirrors should enter the apartment, the person would be greatly startled to see flames apparently playing over the surface of the table, whilst another spectator might enter from another door and see nothing but a long beam of light, rendered visible by the floating particles of dust. To give proper effect to this experiment the concave mirror should be large, and no other light must illuminate the room except that from the fire.

On the same polished table the appearance of a planet with a revolving satellite may be prettily shown by darkening the fire with a screen, and placing a lighted candle before it, which will be reflected by the concave mirror, and appear on the table as a brilliant star of light, and the satellite may be represented by the flame of a small wax taper moved around the large burning candle. The following is the arrangement used by the author at the Polytechnic Institution for the purpose of exhibiting the properties of the concave mirror. A lantern enclosing a very brilliant light, such as the electric or lime light, is required for the illumination of the objects which are to be projected on to the screw. The lantern and electric lamp of Duboscq was preferred, although, of course, any bright light enclosed in a box, with a plain convex lens to project the beam of light when required, will answer the purpose. (Fig. 274.)

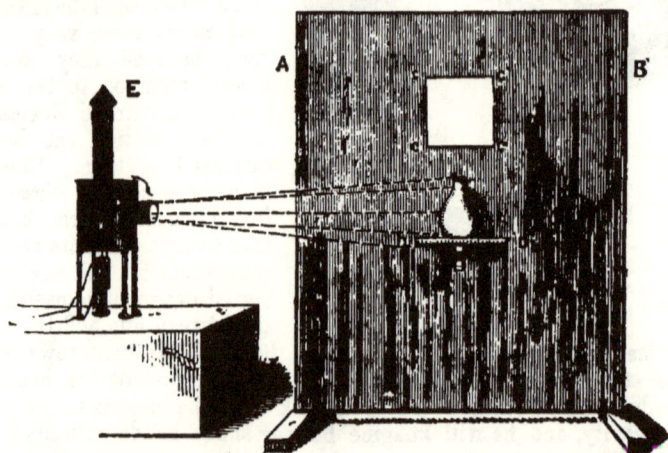

Fig. 274. A B. Portable screen of light framework, covered with black calico. c c c c. Square aperture just above the shelf, D D, upon which the object—viz., a bottle half full of water—is placed. E. Duboscq lantern to illuminate the object at D D.

By removing the diaphragm required to project the picture of the charcoal points on to the screen, a very intense beam of light is obtained, which may be focussed or concentrated on any opaque object by another double convex lens, conveniently mounted with a telescope stand, so that it may be raised or lowered at pleasure. This lens is independent of the lantern, and may be used or not at the pleasure of the operator.

The object is now placed on a shelf fixed to the screen, with a square aperture just above it. The object of the screen is to cut off all extraneous rays of light reflected from the mirror, or to increase the sharpness of the outline of the picture of the object. The screen and object being arranged, and the light thrown on from the lantern, the next step is to adjust the concave mirror, and by moving it towards the object, or backwards, as the case requires, a good image, solid and quasi-stereoscopic, is projected on to the screen. (Fig. 275.)

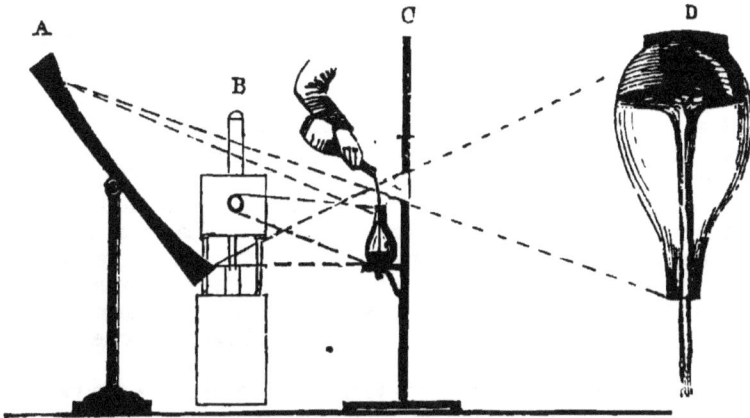

Fig. 275. A. The concave mirror. B. The lantern. C. The portable screen, shelf, and object. D. The inverted image of the bottle filling with water, with the neck downwards, and when thrown on the disc at D producing a most curious illusion.

The act of filling the bottle with water, or better still with mercury, is one of the most singular effects that can be shown; and if all the apparatus is enclosed in a box, so that the picture on the screen only is apparent, the illusion of a bottle being filled in an inverted position is quite magical, and invariably provokes the inquiry, how can it be done? The study of numismatics, the science of coins and medals, is generally considered to be limited to the taste of a very few persons, and any description of a collection of coins at a lecture would be voted a great bore, unless, of course, the members of the audience happened to be antiquaries; great light, however, may be thrown on history by a study of these interesting remains of bygone times, and a lecture on this subject, illustrated with pictures of coins thrown on to the disc by a concave mirror in the manner described, might be made very pleasing and instructive.

Coins, or plaster casts of coins *gilt*, flowers, birds, white mice, the human face and hands, may all, when fully illuminated, be reflected by the concave mirror on to the disc. A Daguerreotype picture at a certain angle appears, when reflected by the concave mirror, to be like any ordinary collodion negative, and all the lights and shadows are reversed, so that the face of the portrait appears black, whilst the black coat is white. On placing the Daguerreotype in another position, easily found by experiment, it is now reflected in the ordinary manner, showing an enlarged and perfect portrait on the disc. In using the Daguerreotype the glass in front of it must be removed. The pictures from the concave mirror may be also projected on thick smoke procured from smouldering damped brown paper, or from a mixture of pitch and a little chlorate of potash laid on paper, and allowed to burn slowly by wetting it with water.

An image reflected from smoke would be visible to a number of spectators, just as the light from the furnace fires of the locomotive is frequently visible at night, being reflected on the escaping column of steam.

It was probably with the help of some kind of smoke and the concave speculum that the deception practised on the worshippers at the temple of Hercules at Tyre was carried out, as it is mentioned by Pliny that a consecrated stone existed there "from which the gods easily rose." At the temple of Esculapius at Tarsus, and that of Enguinum in Sicily, the same kind of optical delusions were exhibited as a portion of the religious ceremonies, from which no doubt the priests obtained a very handsome revenue, much more than could be obtained in modern times by the mere exhibition of such wonders at Adelaide Galleries, Polytechnics, or Panopticons.

The smoke from brown paper is very useful in showing the various directions of the rays of light when reflected from plane, convex, and concave surfaces. The equal angles of the incident and reflected rays may be perfectly shown by using the next arrangement of apparatus. (Fig. 276.)

Fig. 276. A. Rays of light slightly divergent issuing from the lantern, and received on a little concave mirror, which brings the rays almost parallel, and reflects them to B, a piece of looking-glass, from which they are again reflected. c is the incident, and D the reflected rays. F. Smoke from brown paper.

A very dense white smoke is obtained by boiling in separate flasks (the necks of which are brought close together) solutions of ammonia and hydrochloric acid.

The opposite properties of convex and concave mirrors—the former scattering and the latter collecting the rays of light which fall upon them—are also effectively demonstrated by the help of the same illuminating source and proper mirrors, the smoke tracing out perfectly the direction of the rays of light. (Fig. 277.)

Fig. 277.   The smoke shows the rays of light falling on a convex mirror, and rendered still more divergent.

The smoke developes the cone of rays reflected from a concave mirror in the most beautiful manner, and by producing plenty of

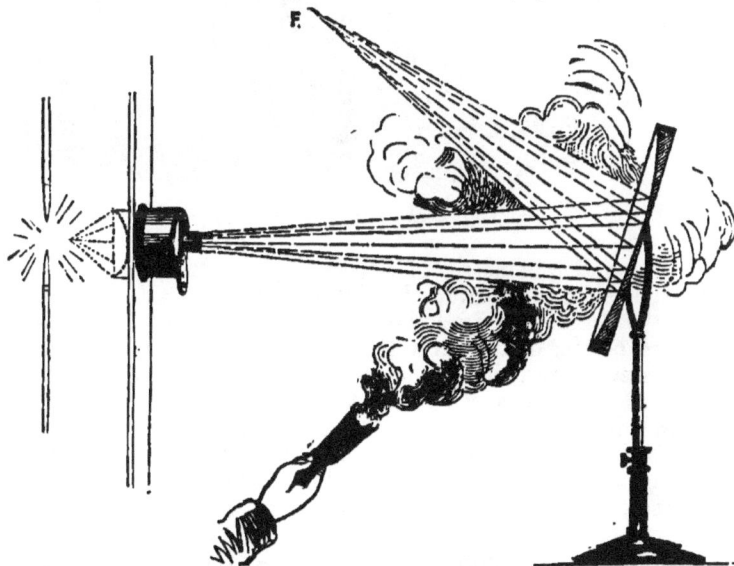

Fig. 278.   The smoke shows rays of light falling on the concave mirror. In this experiment attention should be directed to the bright point, x, the focus where the convergent rays meet.

smoke, and turning the mirror about—the position of the focus (*focus*, a fire-place), is indicated by a brilliant spot of light, and the reason the images of objects reflected by the concave mirror are reversed, may be better understood by observing how the rays cross each other at that point. (Fig. 278.)

One of the most perfect applications of the reflection of light is shown in the "Gregorian reflecting telescope," or in that magnificent instrument constructed by Lord Rosse, at Parsonstown, in Ireland. (Fig. 279.)

Fig. 279. Lord Rosse's gigantic telescope.

The description of nearly all elaborate optical instruments is somewhat tedious, but we venture to give one diagram, with the explanation of the Gregorian reflecting telescope. (Fig. 280.)

At the bottom of the great tube T T T T, (Fig. 280), is placed the large concave mirror D U V F, whose principal focus is at M; and in its middle is a round hole P, opposite to which is placed the small mirror L, concave towards the greater one, and so fixed to a strong wire M, that it may be moved farther from the great mirror or nearer to it, by means of a long screw on the outside of the tube, keeping its axis still in the same line P m n with that of the great one. Now since in viewing a very remote object we can scarcely see a point of it but what is at least as broad as the great mirror, we may consider the rays of each pencil, which flow from every point of the object, to be parallel to each other,

U

and to cover the whole reflecting surface D U V F. But to avoid confusion in the figure, we shall only draw two rays of a pencil flowing from each

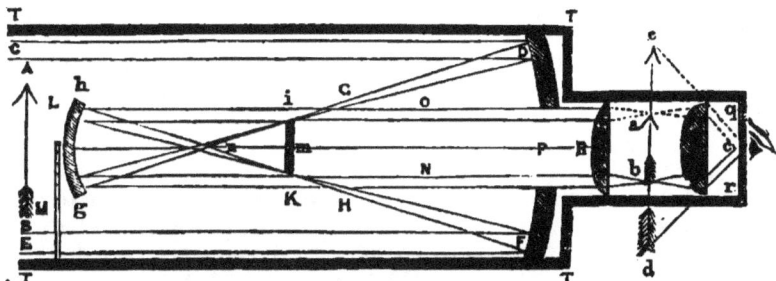

Fig. 280. The Gregorian reflecting telescope.

extremity of the object into the great tube, and trace their progress through all their reflections and refractions to the eye f, at the end of the small tube t t, which is joined to the great one.

Let us then suppose the object A B to be at such a distance, that the rays E flow from its lower extremity B, and the rays C from its upper extremity A. Then the rays C falling parallel upon the great mirror at D, will be thence reflected by converging in the direction D G; and by crossing at i in the principal focus of the mirror, they will form the upper extremity i of the inverted image i K, similar to the lower extremity B of the object A B; and passing on the concave mirror L (whose focus is at N) they will fall upon it at g and be thence reflected, converging in the direction N, because g m is longer than g n; and passing through the hole P in the large mirror, they would meet somewhere about r, and form the lower extremity d of the erect image a d, similar to the lower extremity B of the object A B. But by passing through the plano-convex glass R in their way they form that extremity of the image at b. In like manner the rays E which come from the top of the object A B and fall parallel upon the great mirror at F, are thence reflected converging to its focus, where they form the lower extremity K of the inverted image i K, similar to the upper extremity A, of the object A B; and passing on to the smaller mirror L and falling upon it at h, they are thence reflected in the converging state h o; and going on through the hole P of the great mirror, they would meet somewhere about q, and form there the upper extremity a of the erect image a d, similar to the upper extremity A of the object A B; but by passing through the convex glass R in their way, they meet and cross sooner, as at a, where that point of the erect image is formed. The like being understood of all those rays which flow from the intermediate points of the object, between A and B, and enter the tube T T, all the intermediate points of the image between a and b will be formed; and the rays passing on from the image through the eye-glass s, and through a small hole e in the end of the lesser tube t t, they enter the eye f which sees the image a d (by means of the eye-glass), under the large angle c e d, and magnified in length, under that angle, from c to d.

To find the magnifying power of this telescope, multiply the focal distance of the great mirror by the distance of the small mirror, from the image next the eye, and multiply the focal distance of the small mirror by the focal distance of the eye-glass; then divide the product of the latter, and the quotient will express the magnifying power. (Fig. 280.)

We now come to that much disputed and often quoted experiment of Archimedes, who is stated to have employed metallic concave specula or some other reflecting surface by which he was enabled to set fire to the Roman fleet anchored in the harbour of Syracuse, and at that time besieging their city, in which the great and learned philosopher was shut up with the other inhabitants. The story handed down to posterity was not disputed till about the seventeenth century, when Descartes boldly attacked the truth of it on philosophical grounds, and for the time silenced those who supported the veracity of this ancient Joe Miller. Nearly a hundred years after this time, the neglected Archimedes fiction was again examined by the celebrated naturalist Buffon, and the account of his experiments detailed by the author of "Adversaria," in Chambers' Journal, is so logical and conclusive, that we give a portion of it verbatim.

"For some years prior to 1747, the French naturalist Buffon had been engaged in the prosecution of those researches upon heat which he afterwards published in the first volume of the Supplement to his 'Natural History.' Without any previous knowledge, as it would seem, of the mathematical treatise of Anthemius (περι παραδοξων μη-χανηματων), in which a similar invention of the sixth century is described,* Buffon was led, in spite of the reasonings of Descartes, to conclude that a speculum or series of specula might be constructed sufficient to obtain results little, if at all, inferior to those attributed to the invention of Archimedes.

"This, after encountering many difficulties, which he had foreseen with great acuteness, and obviated with equal ingenuity, he at length succeeded in effecting. In the spring of 1747, he laid before the French Academy a memoir which, in his collected works, extends over upwards of eighty pages. In this paper, he describes himself as in possession of an apparatus by means of which he could set fire to planks at the distance of 200, and even 210 feet, and melt metals and metallic minerals at distances varying from twenty-five to forty feet. This apparatus he describes as composed of 168 plain glasses, silvered on the back, each six inches broad by eight inches long. These, he says, were ranged in a large wooden frame, at intervals not exceeding the third of an inch; so that, by means of an adjustment behind, each should be moveable in all directions independently of the rest—the spaces between the glasses being further of use in allowing the operator to see from behind the point on which it behoved the various disks to be converged.

* See Gibbon's "Decline and Fall," chap. xl., section v., note g.

U 2

"These results ascertained, Buffon's next inquiry was how far they corresponded with those ascribed to the mirrors of Archimedes—the most particular account of which is given by the historians Zonaras and Tzetzes, both of the twelfth century.* 'Archimedes,' says the first of these writers, 'having received the rays of the sun on a mirror, by the thickness and polish of which they were reflected and united, kindled a flame in the air, and darted it with full violence on the ships which were anchored within a certain distance, and which were accordingly reduced to ashes.' The same Zonaras relates that Proclus, a celebrated mathematician of the sixth century, at the siege of Constantinople, set on fire the Thracian fleet by means of brass mirrors. Tzetzes is yet more particular. He tells us, that when the Roman galleys were within a bow-shot of the city-walls, Archimedes caused a kind of hexagonal speculum, with other smaller ones of twenty-four facets each, to be placed at a proper distance; that he moved these by means of hinges and plates of metal; that the hexagon was bisected by 'the meridian of summer and winter;' that it was placed opposite the sun; and that a great fire was thus kindled, which consumed the Roman fleet.

"From these accounts, we may conclude that the mirrors of Archimedes and Buffon were not very different either in their construction or effects. No question, therefore, could remain of the latter having revived one of the most beautiful inventions of former times, were there not one circumstance which still renders the antiquity of it doubtful: the writers contemporary with Archimedes, or nearest his time, make no mention of these mirrors. Livy, who is so fond of the marvellous, and Polybius, whose accuracy so great an invention could scarcely have escaped, are altogether silent on the subject. Plutarch, who has collected so many particulars relative to Archimedes, speaks no more of it than the former two; and Galen, who lived in the second century, is the first writer by whom we find it mentioned. It is, however, difficult to conceive how the notion of such mirrors having ever existed could have occurred, if they never had been actually employed. The idea is greatly above the reach of those minds which are usually occupied in inventing falsehoods; and if the mirrors of Archimedes are a fiction, it must be granted that they are the fiction of a philosopher."

Supposing that Archimedes really did project the concentrated rays of the sun on the Roman vessels, one cannot help pitying the ignorance of the Admiral Marcellus. Had this officer been acquainted with the laws of the reflection of light, he might have laughed to scorn the power of Archimedes, and by receiving the unfriendly rays on one of the bright brazen convex shields of his soldiers, Marcellus could have scattered the concentrated rays, and prevented the burning of his vessels.

In these days of learning it therefore appears strange to find any one advocating the possible use of specula or reflecting mirrors for the purposes of offence or defence, but M. Peyrard a few years ago proposed

* Quoted by Fabricius in his "Biblioth. Græc.," vol. ii., pp. 551, 552.

to produce great effects by mounting each mirror in a distinct frame, carrying a telescope so that one person could direct the rays to the object intended to be set on fire, and he gravely calculated, presuming on the ignorance of the attacked, that with 590 glasses of about twenty inches in diameter, he could reduce a fleet to ashes at the distance of a quarter of a league! and with glasses of double that size at the distance of half a mile! What effect a shell or shot would produce upon this ancient weapon is not stated; this we may safely leave our readers to determine for themselves. The experiment of Archimedes has long been a favourite one with the boys. (Fig. 281.)

Fig. 281. One of the " miseries of *reflection*."

The total internal reflection of light by a column of water is an experiment that admits of great variety so far as colour is concerned, and is one of the most novel and beautiful experiments with light presented to the public within the last few years. The author had the pleasure of introducing it in the first place at the Polytechnic Institution, where the optical novelty excited the greatest attention, and received the approbation of her Most Gracious Majesty, and his Royal Highness the Prince Consort, with the Royal Family, who were pleased to pay a private evening visit to the Polytechnic, and amongst other things minutely examined the "Illuminated Cascade," which had been erected by Mons. Duboscq of Paris.

The illumination of the descending columns of water was obtained by converging the rays from a powerful electric light upon the orifice from which the water escaped, the Duboscq lantern already explained being employed, and in front of it were placed three cylinders, each having a circular window behind and opposite the lens, and an aperture of about one inch in diameter on the opposite side for the escape of water. The

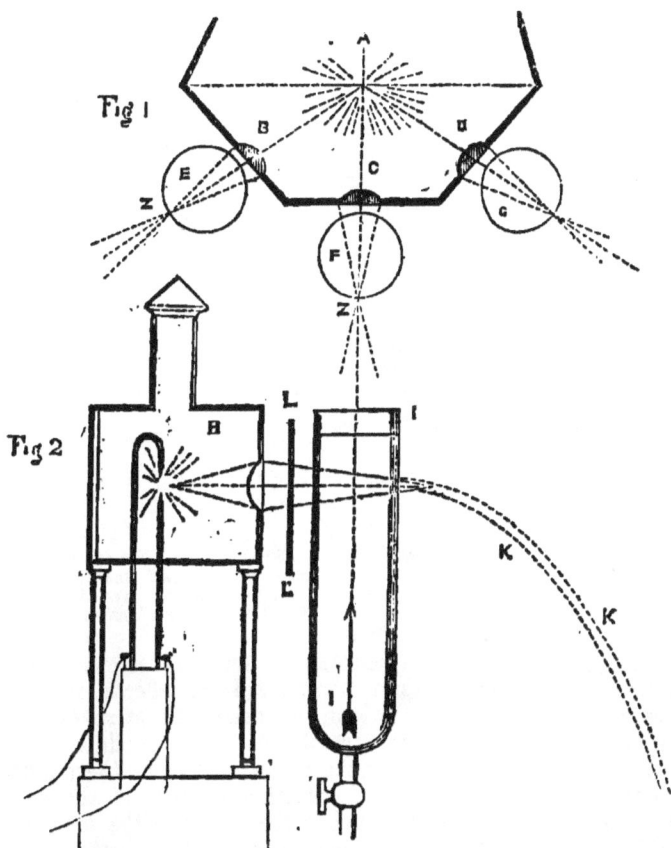

Fig. 282.—Fig. 1. A. The electric light. B C D. The three sides and lenses of the lantern. E F G. The three cylinders of water, each with a circular glass window and orifices at z z z, from which the water and rays of light pass out.—Fig. 2. H. Section of one side of the Duboscq lantern. I I. Cylinder of water, which enters from below. K K. The stream of illuminated water. L L. Bit of coloured glass held between the lantern and the cistern of water.

lantern used was of a peculiar shape, and had three sides, the electric light being in the centre of them, and passing through three separate plano-convex lenses to the three cylinders from which the water escaped.

Attention may be directed to the fact that the light merely passes out of the orifices as a diverging beam of light until the flow of water commences, when the rays are immediately taken up and reflected from point to point inside the arched column of water, and illuminating the latter in the most lovely manner, it appears sometimes like a stream of liquid metal from the iron furnace, or like liquid ruby glass, or of an amethyst or topaz colour, according to the colours of the plates of glass held between the mouths of the lantern and the circular windows in the cylinders of water. The same experiment created quite a *furore* at the Crystal Palace when it was introduced in one of the author's lectures delivered in that noble place of amusement. In order that our readers may understand the arrangement of the apparatus, we have given at page 294 a ground plan view of it, as also the appearance of the cascade when exhibited at the Polytechnic to the Royal party. (Fig. 284.)

Another curious effect observed with the illuminated cascade, is the descent of balls of light as the reflection is cut off for a moment by passing the finger through the stream of water, showing that a certain time is occupied in the reflection of light from one end of the cylinder of water to the other; indeed the best idea of the *rationale* of the experiment is formed by substituting in imagination a silver tube highly polished in the interior, for the

Fig. 283. A B. The sides of the cascade. The dotted lines show the reflection of only two rays of the beam of light passing down inside the water.

descending jet of water. The reflection of sound takes place precisely in the same manner, and the vibrations of the air are reflected from plane, concave, and convex surfaces. It is on this principle that waves of sound thrown off from different surfaces (as of hard rocks), produce the effect of the *echo*. The sounds arrive at the ear in succession, those reflected nearest the ear being first, and the reflecting surfaces at the greatest distance sending the waves of sound to the ear after the former. At Lurley Falls on the Rhine, there is an echo which repeats seventeen times. Whispering galleries, again, illustrate the reflection of sound from continuous curved surfaces, just as the archèd column of water reflects from its interior curved surfaces the rays of light.

Speaking-tubes are well known in which the waves of sound are successively reflected from the sides, exactly like the "Illuminated Cascade" (Fig. 283). The speaking-trumpet is also another and familiar example of the same principle. Probably when Albertus Magnus constructed the brazen head, which had the power of talking, it was nothing more than a metallic head with a few wheels and *visible* mechanism inside, but connected with a lower apartment by a hollow metal tube, where Albertus Magnus descended, and astonished the ignorant with

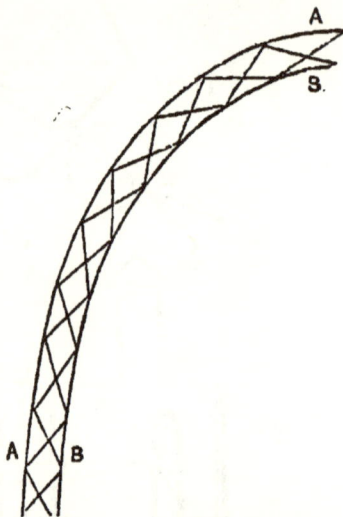

the then unknown principle of the speaking tube. Light entering at one end of a bright metallic tube is reflected from the sides of the tube till it reaches the other, and precisely the same effect occurs in the interior of the cascade of water. (Fig. 284).

Fig. 284. End of Polytechnic Hall, where the illuminated cascade was displayed to her Majesty, H.R.H. the Prince Consort, and Royal party. The cascades issued from behind some artificial rock-work.

THE KALEIDOSCOPE.

If this article on light and optics had gone minutely into the mathematical and purely scientific portion of the subject, we should have had frequent occasion to mention the name of Sir David Brewster, a distinguished philosopher, whose name is peculiarly identified with this interesting branch of physics. It is always pleasing to find men of such standing not only devoting themselves to arguments which college wranglers would study with pleasure, but also descending to a lower level, and inventing optical instruments that delight and amuse the non-scientific and juvenile part of the community. The names of Sir David Brewster and Professor Wheatstone have been connected during the last few years with the invention of the stereoscope, an instrument that will be noticed in another part of this book, but here we shall describe one of the most original optical instruments ever devised, and although it is now regarded as a mere toy, its merits are very great. The title of the instrument is borrowed from the Greek καλος, beautiful, ειδος, a form or appearance, σκοπεω, to see; and the public certainly endorsed the name when they purchased 200,000 of these instruments in London and Paris during the space of three months. It is said that the sensation it excited in London, throughout all ranks of the community, was astonishing, and people were everywhere seen, even at the corners of the streets, looking through the kaleidoscope. The essential parts of this instrument are two mirrors of unsilvered black parallel glass, or plate glass painted black on one side, which should be from six to ten inches in length, and from one inch to an inch and a half in breadth at the object end, while they are made narrower at the other end, to which the eye is applied. The mirrors are united at their lower edges by a strip of black calico fixed with common glue, and are left open at the upper edges, and retained at the proper angle by a bit of cork properly blackened. The angles are 36°, 30°, 25°$\frac{5}{7}$, 22°$\frac{1}{2}$, 20°, 18°, which divide the circumference into 10, 12, 14, 16, 18, 20 parts, thus 36 × 10 = 360, or 18 × 20 = 360, and the strictest attention must be paid to this part of the adjustment, or the figures produced will not be symmetrical. After the mirrors are adjusted to

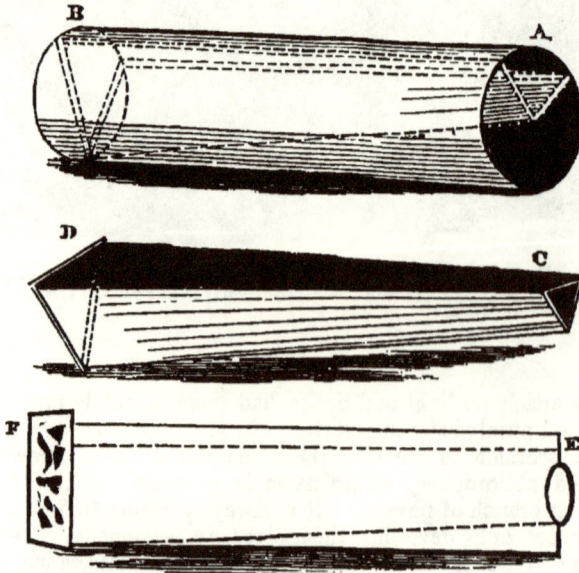

Fig. 285. A B. The tube containing the two mirrors, shown by dotted lines. A is the small end where the eye is placed. B. The object end. C D. Another view of the mirrors arranged to place in the tube; the shaded portion represents the black velvet. E. Double convex lens. F. Box to contain objects, and usually fitted with ground glass outside.

the proper angle, the space between the two upper edges should be covered across with black velvet and the mirrors placed in a tin or brass tube, so that the broad ends shall barely project beyond the end, while the narrow end is placed so that the angle formed by the junction of the mirrors shall be a little below the middle of that end of the tube. A cover with a circular aperture in the centre is then to be fitted to the narrow end of the mirrors, which should in general be furnished with a convex lens whose focal length is an inch or two greater than the length of the mirrors. A case for holding the objects, and for communicating to them a revolving motion, is fitted to the object end of the tube. The objects best suited for producing pleasing effects are small fragments of coloured glass, wires of glass, both spun and twisted, and of different colours and shades of colours, and of various shapes, in curves, angles, circles; also, beads, bugles, fine needles, small pieces of lace, and fragments of fine sea-weed are very beautiful. M. Sturm, of Prague, has lately fixed the images of the kaleidoscope, so that they are available for the production of patterns in every branch of silk, cotton, and mixed fabrics. Photographs could be taken of the most beautiful of these accidental designs, which only occur once, and if not copied are lost.

CHAPTER XXII.

THE REFRACTION OF LIGHT.

THIS term appears to be often confounded with that of reflection, and signifies the bending or breaking back of a ray of light (*re*, back, and *frango*, to break); and it will be remembered that when light falls on the surface of a solid (either liquid or gaseous) body, it may be reflected (*re*, back, and *flecto*, to bend), refracted, polarized, or absorbed. In the previous chapter the property of the reflection of light has been fully investigated, and in this one refraction only will be considered. It is a property which has been, and will continue to be, of the greatest practical utility in its application to the construction of all magnifying glasses, whether belonging to the telescope, microscope, magic lantern, or the dissolving views; or the minor refracting instruments—such as spectacles, opera-glasses, &c.; and it should be remembered that their magnifying power depends solely on the property of refraction.

If substances such as glass had not been endowed with this property, it would be difficult to understand how the great discoveries in the science of astronomy could have been made, or what information we could have gained respecting those interesting truths so constantly revealed by the aid of the microscope. Numerous instances might be quoted of the value of this latter instrument in the detection of adulteration, and the examination of organic structures. When so many talented and industrious scientific men are at work with this instrument, it is perhaps invidious to point to one singly, though we must make an exception in favour of Professor Ehrenberg, of Berlin, whose microscope did such good service in procuring undeniable proof of the Simonides' fraud; he has made use of it again to detect the thief that stole a barrel of specie, which had been purloined on one of the railways. One of a number of barrels, that should have contained coin, was found on arrival at its destination to have been emptied of its precious contents, and refilled with sand. On Professor Ehrenberg being consulted, he sent for samples of sand from all the stations along the different lines of railway that the specie had passed, and by means of his microscope identified the station from which the sand must have been taken. The station once discovered, it was not difficult to hit upon the culprit in the small number of *employés* on duty there.

The simplest case of refraction occurs in tracing the course of a ray of light through the air, and into the medium water; in this case it passes from a rare to a dense medium, and the fact itself is well illustrated by the next diagram, in which the shaded portion represents water, and the paper that it is drawn upon the air. The line A B is a perpendicular ray of light, which passes straight from the air into and through the water, without being changed in its direction. The line C D is another ray, inclined from the perpendicular, and entering the water at an angle, does not pass in the straight line indicated by the dotted line, but is refracted or bent towards the perpendicular at D E.

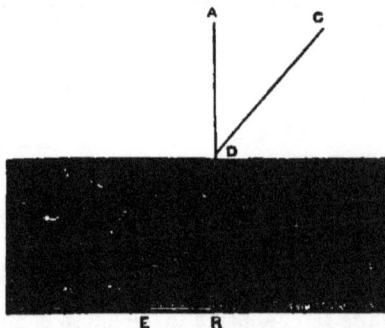

Fig. 286.

This fact reduced to the brevity of scientific laws is thus expressed:— When a ray of light falls perpendicularly on a refracting surface, *it does not experience any refraction or change of direction.* When light passes out of a rare into a dense medium, as from air into water, *the angle of incidence is greater than the angle of refraction.* And when light passes from a dense into a rare medium, as out of water into air, *the reverse takes place,* and *the angle of incidence is smaller than the angle of refraction.*

In order to illustrate these laws, a zinc-worker or tinman may construct a little tank, with glass windows in the front and sides, the latter being as deep as the half-circle described on the back metal plate of the tank, which of course rises higher, in order to show the full circle; this should be japanned white, and a perpendicular and horizontal black line described upon it—the whole, with the exception of the circle, being japanned black. If the Duboscq lantern is arranged with the little mirror, as described in fig. 276, page 287, the ray of light may be thrown perpendicularly, or at an angle, through the water, and the actual breaking back of the ray of light is rendered distinctly apparent. (Fig. 287.)

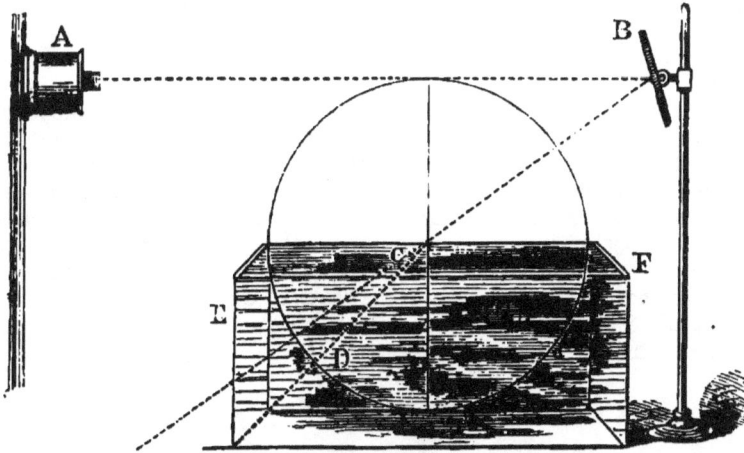

Fig. 287. A. Duboscq lantern. B. The mirror. B C. The incident ray. C D. The refracted ray. E F. Tank, containing water up to the horizontal line of the circle.

The refraction of light is also well displayed by Duboscq's apparatus, with the plano-convex lens, and a brass arrow as an object, with another double convex lens to focus it. When a good sharp outline of the arrow is obtained on the disc, a portion of the rays of light producing it may then be truly broken out or refracted by laying across the brass arrow a square bar of plate glass. (Fig. 288).

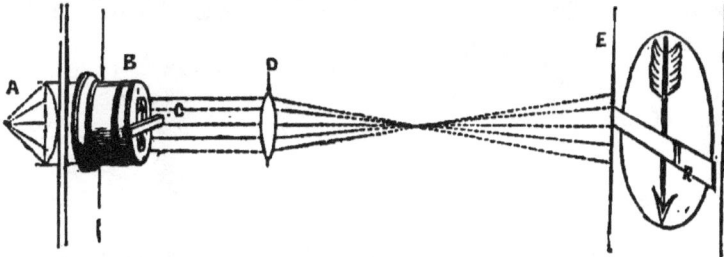

Fig. 288. A. Rays of light from the electric light. B. The cap, with figure of arrow cut out. C. The bar of plate glass. D. The double convex glass to focus E, the image on the disc, and portion refracted at E.

There are many simple ways in which the refraction of light is displayed, such as the apparent breaking of an oar where it enters the water, or the remarkable manner in which the bottom is lifted up when we look, at any angle, through the clear water of a deep river or lake; the latter circumstance has unhappily led to most serious accidents, in consequence of children being induced by the apparent shallowness of the water to get in and bathe. Fish, again, unless seen perpendicularly from a boat, always appear nearer than their true position, and the Indians, when they spear fish, always take care to strike as near the perpendicular as possible; experienced shots know they must aim a little lower and nearer than the apparent position of a fish in order to hit it.

Having learnt that light is bent from its course, it might be supposed that all objects looked at through plate glass should appear distorted; but it must be remembered that the sides of the glass being nearly parallel, an equal amount of refraction occurs in every direction—so that, unless the window is glazed with uneven wavy glass, the object, for all practical purposes, does not apparently change its position, being neither moved to the right or the left, or upward or downward. In order to bend the rays of light in the required direction, the glass must be cut into certain figures called prisms, plane glasses, spheres, and lenses, some of which are shown in the annexed cut. (Fig. 289.)

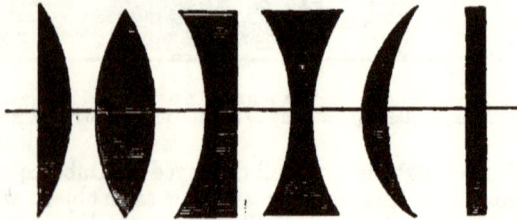

Fig. 289.

It would be tedious to trace out, by a regular series of diagrams, the passage of light through the variety of combinations of lenses; and as the plane, convex, and concave surfaces have been examined with respect to their effect on the reflection of light, they may be referred to again with regard to their influence in refracting light. In the latter it will be found that convex and concave lenses have just the opposite properties of mirrors; thus, a convex lens receiving parallel rays will cause them to converge to a focus.

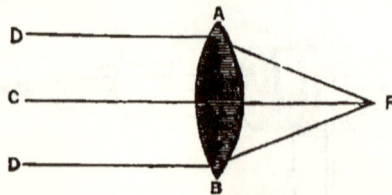

Fig. 290. A B. A double convex lens. C is a ray of light, which falls perpendicularly on A B, and therefore passes on straight to F, the focus. D D. Rays falling at an angle on A B, refracted to focus, F.

(Fig. 290.) The case of *short-sighted* persons arises from too great a convexity of the eye, which makes a very near focus; and that of old people is a flattening of the eye, by which the focus is thrown to a greater distance. The remedy for the latter is a convex spectacle-glass, whilst a concave lens is required for the former, to scatter the rays and prevent their coming to a point too soon.

The action of a concave refracting surface is again the opposite to a concave reflecting surface — the former disperses the rays of light, whilst the latter collects them. A concave lens, as might be expected, produces exactly the contrary effect on light to that of a concave mirror. (Fig. 291.)

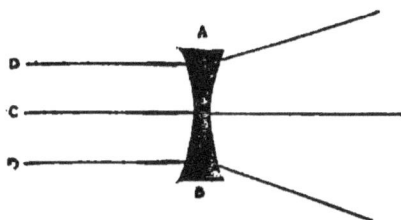

Fig. 291. A B. A double concave lens. C Is a ray of light which falls perpendicularly on A B, and passes through without any alteration of its course. D D. Rays falling at an angle on A B, are refracted and diverged.

These facts are well shown with the aid of the lantern and electric light. The rays of light are refracted in a visible manner when received on a concave or convex lens, provided a little smoke from paper is employed, as in the mirror experiments. (Fig. 292.)

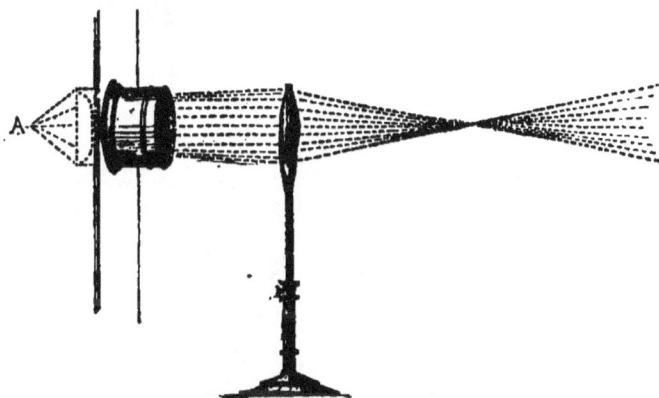

Fig. 202. A. The electric light. B. The lens.

Bearing these elementary truths in mind, it will not be difficult to follow out a complete set of illustrations explanatory of the construction and use of various popular optical contrivances.

CHAPTER XXIII.

I. *The Magic Lantern.*

No other optical instrument has ever caused so much wonderment and delight, from its origin to the present time, as this simple contrivance. For a long time its true value was overlooked, and only ridiculous or comic slides painted, but its educational importance is now being thoroughly appreciated, not only on account of the size of the diagrams that may be represented on the disc, but also from the fact that the attention of an audience is better secured in a room when the only object visible is the diagram under explanation. The lenses it contains are a "bull's eye" or plano-convex, nearest the light, and a double convex glass, for the purpose of focussing the picture which is inverted and placed between the two lenses. (Fig. 293.)

In many books full directions are given for painting the glass slides, but this is an art that requires very great practice and experience. A person may know how to draw and paint on paper or canvas, but it is quite a different thing where glass is concerned, and unless the juvenile artist has taken lessons from a regular painter on glass, his or her efforts are likely to be very unsatisfactory. In many popular works embracing the subject of optics, full directions are given on the mode of painting the slides for the magic lantern, or dissolving views; a new era, however, has dawned upon this mode of illustration, in the preparation of photographs on glass of the most lovely description, and now instead of exhibiting mere daubs of weak colouring, photographic pictures of singular perfection can be procured of Messrs. Negretti and Zambra, Holborn, who have turned their attention especially to this branch, and supply slides of all sizes.

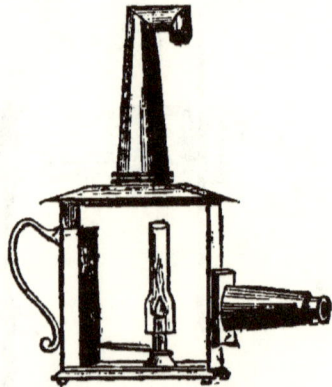

Fig. 293. The magic lantern.

II. *The Dissolving Views.*

This very pleasing modification of the ordinary magic lantern is displayed with the assistance of two lanterns of the same size, provided with lamps and lenses which are exactly alike. They are best arranged on one board, side by side, and if kept parallel with each other, the circles of light thrown from the two lanterns would not coincide on the screen; it is therefore necessary to place one of them at an angle which will vary according to the distance from the screen. The task of making the two circles of light overlap each other precisely on the disc, is called centering the lanterns, and is the first thing that must be attended to before exhibiting the slides. The slides for the dissolving views are all painted of the same size, and supposing a scene such as a church with a bridal procession and the trees in full foliage, to represent summer, is first thrown on to the disc, it may be changed to winter by putting another picture of the same subject, but painted to represent bare trees, and the church and ground covered with snow, and a grave open, with a funeral procession. The two pictures must not be projected on the screen at the same time, and here the dissolving mechanism is required; it consists of two fans so arranged that they may be raised or lowered by a rack-work and handle; one fan in descending covers one of the nozzles of the lanterns, and the other leaves the second lantern open, and free to project the picture; the dissolving is managed by slowly moving the handle of the rack-work, so that one quarter of the picture already on the disc is cut off, and one quarter of the new one thrown on. As the movement proceeds, one half of the old picture is shut out, and one half of the new slide takes its place, and so on, till the whole of the original picture is cut off by the fan and the new one comes into view, and it is in this way the effect of the change from summer to winter is produced. (Fig. 294.)

Fig. 294. Nozzle of one lantern, with the fan, A, raised, and in the position to throw a picture on the disc. B. The other fan shutting off the second lantern.

When two pictures such as those already described, dissolve one into the other, of course the same building or other marked portion of the subject, must strictly coincide in each picture on the disc, or else the two pictures are apparent, and the illusion is destroyed. The pictures must all be centered before the exhibition commences. By the arrangement of Mons. Duboscq, one electric light serves to illuminate both lanterns by making use of mirrors. The dissolving apparatus is likewise very simple, and consists of two diamond-shaped openings in a brass frame, which open and shut alternately by a slide worked with a handle. The single light is not to be recommended, as it is somewhat troublesome to manage properly. (Fig. 295.)

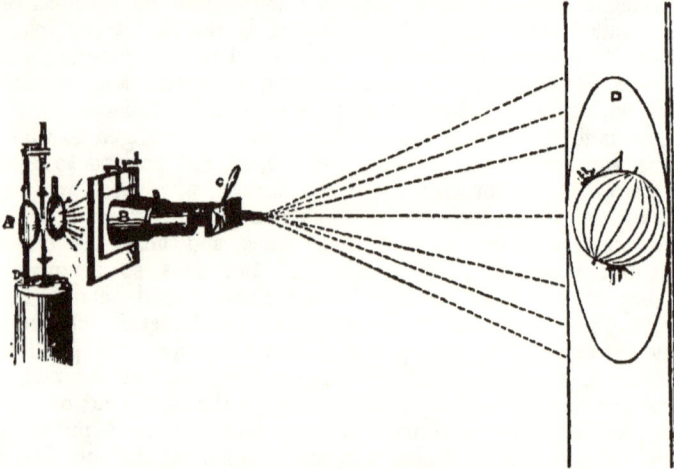

Fig. 295. A. The electric light. B B. The two sets of lenses for the two pictures. c. The dissolving mechanism. D. The picture on screen.

When dissolving views are required on a grand scale, the lenses must be exceedingly large, and the condenser (corresponding with the "bull's-eye" of the simple magic lantern) should be at least nine or eleven inches in diameter, and the front glasses must be of a superior make. The lenses for a large lantern lit by the oxy-hydrogen light, are arranged as in the next cut. (Fig. 296.)

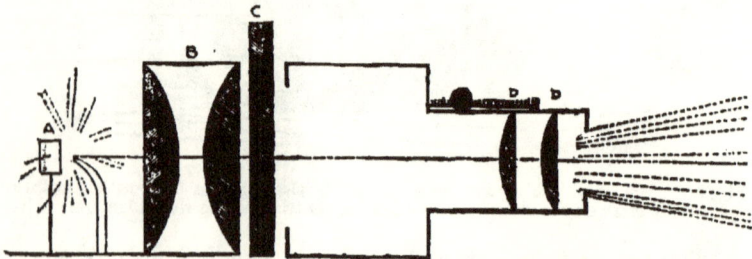

Fig. 296. A. The lime light. B. The condensers. c. The picture. D D. The front lenses for focussing, with rack-work.

At the Polytechnic the author had no less than six lanterns working at or about the same time, to produce effects, in the views illustrating the voyages of Sinbad the Sailor; and in order to obtain the increased

x

results required for dioramic effects, such for instance as the Siege of
Delhi, showing the bursting of the shells, &c., the four fixed lanterns (the
fronts of which are shown in the next cut) were always employed. The
two upper lanterns are dissolved by discs of brass worked by the hand,
and the lower ones with the fans. (Fig. 297.)

Fig. 297. Fronts of the four lanterns, showing how the dissolving mechanism is arranged.

"Behind the scenes" always has a great attraction for young people; we
have, therefore, in the frontispiece, with the help of Mr. Hine (who painted
a great number of the photographs shown at the Polytechnic during the
author's management), given a section of the large theatre taken whilst
the effective scene of the Siege of Delhi was in progress. The optical
effects were assisted by various sounds in imitation of war's alarms, for
the production of which, more *volunteers* than were required would
occasionally trespass behind the screen, and produce those terrific sounds
that some persons of a nervous temperament said were really stunning.
In a page picture, we have also given a correct drawing of the interior
of the optical box at the Polytechnic, with the four fixed lanterns, and
side cupboards to hold the glass pictures. The four lanterns worked on
a railway, with wheels and a circular turn-table; they could be removed,
and the microscope arranged in their places.

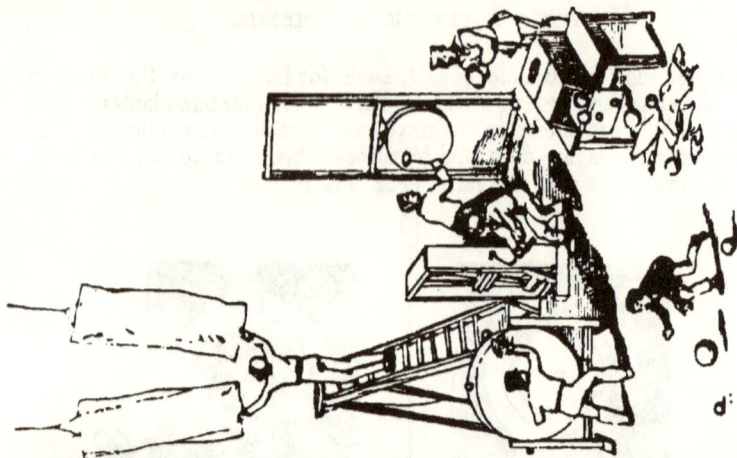

Before and behind the screen at the Polytechnic during the exhibition of the dioramic effects of the siege of Delhi.　　p. 306.

III. *The Oxy-Hydrogen Microscope.*

Many persons will recollect the first exhibition of this instrument in Bond-street, by Mr. J. T. Cooper, and Mr. Cary, succeeded by the Adelaide Gallery exhibition of scientific wonders and an oxy-hydrogen microscope. The apparatus for this purpose consists of three condensing lenses and an object glass. The objects, such as live aquatic insects, are placed in glass troughs containing water; the other objects, ferns, feathers, butterflies, algæ, &c. &c., being mounted on slides in the ordinary way with Canada balsam. (Fig. 298.)

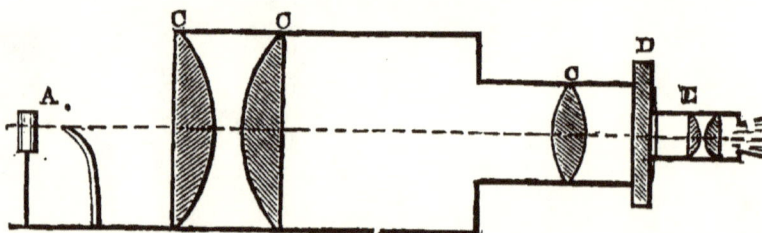

Fig. 298. A. The lime light. c c c. Condensers. D. The object, such as a tank of water containing live insects. E. The object glasses.

IV. *The Physioscope.*

This instrument, brought out at the Polytechnic during the time that Mr. J. F. Goddard managed the optical department of the institution, always excited the greatest mirth and astonishment amongst the numerous visitors; and *habitués* of the old place may remember the good-natured inimitable maudlin simper with which poor Mr. Tait (who was one of the living objects shown on the disc) used to drink off the glass of wine and then wink at the audience. When we say Mr. Tait used to wink, of course it is understood that he was personally invisible, and his apparition or image only appeared on the disc. The countenance is brilliantly illuminated by the oxy-hydrogen light, and being placed near the lenses, the rays are reflected from the face into the physioscope, and being properly focused, and the inversion of the image corrected, the perfect representation of the human countenance is apparent on the disc. The lenses and concave reflectors required are shown in the section of the physioscope. Messrs. Carpenter and Westley, of Regent-street, have brought the manufacture of magic lanterns to great perfection; and Mr. Collins, of the Polytechnic, constructs every kind of dissolving view apparatus, oxy-hydrogen microscopes, physioscopes, &c. (Fig. 299.) With this instrument any opaque objects (provided they reflect light properly) may be displayed to a large audience. Plaster casts appear with singular beauty and softness, whilst flowers, stuffed birds, and especially humming birds, are excellent objects for the physioscope.

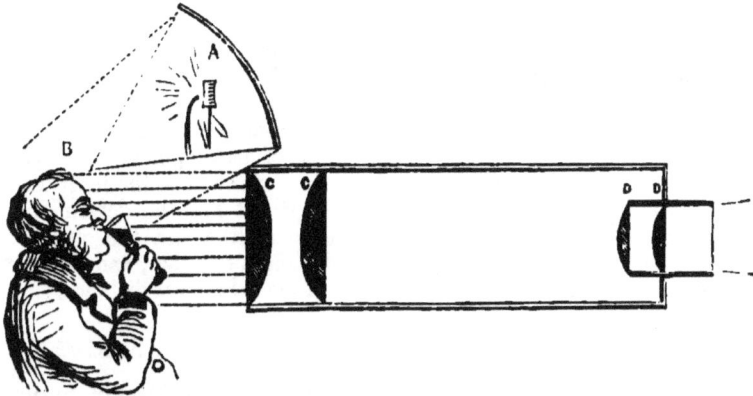

Fig. 209. A. One or more lime lights, throwing rays reflected by concave mirrors on to the face B, from whence they are reflected to C C, the first condensers. D D. Object glasses. This instrument is made by Mr. Collins, who has the tools for making the reflectors with correct curves. The picture of the face on the disc is covered with black spots if the reflectors are not perfect.

V. *The Camera Obscura.*

A "dark chamber" is the name of a most amusing, and now, in the improved form, extremely valuable instrument for photographic purposes. It is occasionally to be met with in public gardens, and there is a very good one on the Hoe at Plymouth. The construction of the camera for observing the surrounding country is very simple, and merely consists of a flat mirror placed at an angle, by which the picture is reflected through a double-convex lens on to a white table beneath. (Fig. 300.)

Fig. 300. A. The mirror. B. The convex lens. C. The white table.

The term "focusing," or the art of moving the lenses so that a sharp image may be obtained, has been frequently mentioned in this article, and perhaps it may be as well to describe the mode of ascertaining the focal distance of a lens by experiment.

Hold the lens opposite the window so that a bright picture of the window-sash may be obtained on a sheet of paper pinned against the wall, and the distance of the lens from the paper will be the focal length.

If the lens has a very long focal length, it may be determined as follows:—Measure the distance between the lens and the object, and also from the image; multiply these distances together, and divide the product by their sums; the quotient will give the focal distance.

VI. *The Decomposition of Light—"its Analysis and Synthesis."*

It is in the Italian language that the bride, the emblem of purity, is called Lucia (*Lux*, light); and surely if an illustration were required of beauty and singleness, light would be named poetically as appropriate; but physically it is not of a single nature, it is composite, and made up of seven colours. The instrument required to refract a ray of light sufficiently to break it into its elementary colours is called the prism, and is a solid having two plane surfaces, called its refracting surfaces, with a base equally inclined to them. (Fig. 301.)

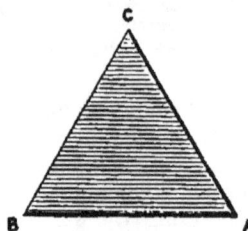

It was in 1672 that Sir Isaac Newton made his celebrated analysis of light, by receiving a sunbeam (as it passed through a hole in a shutter) on to the refracting surface of a prism, and throwing the image or spectrum on to a screen, where he observed the seven colours, red, orange, yellow, green, blue, indigo, and violet, and thus proved " *that there are different species of light, and that each species is disposed both to suffer a different degree of refrangibility in passing out of one medium into another, and to excite in us the idea of a different colour from the rest ; and that bodies appear of that colour which arises from the composition of those colours the several species they reflect are disposed to excite.*"

Sir Isaac Newton's name would have been immortalized by this discovery alone, even if he had not possessed that transcendent ability which raised him above all other mathematicians and physicists. It is at the same time interesting to know that the ancient author Claudian (A.D. 420) inquires "whether colour really belongs to the substances themselves, or whether by the reflection of light they cheat the eye—*enquires silve color proprius rerum, lucisne repulsa eludant aciem.*"

Sir Isaac Newton determined that the spectrum could be divided into 360 equal parts, of which red occupied 45, orange 27, yellow 48, green 60, blue 60, indigo 40, violet 80. He also discovered that if the highly refracted rays, the seven colours, or spectrum were received into

Fig. 301. The prism. The base, A B, is equally inclined to the refracting surfaces, C A, C B.

a concave mirror or a double-convex lens, that they again united and formed white light. In order to demonstrate the properties of the prism in various positions, the next diagram may be adduced. (Fig. 302.)

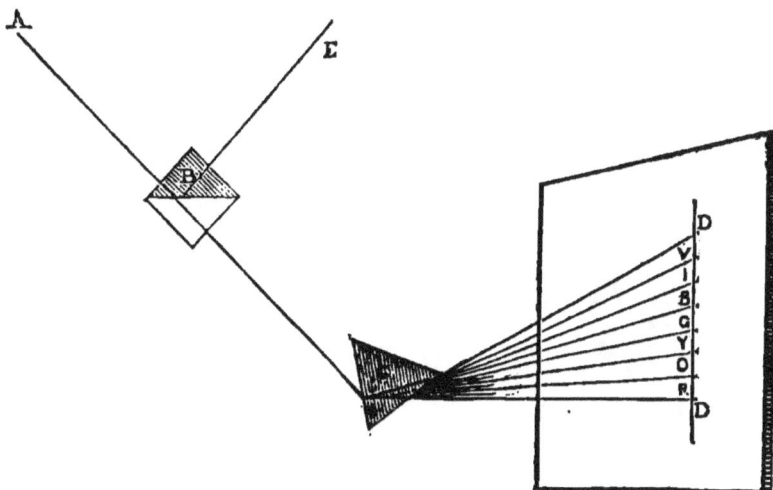

Fig. 302. A. The ray of light passing through two prisms B placed base to base. In this position the light passes through to the second prism, c, without alteration. At c the decomposition of light occurs, and the spectrum is shown at D D. The top prism at B used singly would reflect the ray to B without decomposing it into the coloured rays.

The rainbow is the most beautiful natural optical phenomenon with which we are acquainted; it is only seen in rainy weather when the sun illuminates the falling rain, and the spectator has the sun at his back. There are frequently two bows seen, the interior and exterior bow, or the primary and secondary, and even within the primary rainbow, and in contact with it, and outside the secondary one, there have been seen other bows beyond the number stated.

The primary or inner rainbow consists of seven different coloured bows, and is usually the brightest, being formed by the rays of light falling on the upper parts of the drops of rain. The exterior bow is formed by the rays of light falling on the lower parts of the drops of rain; and in both cases the rays of light undergo refraction and reflection, hence the opinion of Aristotle, that the rainbow is caused only by the reflection of light, is not correct.

The first refraction occurs when the rays of light enter, and the second when they emerge from the spheroids of water in the first bow; the refracted rays undergo only one reflection, whereas in the second the brilliancy of the colours is impaired by two reflections.

The spectrum from the electric light is one of the most gorgeous exhibitions of colour that can be conceived; and the instruments required for the purpose are illustrated in No. 1 (Fig. 303), whilst the synthesis of the coloured rays and production of white light is shown at No. 2 of the same figure. (Fig. 303.)

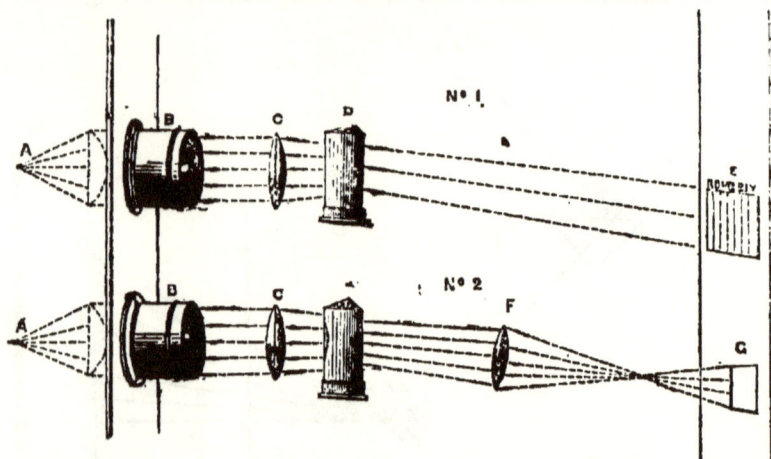

Fig. 303. No. 1. A. The electric light. B. The narrow slit through which the light passes to the convex lens, C. D. The prism. E. The spectrum. No. 2 is the same for A B C D; but F is the convex lens collecting the scattered rays, and forming white light at G.

VII. *Duration of the Impression of Light.*

If a circular disc is painted with the prismatic colours taken in the same proportion with respect to each other in which they are exhibited in the spectrum made by the prism, and the wheel is turned swiftly, then the individual colours disappear, and nearly white light is apparent. The cause is due to the same principle that creates the appearance of a complete circle of fire when a burning squib is moved quickly round before it is thrown away to burst, and as it is evident that the burning squib cannot be in every part of the circle at the same moment, there must be some inherent faculty belonging to the human eye which enables it to retain for a definite period the impression of images that may fall upon it; and this principle has been so far pressed, as it were, beyond its limits, that it is gravely asserted the image of a man's murderer "might be discovered on the retina of the eye-ball if that could be examined sufficiently quick after death." The fixture of the picture is said to be due to a sort of natural photographic process; but such fanciful statements often lead the mind into dream-land only, and so we will return to the fact of the duration of the impression of light on the eye as evidenced by several ingenious optical instruments, and especially by the scientific inventions of Dr. Faraday, Dr. Paris, and of Mr. Thomas Rose of Glasgow.

By careful experiment M. D'Arcy found that the light of a live coal moving at the distance of 165 feet, maintained its impression on the retina during the seventh part of a second. Hence the cause of the recomposition of white light when the colours on the disc are quickly rotated. Each colour at any point succeeds the other before the impression of the last is gone from the eye, and provided the colours move round within the seventh part of a second, they are all impressed together on the eye, and meeting on the retina, produce the effect of white light.

VIII. *The Phenakistiscope.*

This amusing instrument consists of a turning wheel upon which figures appear to jump, walk, or dance. The disc or wheel is of cardboard, upon which are painted (towards the periphery) figures in eight, ten, or twelve postures. Thus, if it is desired to represent clowns turning round in a circle, twelve different positions of the figure in the act of turning are painted on the disc, and above each of the figures on the wheel a slit is cut about one inch long, and a quarter of an inch wide in a direction corresponding with the radii of the circle. This simple form of the instrument is used by placing the figured side towards a looking-glass and then causing it to revolve at a certain speed, which is ascertained by experiment; and as the spectator looks through the slits into the looking-glass, the clowns appear to turn round. At the Polytechnic Institution there are two of these wheels with looking-glasses, and although the same designs have done duty for many years, they still attract the public attention. (Fig. 304.)

Fig. 304. Design for the phenakistiscope. The spectator is supposed to be looking towards a mirror through the slits. It is supported by a handle through the centre, round which it is twirled by the other hand.

In the "Journal of the Royal Institution" Mr. Faraday has described some very interesting experiments and optical illusions produced by the revolution of wheels in different directions and velocities. The wheels are made of cardboard, and by cutting out two cog wheels of an equal size, and placing one above the other on a pin, the usual hazy tint when the cogs are acting is apparent when they are whirled round; but if the two cog wheels are made to move in opposite directions, there will be the extraordinary appearance of a fixed spectral wheel. If the cogs are cut in a slanting direction on both wheels, the spectral wheel will exhibit slanting cogs; but if one wheel is turned so that the cogs shall point in opposite directions, then the spectral wheel will have straight cogs. A number of such wheels set in motion in a darkened room, and illuminated suddenly with the light from the electric spark, appear to stand perfectly still, although moving with a great velocity. An expensive instrument has been constructed by Duboscq, for the purpose of showing the usual phenakistiscope effects on the screen with the magic lantern; a very limited picture, however, is shown, and there is still great room for the improvement of the apparatus. (Fig. 305.)

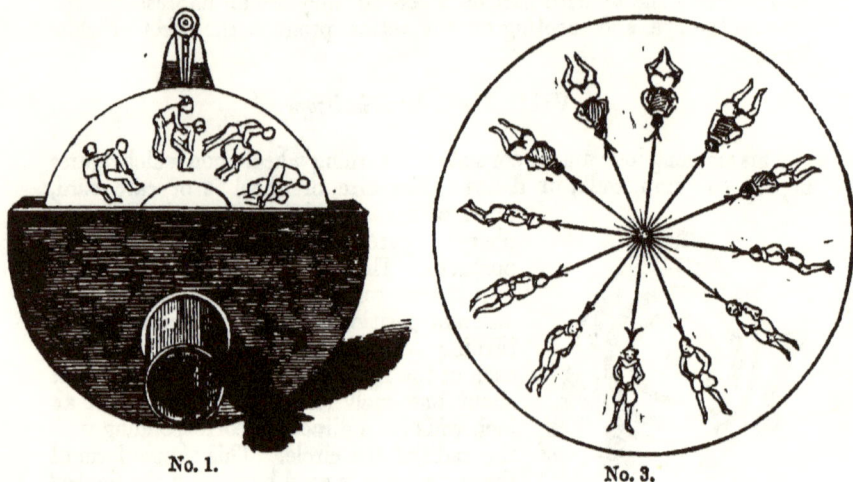

No. 1.

No. 3.

No. 2.

Fig. 305. Phenakistiscope made by Duboscq, of Paris. No. 1. Apparatus in elevation with the condensers. No. 2. Section of the apparatus. A. The light. B. Condenser, or plano-convex lens. C. Round glass disc with design painted on it. D. Wooden disc with four double-convex lenses placed at equal distances from each other, so as to coincide with C, whilst rotating. Both the latter and C rotate, and the picture is focussed on the disc by the lenses F. No. 3. Glass plate, with device painted thereon.

IX. *The Thaumatrope.*

This very simple toy was invented by the late Dr. Paris, who gave it an appropriate name, compounded of the Greek words, θαῦμα, wonder, τρέπω, to turn. The duration of the impressions of light on the eye is very apparent whilst using this toy, which is usually made of a circular piece of cardboard, having on one side a painting of a man's head, and on the other a hat; or a picture of a lighted candle on one face of the cardboard, and an extinguisher on the other; or a gate, and a horseman leaping it. Each pair of designs painted on opposite sides of the cardboard appear to be one when twisted round by strings tied to the opposite edges of the cardboard circle. The *rationale* of this experiment being, that the picture of one design—such as the head and face—is retained by the eye until the hat appears, and being mutually impressed upon the nerve of vision at very nearly the same instant of time, they appear as one picture.

X. *The Kalotrope.*

This is an optical arrangement by Mr. Thomas Rose, of Glasgow, primarily designed for showing the illusions of the phenakistiscope and kindred devices to a numerous audience; but more remarkable for its presentations of very beautiful spectra, composed of the multiplication, combination, and involution of simple figures disposed around a disc. The arrangement consists of a movement for giving considerable velocity to two concentric wheels, working nearly in contact, and moving in contrary directions. But the only part of the apparatus that requires special explanation and illustration is the device disc and the disc of apertures; the first of which is placed on the hinder wheel, and the second on the front wheel. We give figures of the two discs, premising, however, that each is capable of an almost infinite variety of characters. No. 1 (Fig. 306) presents in its four quadrants the perforations for four distinct discs of apertures; and No. 2 is a device disc, consisting of twelve equidistant black balls. Under *a* the balls will be presented as twenty-four ovals; under *b*, as forty-eight involved figures, beautifully variegated; under *c*, as an elaborate lacework; and under *d*, as a rich variegation of form and colour. Every fresh disc of devices and disc of apertures of course opens up a new field of effect. Thus, if we take a disc bearing twelve repeats of a ball in the interior of a ring, each repeat being so painted that its position is advanced in the ring until it reaches in the twelfth ring the point whence it started, and place this on the back disc of the kalotrope, having previously removed the first one, no effect is observed when the wheel is rotated beyond the spreading out of the design and general appearance of hazy black circles. When, however, the disc, with twelve slits or apertures, is now placed on the front wheel, and the two rotated in opposite directions, then the whole figure starts as it were into existence, and each ball apparently moves round the interior of its circle.

MR. ROSE'S KALOTROPE AND PHOTODROME. 315

The apparatus was produced at the Royal Polytechnic Institution by the author, and excited much interest. (Fig. 306.)

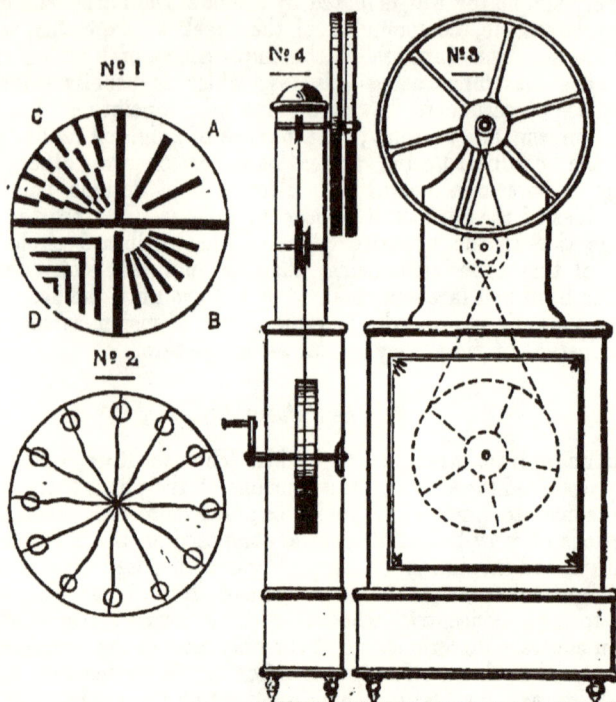

Fig. 306. Nos. 1 and 2 are the discs. No 3. Kalotrope in elevation. No. 4. Side view of kalotrope, showing the multiplying wheels and the perforated and painted discs moving in opposite directions.

XI. *The Photodrome.*

This is a second optical arrangement by Mr. Rose for showing spectral illusions; and it is superior to the last, inasmuch as it offers to the public lecturer a most effective means of presenting these deceptions to a large audience. It differs from the kalotrope in several important points. It dispenses with the discs of apertures, and leaves the device disc with its face fully exposed to the spectators. The effects are produced by a powerful light, thrown through the tube of a lantern, and broken by a wheel working across it. The apparatus, as it at present stands in the inventor's possession, consists of two distinct parts; the one a movement for the device discs, and the other for the light. A wheel four feet in diameter is connected with a train of movement capable of giving it five hundred or six hundred revolutions per minute. On this wheel the device disc is placed, in full view of the spectators, and set in motion. From an opposite gallery the light is thrown, and broken by a wheel of such diameter and number of apertures as will admit the velocity of the *photodrome* (or light-runner) to be at least *six* times the velocity of the device disc; whilst the apertures are of such width as to restrict the duration of the light-flash to about one-two-thousandth of a second. The wheel working across the light has a train of movement for raising the velocity to two thousand revolutions per second. The management of the apparatus is very simple. The device-wheel is brought to a steady, rapid rotation, and the operator on the light then works his wheel with gradually increasing velocity, until he overtakes the figures of the device, where, by mere delicacy of touch, he is able to hold them stationary or give them motion, at pleasure.

Theories of light and colour still agitate the scientific world, although that man must be bold who will assert that his hypothesis is fitted to explain every difficult point that arises as our experimental knowledge increases. Mr. G. J. Smith, of the Perth Academy, has propounded a very ingenious theory of light and colour, supported by some clever experiments. But, as Solomon says, "there is nothing *new* under the sun," and in an able paper Mr. Rose, of Glasgow, lays claim to the anticipations of Mr. Smith's theory as follows:—

"My attention has been directed to a paper entitled 'The Theory of Light,' by G. John Smith, Esq., M.A., of Perth Academy. I think it is now nearly two years since I communicated an interesting fact to Professor Faraday, and to a member of our local Philosophical Institution, which may fairly claim to have anticipated Mr. Smith's theory. The fact was this : that if a piece of intensely white card be held in one hand, with the light of a powerful gas-jet falling upon it, and if the other hand has command of the gas-tap, as the light is gradually reduced, the card will assume the prismatic colours down to intense blue, and as the light is restored the colours will present themselves in inverse order. The experiment showed, very conclusively to my mind, that light is homogeneous, and that what we name colour is only the various affection of the optic nerve by a greater or lesser radiation of light from a focal point in an imperfect reflector—say, in the instance, a white card. I apprehend that Mr. Smith confuses his theory when he speaks of alternations of light and shadow producing colour. Shadow, or darkness, is mere negation of light. We do not see mixtures of light and darkness, or blackness and whiteness, but light in its several degrees of intensity. Mr. Smith's experiments present only what my kalotrope has done, and what my later device, the photodrome (now nearly three years old) is doing in a much more perfect manner. It is one of the mysteries intelligible only to the initiated, that whilst Mr. Smith's paper seems to have been received with great favour by the British Association, my communication relative to the photodrome was voted 'not *sufficiently practical.*' . . . . .

"Since I have come before the public with an experiment, which in any view is an interesting one, permit me to reproduce it under several distinct conditions, and to add a brief narrative of remarkable presentations of colour that have come before me, and which, so far as I am aware, are perfectly novel, or known only through the more recent experiments of Mr. Smith. Professor Faraday very courteously acknowledged my communication of the experiment with the card, but said that it only partially succeeded with him, and added that probably this was owing to some decay of sensitiveness in his eyes. More likely I failed to state with sufficient clearness the conditions of the experiment, since I have always found nine persons out of ten perfectly agreed as to the effects produced when they have been at my side. The transitions from white to yellow, orange, red, and thence to intense blue, are, I may say, invariably admitted. Success depends on a very slow and regular reduction and restoration of the light. I have given one method of performing the experiment, and will add other two. Allow the light to remain undisturbed, and begin by holding the card near to it; then keep the hand steady and the eye intently fixed upon the card, and retire gradually with your back to the light, and the colours will change in the order of the prismatic spectrum from yellow to intense blue. On returning backwards towards the light the colours will again present themselves, but in inverse order. In this form of the experiment we are certain that the light remains precisely the same throughout. The third method is this: Place a circle of white card, about three inches in diameter, in the centre of a black board, and let a spectator stand within twelve inches of the board, with his eyes fixed upon the card. Let an operator be provided with a light so covered that it shall not fall on the eye of the spectator; then, as he retires with the light or returns with it, the spectator will see the colours as before. This arrangement evidently subjects the experiment to a severe test, since the black board enhances the whiteness of the card, and tends to preserve it. . . . . Whilst pursuing my principal object, I frequently noticed most remarkable presentations of colour; but, as the conditions were for the most part unsuitable to the lecture-room, I gave them only a passing regard. Allow me to instance a few of the experiments.

"The first refers to the kalotrope, which may be briefly described as an arrangement of two concentric wheels, working nearly in contact and in contrary directions. Discs of various devices are provided for the hinder wheel, and a number of perforated black discs for the one in front. When a disc charged with twelve black radii is placed on the hinder wheel, the six spokes of the front wheel, in passing rapidly across it, convert the twelve *black* radii into twenty-four apparently stationary *white* radii upon a tinted ground. Here is a remarkable presentation of the complementary, inasmuch as it is placed permanently before the eye by persistence.

"The second experiment is performed with the photodrome, which consists of an independent wheel to receive the device discs, and an apparatus (altogether apart, and, if desired, out of sight) by which flashes of light are thrown upon the disc in rapid and regular succession. Now, if a disc charged with twelve dark blue balls, nearly in contact, be placed upon the wheel, and a little natural light be allowed to fall upon it, so soon as it is thrown into rapid revolution, and flashes of artificial light (insulated in a lantern) are duly measured out upon it, we see twelve apparently stationary light-blue balls upon a zone of bright orange. Here, again, there is nothing for which we are not prepared; the complementary is suddenly presented, and it is maintained permanently before the eye by persistence.

"A third experiment may prove interesting in its relation to Mr. Smith's ingenious theory. Place the kalotrope opposite a bright northern noonday sky, remove the front wheel, and affix to the hinder wheel one of the perforated black discs used for the kalotropic effects. The experimentalist stands at the back of the instrument, and can see the sky only through the apertures in the black disc. Cause these apertures to pass the eye at intervals varying from one-half to one-sixth of a second, and very remarkable presentations of colour are seen. Under the lower velocities the sky flashes, and assumes an unnatural brilliancy, and the intervals of the fourth and fifth of a second give it sometimes a crimson, at others a deep purple colour. Now, what are we to infer from this experiment? Certainly *not* that the pulsations have absolutely produced variety of colour. At every pulsation the full natural light falls upon the eye, and the intervals between the pulsations give time for the reaction necessary to the suggestion of complementary colour, and that under manifold modifications arising out of the ever-changing condition of the eye during the experiment. If the apertures pass the eye with a velocity exceeding one-sixth of a second, the effect ceases. There is then perfect persistence, and the eye apprehends nothing but the ordinary light of the sky, reduced in intensity, with nothing to break its uniformity or give it a chromatic character.

"A fourth experiment is kindred to the last. Place the kalotrope under the same adjustment and management as before, in front of a brilliant sunset, and the spectator will see, with more than a poet's vision,

'The rich hues of all glorious things.'"

XII. *The Kaleidoscopic Colour-top.*

This invention by John Graham, of Tunbridge, is designed to show that when white or coloured light is transmitted to the eye through small openings cut into patterns or devices, and when such openings are made to pass before the eye in rapid successive jerks, both form and colour are retained upon the nerve of the visual organ sufficiently long to produce a compound pattern, all the parts of which appear simultaneously, although presented in succession. The instrument forms, therefore, a pleasing illustration of the law that the eye requires an almost inappreciably short space of time to receive an impression, and that such impression is not directly effaced, but remains for an assignable though very limited period. The results are obtained by rotating two discs on a wheel, the lower disc containing colours, and the upper one the openings; this latter disc is made to vibrate as well as to rotate, thus allowing the eye to receive the coloured light reflected from below, which light assumes, at the same time, the forms of the patterns through which it has been transmitted. The instrument serves also to illustrate most of the important phenomena of colour.

XIII. *Simple Microscopes and Telescopes.* ·

The Stanhope lenses are now sold at such a cheap rate, and are so useful as simple portable microscopes, that it is hardly worth while to detail any plan by which a cheap single-lens magnifier may be obtained. Eloquent vendors of cheap microscopes are to be found in the streets, who make their instrument of a pill-box perforated with a pin-hole, in which a globule of glass fixed with Canada balsam is placed; and the spherical form of the drop affords the magnifying power: or a thin platinum wire may be bent into a small circular loop, and into this may be placed a splinter of flint-glass; if the flame of a spirit-lamp is urged upon the loop of platinum wire and glass by the blowpipe until it melts, a small double-convex lens may be obtained, which will answer very well as a magnifying-glass. Practice makes perfect, and after two or three trials, a good single lens may be obtained, which can be mounted between two small pieces of lead, brass, or cardboard, properly fixed together, with holes through them just large enough to retain the edge of the tiny lens. A prism can be made of two small pieces of window-glass stuck together with a lump of soft beeswax, and if a few drops of water are placed in the angle, they are retained by capillary attraction. The prism is used by holding it against a large pin-hole or small slit in a bit of card, and directing them towards the sky, when the beautiful colours of the spectrum will be apparent if the card and prism are brought close to the eye.

The most simple form of the refracting telescope is made with a lens of any focal length exceeding six inches, placed at one end of a tin or cardboard tube, which must be six inches longer than the focal length of the lens; the tube may be in two parts, sliding one within the other, and when the eye is placed at the other end, an inverted image of the object looked at, is apparent. By using two double-convex lenses, a more perfect simple astronomical telescope is obtained. The object-glass, *i.e.*, the lens next the object looked at, must be placed at the end of a tin or pasteboard tube larger than its focus, and the second lens, called the eye-glass, because next the eye, is a smaller tube, termed the eye-tube; and if the focal length of the object-glass is three feet, the eye-glass must have a one-inch focus, and of course the eye-tube and glass must slide freely in the tube containing the object-glass. An object-glass of forty feet focus will admit of an eye-glass of only a four-inch focus, and will, therefore, magnify one hundred and twenty times. A tube of forty feet in length would of course be very troublesome to manage, and therefore it is usual to adopt the plan originally devised by Huygens, viz., that of placing the object-glass in a short tube on the top of a high pole with a ball-and-socket joint, whilst the eye-glass is brought into the same line as the object-glass, and focused with a tube and rack-work properly supported. In an ordinary terrestrial telescope there are four lenses, in order that the objects seen by its assistance shall not be inverted; and whenever objects are examined by a common telescope, they are found to be fringed, or surrounded with prismatic colours. This disagreeable effect is corrected by the use of *achromatic* lenses, in which two kinds of glass are united; and the light decomposed by one glass, uniting with the colours produced by the other form white light, thus a double convex lens of crown glass, c c, may be united with a plano-convex lens of flint glass, F F, which must have a focus about double the length of that of the crown-glass lens. The concave lens corrects the colour or chromatic aberration of the other, and leaves about one-half of the refracting power of the convex lens as the effective magnifying power of the compound lens. The French opticians cement the lenses very neatly together, and use them in ordinary spy and opera glasses. (Fig. 307.)

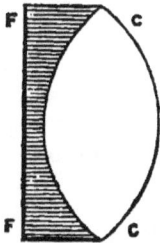

Fig. 307. A compound achromatic lens, composed of c c, the double-convex lens of crown-glass, and F F, the plano-concave lens of flint-glass.

XIV. *The Stereoscope.*

This instrument has now attained a popularity quite equal to, if it does not surpass, that formerly enjoyed by the kaleidoscope, and without entering upon the much-vexed question of priority of discovery, it is sufficient again to mention with the highest respect the names of Sir David Brewster and Professor Wheatstone as identified with the discovery and use of this most pleasing optical instrument.

The principle of the stereoscope (meaning, *solid I see*) is copied from nature: *i.e.*, when both eyes are employed in the examination of an object, two separate pictures, embracing dissimilar forms, are impressed upon the retinæ, and produce the effect of solidity; if the pictures formed at the back of the eyes could be examined by another person with a stereoscope, they would come together, and also produce the effect of solidity.

Stereoscopic pictures are obtained by exposing sensitized paper in the camera to the picture of an object taken in two positions, or two cameras are employed to obtain the same result. If the latter mode is adopted, the stereoscopic pictures must not be taken from positions too widely separated from each other; or else, when the two pictures are placed in the stereoscope, they will stand out with a relief that is quite unnatural, and the object will appear like a very reduced solid model, instead of having the natural appearance presented by pictures which have been taken at positions too distant from each other.

Sir David Brewster says, "In order to obtain photographic pictures mathematically exact, we must construct a binocular camera which will take the pictures simultaneously, and of the same size; that is, by a camera with two lenses of the same aperture and focal length, placed at the same distance as the two eyes. As it is impossible to grind and polish two lenses, whether single or achromatic, of exactly the same focal lengths, even if we had the very same glass for each, I propose to bisect the lenses, and construct the instrument with semi-lenses, which will give us pictures of precisely the same size and definition. These lenses should be placed with their diameters of bisection parallel to one another, and at a distance of 2½ inches, *which is the average distance of the eyes in man;* and when fixed in a box of sufficient size, will form a binocular camera, which will give us at the same instant, with the same lights and shadows, and of the same size, such dissimilar pictures of statues, buildings, landscapes, and living objects, as will reproduce them in relief in the stereoscope." Thus with a single camera provided with semi-lenses, or two lenses of the same focal length, stereoscopic pictures can be obtained.

To bring the images of the two pictures together, and produce the effect of solidity; either of two instruments may be employed. The reflecting stereoscope is the invention of Professor Wheatstone. The refracting or lenticular stereoscope that of Sir David Brewster.

The former is constructed by placing two upright boards on a wooden stand at a moderate distance from each other; the stereoscopic pictures are attached to these boards, which may be made to move up or down, and if the pictures are held in grooves, they may be pulled right or left at pleasure, and thus four movements are secured—viz., upward, downward, right, or left. Between the two stereoscopic pictures are placed two looking-glasses, so adjusted that their backs form an angle of ninety degrees with each other. (Fig. 308.)

Fig. 308. Wheatstone's reflecting stereoscope.

The pictures are illuminated at night by a lamp or gas flame placed at the back of the mirrors, which, when fixed together, have the same shape as a prism; indeed, Professor Wheatstone substituted a prism for the mirrors, and thus paved the way for the invention of the lenticular stereoscope.

Y

The stereoscopic effect is obtained by bringing the eyes close to the inclined mirrors, so that the two reflected images coincide at the intersection of the optic axis; the coincidence of the images is further secured by moving either picture a little to the right or left, and if the upright boards move bodily in grooves to or from the centre mirror, the greatest nicety of adjustment is procured.

During the last three years of the author's directorship of the Polytechnic—viz., in 1856, 1857, 1858—nearly the whole of the pictures shown by the dissolving-view apparatus were coloured photographs from Mr. Hine's original pictures, painted two feet square in blue and white, and reduced on the glass to about six inches square. The collodion film being frequently thick and difficult to penetrate with light, was etched and scratched away where required, and filled in with colour, and when these pictures were looked at with *one* eye only, they appeared to be almost solid or stereoscopic on the disc.

The lenticular stereoscope consists of a box of a pyramidal shape, open at the base, and provided with grooves in which are placed the stereoscopic pictures; if the latter are taken on glass the base of the box is held directly against the light, but if they are daguerreotypes or paper pictures, then a side light is reflected upon them by means of a lid covered in the inside with tinfoil, which is raised or lowered at pleasure from the top part of the box. Two semi-lenses are now fitted into the narrow part of the box, and are placed at such a distance from each other that the centres of the semi-lenses correspond with the pupil of the eyes, and this distance has already been stated to amount to $2\frac{1}{2}$ inches. (Fig. 309.)

The principle of the lenticular stereoscope is perhaps better seen by reference to the next diagram, in which the centres of the semi-lenses (*i.e.*, a lens cut in half) are placed at $2\frac{1}{2}$ inches apart, with their *thin* edges towards each other, and marked, A B, Fig. 310. The centres of the two stereoscopic pictures C D correspond with the centres of the lenses, and the rays of light *diverging* from C D fall upon the semi-lenses, and being refracted nearly *parallel* are, by the prismatic form of the semi-lenses, deflected from their course, and leave the surfaces of the lenses in the same direction as if they actually emanated from E; and as all images of bodies appear to come in a straight line from the point whence they are seen, the two pictures are superimposed on each other, and together produce the appearance of solidity, so that a stereoscopic result is obtained when the *spectral images* of the two stereoscopic pictures are made to overlap each other. By taking one of the semi-lenses in each hand, and looking at the two pictures, the over-lapping

Fig. 309. Brewster's lenticular stereoscope.

of the *spectral images* becomes very apparent, so that the combined *spectral images*, and not the *pictures* themselves, are seen when we look into a stereoscope. (Fig. 310.)

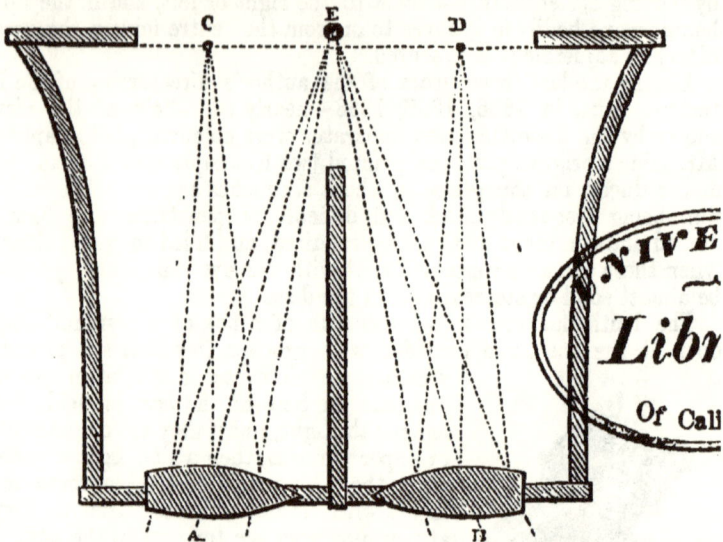

Fig. 310.

Sir David Brewster says, "In order that the two images may coalesce without any effort or strain on the part of the eye, it is necessary that the distance of the similar parts of the two drawings be equal to twice the separation produced by the prism. For this purpose measure the distance at which the semi-lenses give the most distinct view of the stereoscopic pictures, and having ascertained by using one eye the amount of the refraction produced at that distance, or the quantity by which the image of one of the pictures is displaced, place the stereoscopic pictures at a distance equal to twice that quantity—that is, place the pictures so that the average distance of similar parts in each is equal to twice that quantity. If this is not correctly done, the eye of the observer will correct the error by making the images coalesce, without being sensible that it is making any such effort. When the dissimilar stereoscopic pictures are thus united, the solid will appear standing as it were in relief between the two plane representations."

XV. *The Stereomonoscope.*

M. Claudet, whose name has long been celebrated in connexion with the art of photography, has described an instrument by which a single picture is made to simulate the appearance of solidity, and he states that by means of this arrangement a number of persons may observe the effect at the same time. The apparatus required is very simple, consisting of a large double convex lens, and a screen of ground glass. The object A, Fig. 311, is highly illuminated, and placed in the focus of a double convex lens B, when an image of the object is projected, and will

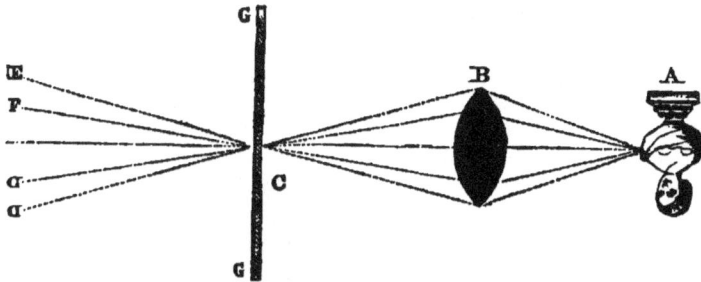

Fig. 311. The stereomonoscope.

be found suspended in the air in the conjugate focus of the lens at C, and from this point the rays of light will diverge as from a real object, which will be seen by separate spectators at D D and E E; and if the screen of ground glass is placed at G G, the image will appear with all the effect of length, breadth, and depth, which belong to solid bodies. (Fig. 311.)

An image formed on ground glass in this manner can be seen only in the direction of the incident rays, and the stereoscopic effect is not apparent when the image is received on a calico or transparent screen, on account of the rays being scattered in all directions.

XVI. *The Stereomoscope.*

This arrangement is an important modification of the other, and consists of a screen of ground glass (A B, Fig. 312), and two convex

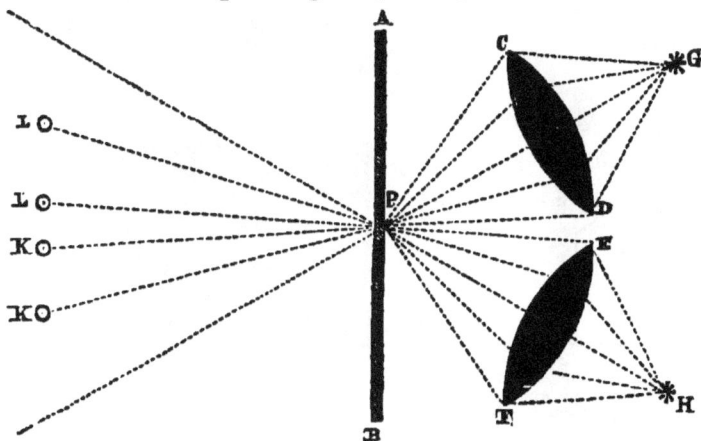

Fig. 312. The stereomoscope.

lenses (C D, and E F) arranged in such a manner that they will project images of the stereoscopic pictures, G H, at the same point on the screen, A B.

It might be thought that a confusion of images would result from projecting two pictures on one point, P—viz., the focus of the two lenses; but as each photograph can be seen only in the direction of its own rays, it follows that if the eyes are so placed that each receives the impression of one stereoscopic picture, the two images must coalesce, and a stereoscopic effect will be the result, as is apparent at K K and L L; so that several persons may look at the stereoscope at one time. (Fig. 312.)

XVII. *The Pseudoscope.*

This curious optical instrument, as its name implies, produces a false image by the refracting power of prisms, and is the invention of Professor Wheatstone. When used with both eyes, the same as the stereoscope, it inverts the relief of a solid body, and makes it appear exactly as if it were an intaglio, or sunk beneath the line surrounding it. For instance, a terrestrial globe when looked at through the pseudo-scope appears to be concave, like Wyld's Globe in Leicester-square, instead of convex. A vase with raised ornaments upon it looks as if it had been turned (to reverse the usual expression) outside in, and

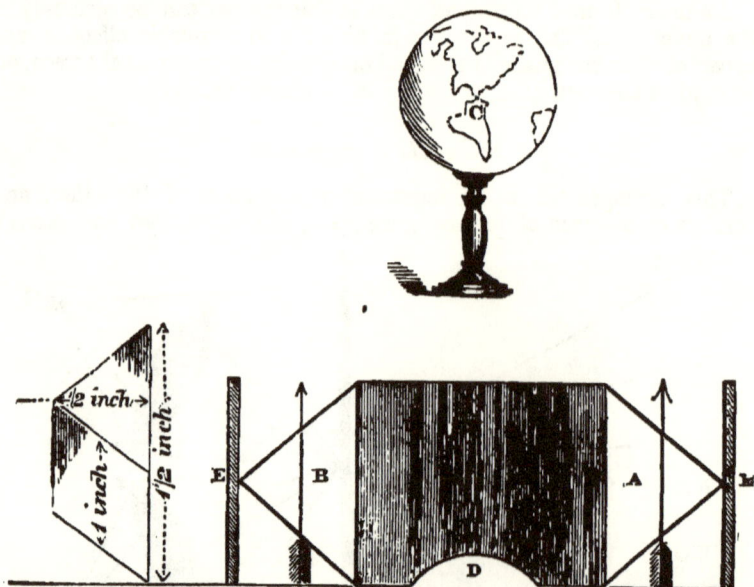

Fig. 313. Horizontal section of the pseudoscope, showing at A B two prisms placed against a block of wood about two inches long and one inch and a half wide, and cut out in the centre to admit the nose at D. The eyes are supposed to be looking at the globe, C, in the direction of the arrows. E E. Brass plates blackened, which shut out the side light, and assist in keeping the prisms in position.

the whole of its convexity is turned to concavity; and of course a face seen under these circumstances looks very curious. (Fig. 313.) The cause is perhaps somewhat difficult to understand; but by taking other and more simple examples of the same effect, the principle may be gradually comprehended.

Sir David Brewster, in his "Letters on Natural Magic," remarks that "one of the most curious phenomena is that *false* perception in vision by which we conceive depressions to be elevations, and elevations depressions—or by which intaglios are converted into cameos, and cameos into intaglios. This curious fact seems to have been observed ·t one of the early meetings of the Royal Society of London, when one of the members, in looking at a guinea through a compound microscope of new construction, was surprised to see the head upon the coin depressed, while other members could only see it embossed, as it really was. . . . . The best method of observing this deception is to view the engraved seal of a watch with the eye-piece of an achromatic telescope, or with a compound microscope, or any combination of lenses which inverts the objects that are viewed through it; a single convex lens will answer the purpose, provided we hold the eye six or eight inches behind the image of the seal formed in its conjugate focus."

After bringing forward various interesting experiments in further explanation of the cause, Sir D. Brewster states it to be his belief that the illusion is the result of an operation of our own minds, whereby we judge of the forms of bodies by the knowledge we have acquired of light and shadow. Hence, the illusion depends on the accuracy and extent of our knowledge on this subject; and while some persons are under its influence, others are entirely insensible to it. This statement is borne out by experience, as the author, whilst Resident Director of the Polytechnic, had four of Wheatstone's pseudoscopes placed in the gallery, with proper objects behind them; and he frequently noticed that some visitors would look through the instrument and see no alteration of the convex objects, whilst others would shout with delight, and call their friends to witness the strange metamorphosis, who in their turn might disappoint the caller by being perfectly insensible to its strange effects.

The pseudo-effects of vision are not confined to the results already explained, but are to be observed especially whilst travelling in a coach, when the eyes may be so fixed as to give the impression of movement to the trees and houses, whilst the coach appears to stand still. In railway carriages, after riding for some time and then coming to a stand still, if another train is set slowly in motion by the one at rest, it frequently happens that the latter appears to be moving instead of the former.

CHAPTER XXIV.

THE ABSORPTION OF LIGHT.

THE analysis of light has been explained in a previous chapter, and it has been shown how the spectrum is produced. Colour, however, may be obtained by other means, and the property enjoyed by certain bodies, of absorbing certain coloured rays in preference to others, offers another mode of decomposing light.

The property of absorption is shown to us in every kind of degree by innumerable natural and artificial substances; and by examining the spectrum through a wedge of blue glass, Sir David Brewster was enabled to separate the seven colours of the spectrum into the three primary colours, red, yellow, and blue, which he proved existed at every point of the spectrum, and by overlapping each other in various proportions, produce the compound colours of orange, green, indigo, and violet.

Connected with this property is the remarkable effect produced by coloured light on ordinary colours, and the sickly hue cast upon the ghost in a melodrama, or the fiery complexion imparted to the hair of Der Freischutz, or the jaundiced appearance presented by every member of a juvenile assembly when illuminated with a yellow light from the salt and burning spirit of "snapdragon," are too well known to require a lengthened description here.

If a number of colours are painted on cardboard, or groups of plants, flowers, flags, and shawls, are illuminated by a mono-chromatic light, and especially the light procured from a large *tow* torch well supplied with salt and spirit, the effect is certainly very remarkable; at the same time it shows how completely substances owe their colour to the light by which they are illuminated, and it also indicates why ladies cannot choose colours by candle-light, unless of course they propose to wear the dress only at night, when it is quite prudent to see the colours in a room lit with gas; and this fact is so well known that with the chief drapers, such as at Messrs. Halling, Pearce, and Stone's, Waterloo House, a darkened room lit with gas is provided during the daytime to enable purchasers of coloured dresses to judge of the effect of artificial light upon them. Whilst the flowers, &c., are lighted up with the yellow light, a magical change is brought about by throwing on suddenly the rays from the oxy-hydrogen light, when the colours are again restored; or if the latter apparatus is not ready, the combustion of phosphorus in a jar of oxygen will answer the same purpose. The light obtained from the combustion of gas affords an excess of the yellow or red rays of light, which causes the difference between candlelight and daylight colours already alluded to.

CHAPTER XXV.

THE INFLECTION OR DIFFRACTION OF LIGHT.

In this part of the subject it is absolutely necessary to return to the theory of undulations with which the present subject was commenced. The inflection of light offers a third method by which rays of light may be decomposed and colour produced. The phenomena are extremely beautiful, although the explanation of them is almost too intricate for a popular work of this kind.

The cases where colour is produced by inflection are more numerous than might at first be supposed; thus, if we look at a gaslight or the setting sun through a wire gauze blind, protecting the eye with a little tank of dilute ink, a most beautiful coloured cross is apparent. An extremely thin film of a transparent matter, such as a little naphtha or varnish dropped on the surface of warm water or soap bubbles, or a very thin film of glass obtained by blowing out a bulb of red-hot glass till it bursts, or an exquisitely thin plate of talc or mica, all present the phenomena of colour, although they are individually transparent, and in ordinary thicknesses quite colourless.

Sir Isaac Newton brought his powerful intellect to bear on these facts, and as a preliminary step invented an instrument for measuring the exact thickness of those transparent substances that afforded colour, and the apparatus displaying Newton's rings is still a favourite optical experiment. It consists of a plano-convex lens, A. (Fig. 314) a slice,

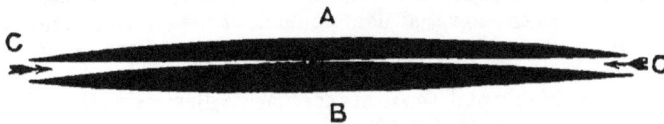

Fig. 314. The two lenses, with the plate or film of air between them, and producing seven coloured rings when the lenses are brought sufficiently close to each other by the screws.

as it were, from a globe of glass twenty-eight feet in diameter, or the radius of whose convex surface is fourteen feet. This plano-convex lens is placed on another double convex lens, B., whose convex surfaces have a radius of fifty feet each, consequently the lenses are very shallow, and the space (c c) included between them being filled with air, can of course be accurately measured. (Fig. 314.) It is usual to mount the lenses in brass rings which are brought together with screws, when the most beautiful coloured rings are apparent, and are produced by the extreme thinness of the film or plate of air enclosed between the two lenses; and the relative thicknesses of the plates of air at which each coloured light is reflected are as follows :—

Red . . . . 133 10 millionths of an inch.
Orange . . . 120 „ „
Yellow . . . . 113½ „ „
Green . . . . 105¼ „ „
Blue . . . . 98 „ „
Indigo . . . . 92½ „ „
Violet . . . . 83¼ „ „

By dividing an inch into ten millions of parts, and by taking 133 of such parts, the thickness of the film of air required to reflect the red ray is obtained, and in like manner the other colours require the minute thicknesses of air recorded in the table above. When the thickness of the film of air is about $\frac{1}{178.000}$dths of an inch, the colours cease to become visible, owing to the union of all the separate colours forming white light, but if the Newton rings are produced in mono-chromatic light, then a greater number of rings are apparent, but of one colour only, and alternating with black rings, i.e., a dark and a yellow succeeding each other; this fact is of great importance as an illustration of the undulatory theory, and demonstrates the important truth, that *two rays of light may interfere with each other in such a manner as to produce darkness.*

Sir David Brewster remarks that, "From his experiments on the colours of thin and of thick plates, Newton inferred that they were produced by a singular property of the particles of light, in virtue of which they possess, at different points of their paths, *fits* or dispositions to be reflected from or transmitted by transparent bodies. Sir Isaac does not pretend to explain the origin of these *fits*, or the cause which produces them, but terms them *fits of transmission* and *fits of reflexion.*"

Sir Isaac Newton objected to the theory of undulations because experiments seemed to show that light could not travel through bent tubes, which it ought to do if propagated by undulations like sound; and it was reserved for the late Dr. Young to prove that light could and would turn a corner, in his highly philosophical experiments illustrating the inflection or bending in of the rays of light.

Dr. Young placed before a hole in a shutter a piece of thick paper perforated with a fine needle, and receiving through it the diverging beams on a paper screen, found that when a slip of cardboard one-thirtieth of an inch in breadth was held in such a beam of light, that the shadow of the card was not merely a dark band, but divided into light and dark parallel bands, and instead of the centre of the shadow being the darkest part, it was actually white. Dr. Young ascertained that if he intercepted the light passing *on one side* of the slip of card with any opaque body, and allowed the light to pass freely on the other side of the slip of cardboard, that all the bands and the white band in the centre disappeared, and hence he concluded that the bands or fringes within the shadow were produced *by the interference of the rays bent into the shadow by one side of the card, with the rays bent into the shadow by the other side.* (Fig. 315).

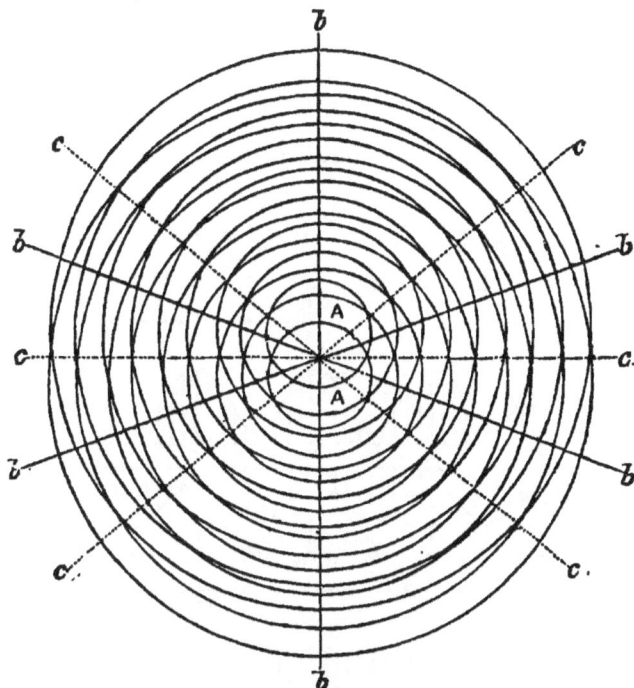

Fig. 315.

In order to show how two waves may interfere so as to exalt or destroy each other, two sets of waves may be propagated on the surface of a still tank or bath of water, from the two points A A (Fig. 315), the black lines or circles representing the tops of the waves. It will be seen that along the lines B B the waves interfere just half way between each other, so that in all these directions there will be a smooth surface, provided each set of waves is produced by precisely the same degree of disturbing force, so as to be perfectly equal and alike in every respect, and the first wave of one set exactly half a wave in advance of the first wave of the other, while at the curve in the direction of all the line C C, the waves coincide, and produce elevations or undulations of double extent; in the intermediate spaces, intermediate effects will, of course, be produced.

Professor Wheatstone has invented some very simple and beautiful acoustic apparatus for the purpose of proving that the same laws of interference exist also in sound, which, as already stated, consists in the vibrations or undulation of the particles of air.

The nature and effects of interference are also admirably illustrated by the following models of Mr. Charles Woodward, President of the Islington Scientific Institution, and to whom we have already alluded.

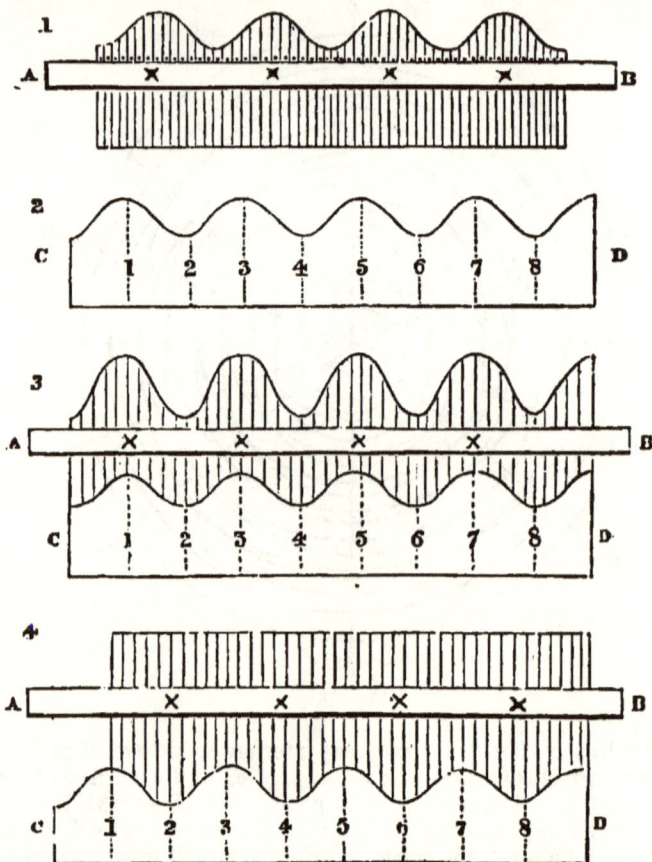

Fig. 316.—No. 1. A model of waves with moveable rods.—No. 2. A model of fixed waves.— No. 3. Intensity of waves doubled by the superposition and coincidence of two equal systems.—No. 4. Waves neutralized by the superposition and interference of two equal systems, the raised part of one wave accurately fitting into and making smooth the hollow of the other, illustrating the fact that two waves of light or sound may destroy each other.

Returning again to the coloured rings, we find that Newton discovered that at whatever thickness of the film of air the coloured ring first appeared, there would be found at twice that thickness the dark ring, at three times the coloured, at four times the dark, and so on, *the coloured rings regularly occurring at the odd numbers, and the dark ones at the even numbers.* This discovery is well illustrated by the models (Fig. 316); and it may be noticed at No. 3 that the highest and the lowest parts of the waves interfere, but coincide and produce a wave of double intensity; the little crosses of the upper model are in a straight line with the numbers 1, 3, 5, 7, and are supposed to represent the coloured rings, whilst in No. 4 the upper series of waves is half an undulation in advance of the lower; and if the eye is again directed from the little crosses downward, the figures 2, 4, 6, 8, even numbers, are apparent, and represent the dark rings, when the waves of light destroy each other. The phenomena of thin plates, such as colours from soap bubbles, and the films of varnish, are well explained by the law of interference. The light reflected from the second surface of the film of air (which must of course, however thin, have two surfaces, viz., a upper and a lower one) interferes with the light reflected from the first, and as they come from different points of space, one set of waves is in advance of the other, No. 4, Fig. 316; they

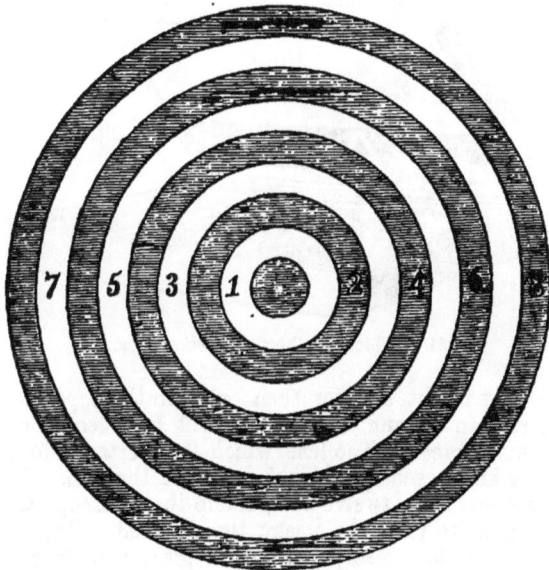

Fig. 317. Appearance of Newton's rings when produced in yellow light, 1, 3, 5, 7, being the yellow rings, and 2, 4, 6, 8, the dark rings. Light by the odd numbers; darkness by the even numbers. The central spot, where the two surfaces are in contact, is dark.

reach the eye with different lengths of paths, and by their *interference* form alternately the luminous and dark fringes, bands, or circles. Bridge's diffraction apparatus, manufactured only by Elliott Brothers, offers itself specially as a most beautiful drawing-room optical instrument. The purpose of this apparatus is to illustrate in great variety, and in the most convenient and compact form, the phenomena of the diffraction or interference of light. This is attained by the assistance of photography. Transparent apertures in an opaque collodion film are produced on glass, and a point of light is viewed through the apertures.

The forms of the apertures are exceedingly various,—triangles, squares, circles, ellipses, parabolas, hyperbolas, and combinations of them, besides many figures of fanciful forms, are included in the set. When an image of the sun is viewed through these apertures, figures of extraordinary beauty, both of form and colour, are produced; and of each of these many variations may be obtained by placing the eye-glass of the telescope at different distances from the object glass. Many of the figures produced, especially when the telescope is out of focus, might suggest very useful hints to those concerned in designing patterns. Although the phenomena are chiefly of interest to the student of science, in consequence of their bearing on theories of light, yet their beauty and variety render them amusing to all. A few words on the mode of using the apparatus may be of service. (Fig. 318.)

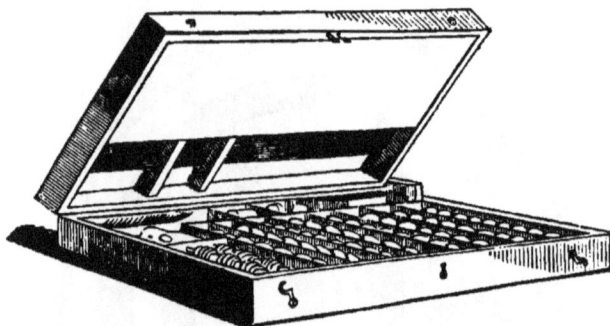

Fig. 318. Elliott Brothers' diffraction apparatus.

Choose a very bright day, for then only can the apparatus be used. Place the mirror in the sun, and let the light be reflected on the back of the blackened screen. The lens which is inserted into this screen will then form an exceedingly bright image of the sun. Then at the distance of not less than twelve feet, clamp the telescope to a table in such a position as to view the image thus formed. Put the eccentric cap on the end of the telescope, clean the glass objects carefully, and attach them to the cap so that they may be turned each in order before the telescope. In this manner, all those which consist of a series of figures may be viewed. Then detach the eccentric cap, and replace it by the other. Into it place any of the single objects. In viewing some of the figures, brightness is advantageous—in others, delicacy; in the former case, let the lens of long focus be inserted in the screen—in the latter case, that of shorter focus. In every case, let the phenomena be observed not only when the telescope is in focus, but also when the eye-glass is pushed in to various distances.

Mr. Warren de la Rue has ingeniously taken advantage of the colours produced by thin films of varnish, and actually *fixed* the lovely iridescent colour produced in that manner on highly polished paper, which is termed "iridescent paper." A tank of warm water at 80° Fahr., about six inches deep, and two feet six inches square, is provided, and a highly glazed sheet of white or black paper being first wetted on a perforated metallic plate, is then sunk with the plate below its surface, care being taken to avoid air bubbles.   A peculiar varnish is then allowed to trickle slowly down a sort of tongue of metal placed in the middle of one of the sides of the tank, and directly the varnish touches the surface of the water it begins to spread out in exquisitely thin films, and by watching the operation close to a window and skimming away all the imperfect films, a perfect one is at last obtained, and at that moment the paper lying on the metal plate is raised from the bottom of the tank, and the delicate film of varnish secured.   When dry, the iridescent colours are apparent, and the paper is employed for many ornamental purposes.

Fig. 319. Reade's iriscope.

An extremely simple and pretty method of producing Newton's rings has been invented by Reade, and is called " Reade's iriscope." A plate of glass of any shape (perhaps circular is the best) is painted on one side with some quickly drying black paint or varnish, and after the other side has been cleaned, it is then rubbed over with a piece of wet soap, and this is rubbed off with a clean soft duster. A tube of about half an inch in diameter, and twelve inches long, is provided, and is held about one inch above the centre of the soaped side of the glass plate, and directly the breath is directed down the tube on the glass, an immense number of minute particles of moisture are deposited on the glass, and these by inflection decompose the light, and all the colours of the rainbow are produced. (Fig. 319.)

The iridescent colours seen upon the surface of *mother-of-pearl*, which Mr. Simonds' excellent commercial dictionary tells us is " the name for the iridescent shell of the pearl oyster, and other molluscs," are referrible to fine parallel lines formed by its texture, and are reproducible, according to Brewster's experiments, by taking impressions of them in soft wax.   The gorgeous colours of certain shells and fish, the feathers of birds, Barton's steel buttons, are not due to any inherent *pigment* or colouring matter that could be extracted from them, but are owing either to the peculiar fibrous, or parallel-lined, or laminated (plate-like) surfaces upon which the light falls, and being reflected in paths of different lengths, interference occurs, and coloured light is produced.

CHAPTER XXVI.

THE POLARIZATION OF LIGHT.

THIS branch of the phenomena of light includes some of the most remarkable and gorgeous chromatic effects; at the same time, regarded philosophically, it is certainly a most difficult subject to place in a purely elementary manner before the youthful minds of juvenile philosophers, and unless the previous chapter on the diffraction of light is carefully examined, the rationale of the illustrations of polarized light will hardly be appreciated. We have first to ask, "What is polarized light?" The answer requires us again to carry our thoughts back to the consideration of the undulatory theory of light, already illustrated and partly explained at pages 262, 330.

After perusing this portion of the subject, it might be considered that waves of light were constituted of one motion only, and that an undulation might be either perpendicular or horizontal, according to circumstances. (Fig. 320.)

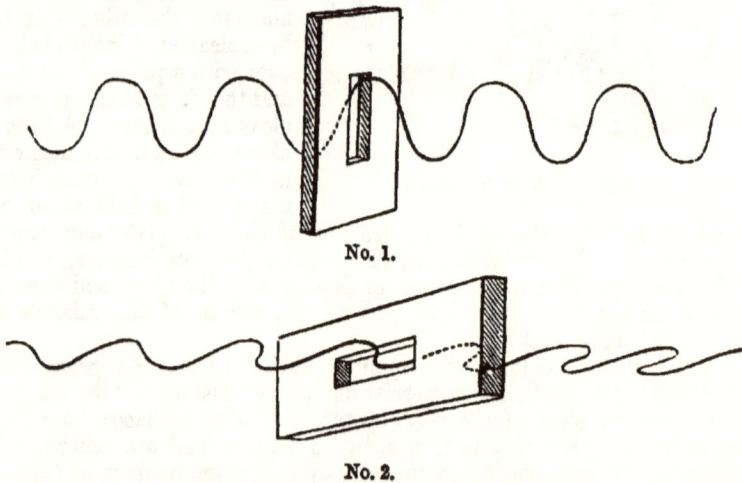

No. 1.

No. 2.

Fig. 320.—No. 1. A wire bent to represent a perpendicular vibration, which if kept in the latter position, will only pass through a perpendicular aperture.—No. 2. A wire bent to represent a horizontal wave which will only pass through a horizontal aperture.

This simple condition of the waves of light could not, however, be reconciled theoretically with the actual facts, and it is necessary in regarding a ray of light, to consider it as a combination of two vibrating motions, one of which, for the sake of simplicity, may be considered as perpendicular, and the other horizontal; and this idea of the nature of an undulation of light originated with the late Dr. Young, who while considering the results of Sir D. Brewster's researches on the laws of double refraction, first proposed the theory of transversal (cross-wise) vibration.   Dr. Young illustrated his theory with a stretched cord, which if agitated or violently shaken perpendicularly, produces a wave that runs along the cord to the other end, and may be often seen illustrated on the banks of a river overhung with high bushes; the bargemen who drive the horses pulling the vessel by a rope, would be continually stopped by the stunted thick bushes, but directly they approach them, they give the horse a lash, and then violently agitate the rope vertically, which is thrown into waves that pass along the rope, and clear the bushes in the most perfect manner. (Fig. 321.)

Fig. 321.   Bargeman throwing his tow-rope into waves to get it over the thick bushes.

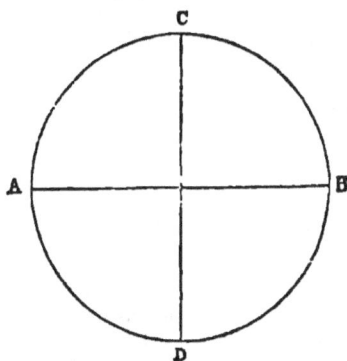

Fig. 322.   A section of a wave of common light made up of the transversal vibration, A B and C D.

Now if a similar movement is made with the stretched rope from right to left, another wave will be produced, which will run along the cord in an horizontal position, and if the latter is compared with the perpendicular undulation, it will be evident that each set of waves will be in planes at right angles to and independent of each other. This is supposed to be the mechanism of a wave of common light, so that if a section is taken of such an undulation, it will be represented by a circle A B C D (Fig. 322), with two diameters A B, and C D; or a better mechanical notion of a wave of common light is acquired from the inspection of another of Mr. Woodward's cardboard models. (Fig. 323.)

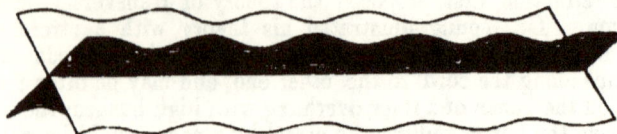

Fig. 323. Model of a wave of common light.

The existence of an *alternating motion of some kind* at minute intervals along a ray is, says Professor Baden Powell, "as real as the motion of translation by which light is propagated through space. *Both* must essentially be *combined* in any correct conception we form of light. That this alternating motion must have reference to certain directions *transverse* to that of the ray is equally established as a consequence of the phenomena; and these *two* principles must form the basis of any explanation which can be attempted." A beam of common light is therefore to be regarded as a rapid succession of systems of waves in which the vibrations take place in different planes.

If the two systems of waves are separated the one from the other, viz., the horizontal from the perpendicular, they each form separately a ray of polarized light, and as Fresnel has remarked, *common light* is merely *polarized light*, having *two planes* of polarization at *right angles* to each other. To follow up the mechanical notion of the nature of polarized light, it is necessary to refer again to Woodward's card wave model (Fig. 323), and by separating the two cards one from the other it may be demonstrated how a wave of common light reduced to its skeleton or primary form is reducible into two waves of polarized light, or how the two cards placed together again in a transversal position form a ray of common light. (Fig. 324.)

No. 1. No. 2. No. 3.

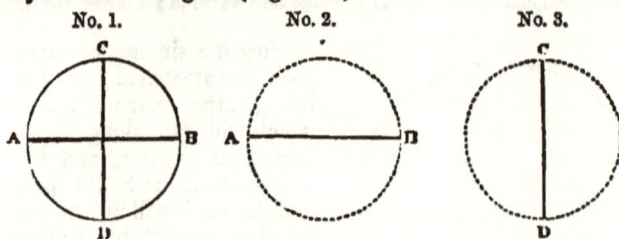

Fig. 324.—No. 1. Common light, made up of the two waves of polarized light, Nos. 2 and 3.

The query with respect to the nature of polarized light being answered, it is necessary, in the next place, to consider how the separation of these transversal vibrations may be effected, and in fact to ask what optical arrangements are necessary to procure a beam of polarized light? Light may be polarized in four different ways—viz., by reflection, single refraction, double refraction, and by the tourmaline—viz., by absorption.

z

Polarization by Reflection, and by Single Refraction.

In the year 1810, the celebrated French philosopher, Mons. Malus, while looking through a prism of Iceland spar, at the light of the setting sun, reflected from the windows of the Luxemburg palace in Paris, discovered that a beam of light reflected from a plate of glass at an angle of 56 degrees, presented precisely the same properties as one of the rays formed by a rhomb of Iceland spar, and that it was in fact polarized. *One* of the transversal waves of polarized light of the common light, being reflected or thrown off from the surface of the glass, whilst the other and second transversal vibration passed *through* the plate of glass, and was likewise polarized in another plane, but by *single refraction*, so that the experiment illustrates two of the modes of polarizing light—viz., by reflection, and by single refraction. This important elementary truth is beautifully illustrated by Mr. J. T. Goddard's new form of the oxy-hydrogen polariscope, by which a beam of common light traverses a long square tin box without change; but directly a bundle of plates of glass composed of ten plates of thin flattened crown glass, or sixteen plates of thin parallel glass plates used for microscopes, are slid into the box at an angle of 56° 45', then the beam of common

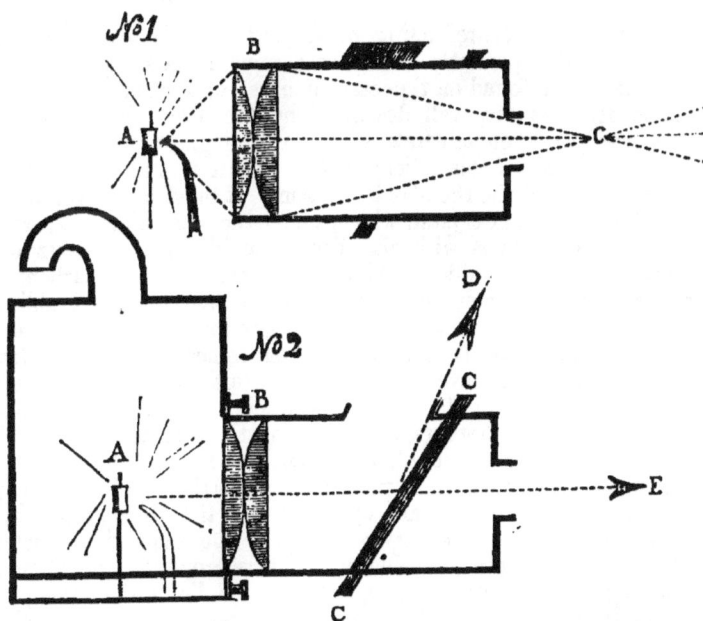

Fig. 325.—No. 1. A is the lime light. B. The condenser lenses. C. The beam of *common* light. Here the glass plates are removed.—No. 2. A. Lime light. B. The condenser lenses. C C. The bundle of plates of glass at an angle of 56° 45'. D is the ray of light polarized by reflection from the glass plates, C C, and E is the beam of polarized light by single refraction, having passed through the bundle of plates of glass, C C.

light is split into two
beams of polarized light,
which pursue their re-
spective paths, one pass-
ing by single refraction
through the glass, and the
other being reflected, and
rendered apparent by
opening an aperture over
the glass plates, and then
again by using a little
smoke from brown paper,
the course of the rays
becomes more apparent.
The same truth is well
illustrated by the card-
board model wave and a wooden plane with horizontal and perpendicular
slits, placed at an angle of 56° 45', as at Fig. 326.

Fig. 326. A A. Model in wood of a bundle of plates of glass at an angle of 56° 45'. B. Beam of common light, with transversal vibration. C. Light polarized by reflection. D. Light polarized by refraction.

POLARIZATION BY DOUBLE REFRACTION.

The name of *Double*-refracting or Iceland Spar is given to a very
clear, limpid, and perfectly transparent mineral, composed of carbonate of lime, and found on the eastern coast of Iceland. Its crystallographic features are well described by the Rev. Walter Mitchell
in his learned work on mineralogy and crystallography, and it is sufficient for the object of this article to state that it crystallizes in rhombs,
and modifications of the rhomboidal system. It must not be confounded
with rock or mountain crystal, which, under the name of quartz, crystallizes in six-sided prisms with six-sided pyramidal tops; quartz being
composed of silica, or silicic acid and calcareous spar of carbonate of
lime. Very large specimens of the latter mineral are rare and valuable, and
the *lion* of specimens of calcareous, or double-refracting spar, is now in the
possession of Professor Tennant, the eminent mineralogist of the Strand.
It is nine inches high, seven and three-quarters inches broad, and five
and a half inches thick; its estimated value being 100*l*. This beautiful
specimen has been photographed, and its stereograph illustrates in a
very striking manner the double refracting properties of the spar.
If a printed slip of paper is placed behind a rhomb of Iceland spar,
two images of the former are apparent, and the stereograph already
alluded to shows this fact very perfectly, at the same time illustrates
the value of the stereoscope. Out of the stereoscope the words " Stereoscopic Magazine " appear doubled, but seem to lie in the same plane;
but directly the picture is placed in the instrument, then it is clearly
seen that one image is evidently in a very different plane from the other.
The double-refracting power of this mineral is illustrated by holding a
small rhomb of Iceland spar, placed in a proper brass tube before
the orifice as at Fig. 327, from which the rays of common light are

z 2

passing; if an opaque screen of brass perforated with a small hole is introduced behind the rhomb, then, instead of one circle of light being apparent on the screen, two are produced, and both the rays issuing in this manner are polarized, one being termed the ordinary and the other the extraordinary ray. (Fig. 327.)

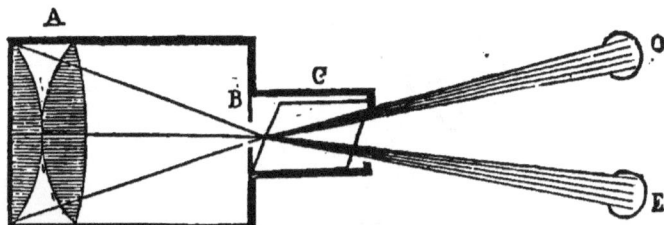

Fig. 327. A. The condensers. B. The hole in the brass screen or stop. C. The rhomb of Iceland spar. O. The ordinary, and E the extraordinary, ray, both of which are polarized light.

The polarizing property of the rhomb is perhaps better shown by the next diagram, where A B represents the obtuse angles of the Iceland spar, and a line drawn from A to B, would be the axis of the crystal. The incidental ray of common light is shown at C, and the oppositely polarized transmitted rays called the ordinary ray O, and extraordinary ray E, emerge from the opposite face of the rhomboid.   If a black line is ruled on a sheet of paper as at K K, and examined by the eye at C, it appears double as at K K and J J. (Fig. 328.)

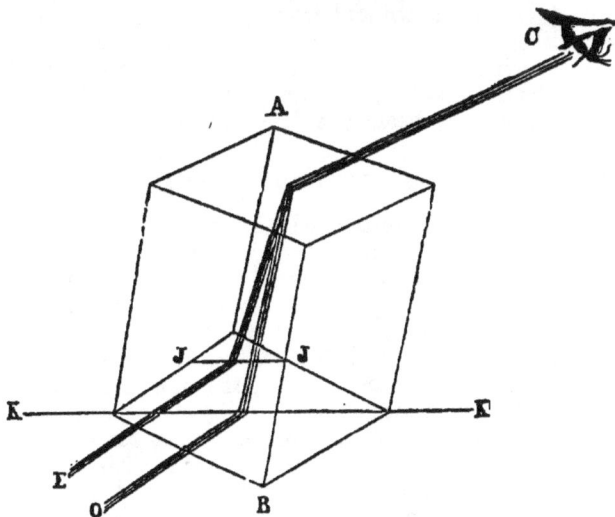

Fig. 328. Rhomb of Iceland spar.

The cardboard model is again useful in demonstrating the polarization of light by double refraction, and if a model of a rhomb of Iceland spar is made of glass plates, one face of which has an aperture like a cross, and the other a horizontal and perpendicular slit, as at Nos. 1 and 2 (Fig. 329), the production of the ordinary and extraordinary rays is demonstrated in a familiar manner, and is easily comprehended.

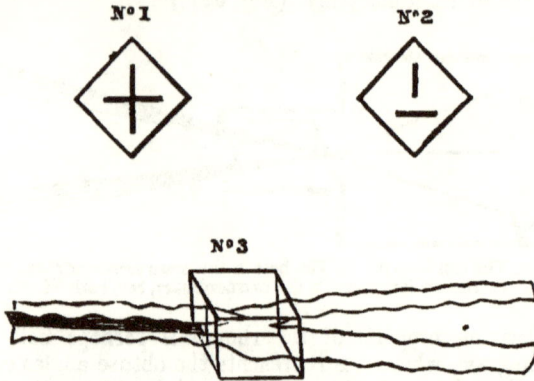

Fig. 329.—No. 1. One face of the model rhomb to admit the transversal vibration, represented by the cardboard model.—No. 2. The opposite face of the rhomb, from which issue the polarized, ordinary, and extraordinary rays.—No. 3. Side view of the model.

In Newton's "Optics" we find the following description of Iceland spar:—"This crystal is a pellucid fissile stone, clear as water or crystal of the rock (quartz), and without colour. . . . . Being rubbed on cloth it attracts pieces of straw and other light things like amber or glass, and with aquafortis it makes an ebullition. . . . . If a piece of this crystalline stone be laid upon a book, every letter of the book seen through it will appear double by means of a double refraction."

POLARIZATION BY THE TOURMALINE.

This mineral was first discovered during the sixteenth century, in the island of Ceylon, afterwards in Brazil, and since that period at various localities in the four quarters of the globe. In the Grevillian collection purchased many years ago by government for the British Museum, there is a fine specimen of red tourmaline valued at 500*l*. The green tourmaline is named Brazilian emerald, and the Berlin blue tourmaline is called Brazilian sapphire; the mineral chiefly consists of sand (silica) and alumina, with a small quantity of lime, or potash, or soda, boracic acid, and sometimes oxide of iron or manganese. When light is passed through a slice of this mineral it is immediately polarized, one of the transversal vibrations being absorbed, stopped, or otherwise disposed of, the other only emerging from the tourmaline, consequently it is one of the most convenient polarizers, although the polarized light partakes of the accidental colour of the mineral. Green, blue, and yellow tourmalines are bad polarizers, but the brown and pink varieties are very good, and it is a most curious fact that white tourmaline does not polarize. (Fig. 330.)

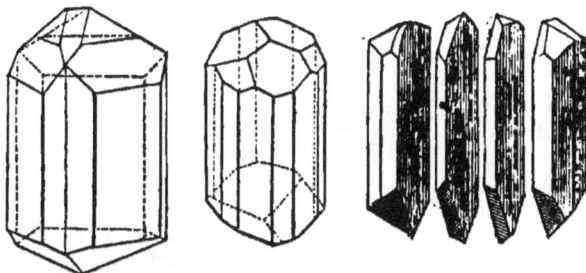

Fig. 330. Crystal of tourmaline slit (parallel to the axis) into four plates, which when ground and polished, may be used for the polarization of light.

The mineral crystallizes in long prisms, whose primitive form is the obtuse rhomboid, having the axis parallel to the axis of the prism. The term axis with reference to the earth, as shown at page 16, is an imaginary *single line* around which the mass rotates, but in a crystal it means a *single direction*, because a crystal is made up of a number of similar crystals, each of which must have its axis, thus the whitest Carrara marble reduced to fine powder, moistened with water and placed under a microscope, is found to consist chiefly of minute rhomboids, similar to calcareous spar. The smallest crystal of this mineral is divisible again and without limit into other rhombs, each of which possesses an axis. (Fig. 331.)

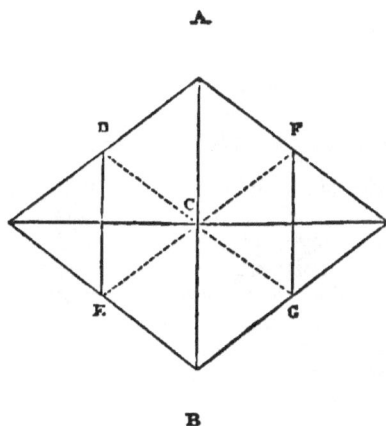

Fig. 331 represents a crystal, the axis of which is the direction A B. The dotted lines show the division of the large crystal into four other and smaller ones, each of which has its axis, A C, C B, D E, F G; and every line within the large crystal parallel to A B is an axis, consequently the term is employed usually in the plural number *axes*.

If a plate of tourmaline is held before the eye whilst looking at the sun (like the gay youth in Hogarth's picture who is being arrested whilst absorbed with the wonders of a tourmaline, which was, in the great painter's time, a popular curiosity,) it may be turned round in all directions without the slightest difference in the appearance of the light, which will be coloured by the accidental tint of the crystal, but if a second slice of tourmaline is placed behind the other, there will be found certain directions in which the light passes through both the slices, whilst in other positions the light is completely cut off.

When the axes of both plates coincide, the light polarized by one tourmaline will pass through the other, but if the axes do not coincide, and are at right angles to each other, then the polarized light is entirely stopped, and the *rationale* of this will be appreciated at once if a tourmaline is regarded (mechanically) as if it were like a grating with perpendicular bars through which the polarized light will pass. Any number of such gratings with the bars parallel would not stop the polarized

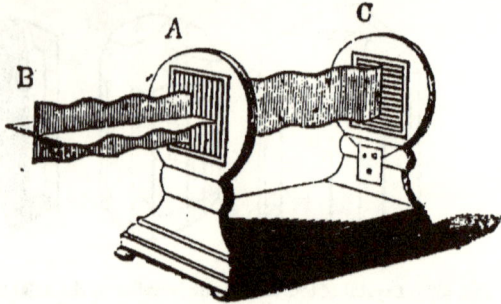

Fig. 332. A. Model of the first slice of tourmaline into which the transversal vibrations, B, are passing; the horizontal wave is absorbed, and the perpendicular polarized one proceeds to the second slice of tourmaline, c, where the bars (the axes) being at right angles to those of A, it is stopped, and cannot pass through until the bars of c are parallel with A.

light, but if the second grating is turned round ninety degrees, the bars will be at right angles to those of the first grating, and the perpendicular wave of polarized light cannot pass. (Fig. 332.)

Splendid Chromatic effects produced by Polarized Light.

Having discussed the various modes of obtaining polarized light, the next step is to arrange an apparatus by which certain double refracting crystals, and other bodies, shall divide a ray of polarized light, and then by subsequent treatment with another polarizing surface, the divided rays are caused to *interfere* with each other, and afford the phenomena of colour. Bodies that refract light singly, such as gases, vapours or liquids, annealed glass, jelly, gums, resins, crystallized bodies of the tessular system, such as the cube and octohedron, do not afford any of the results which will be explained presently, except by the influence of pressure, as in unannealed glass, or a bent cold glass bar. By compression or dilatation, they are changed to double refractors of light. The bodies that possess the property of double refraction (though not to the visible extent of Iceland spar), are all other bodies such as crystallized chemicals, salts, crystallized minerals, animal and vegetable substances possessing a uniform structure, such as horn and quill; all these substances divide the ray of polarized light into two parts, and by placing a thin film of a crystal of selenite (which is one of the best minerals that can be used for the purpose) in the path of the beam of polarized light, coming either from the glass plates, as in No. 2, (Fig. 325), page 338, or from a slice of tourmaline, and then receiving it through the ordinary focusing lenses or object-glasses of the oxy-hydrogen microscope, no colour is yet apparent in the image of the selenite on the screen, until another tourmaline, or a bundle of glass plates, is placed at an angle of 56° 45', and at right angles to the plane of reflection of the first set of plates; then the most gorgeous colours suddenly appear over all parts of the film of selenite as depicted on the screen, like other objects shown by the oxy-hydrogen microscope. (Fig. 333.)

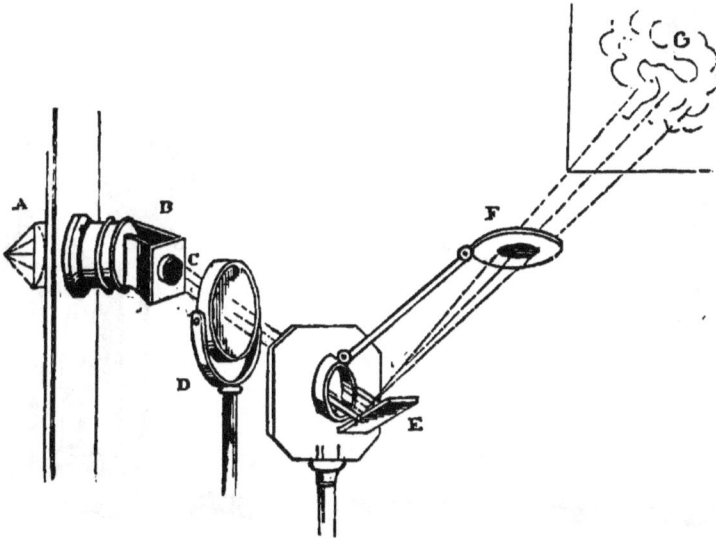

Fig. 333.   Duboscq's polarizing apparatus.   A. The light and the condenser lens. B. The plates of glass at the proper angle.   c. The selenite object.   D. The focusing lens.   E. The second bundle of plates of glass called the analyser.   F. A stop for extraneous rays of light.   G. The image of the film of selenite most beautifully coloured

Goddard's oxy-hydrogen polariscope is one of the most convenient, because either the reflected or refracted polarized rays can be rendered available; it consists of the apparatus shown at Fig. 325, and to this is added a low microscope power, and stage to hold the selenite or other objects, with another bundle of sixteen plates of the thin microscopic glass or mica, called the analyser. A slice of tourmaline, or a Nicol's prism may be employed, instead of the second bundle of reflecting plates. When the ray of polarized light reflected from the first set of glass plates enters the doubly refracting film of selenite, which is about the fortieth or fiftieth part of an inch in thickness, it is split into the ordinary and extraordinary rays, and is said to be *dipolarized*, and forms two planes of polarized light, vibrating at right angles to each other. When the latter are received on another bundle of plates of glass called the analyser, at an angle of 56° 45', but at right angles to the first set of glass plates, they interfere, because in the passage of the two rays from the selenite they have traversed it in different directions, with different velocities; one of these sets of waves will therefore, on emerging from the opposite face of the selenite be retarded, and lie behind the other; but being polarized in different planes, they cannot *interfere* until their planes of polarization are made to coincide, which is

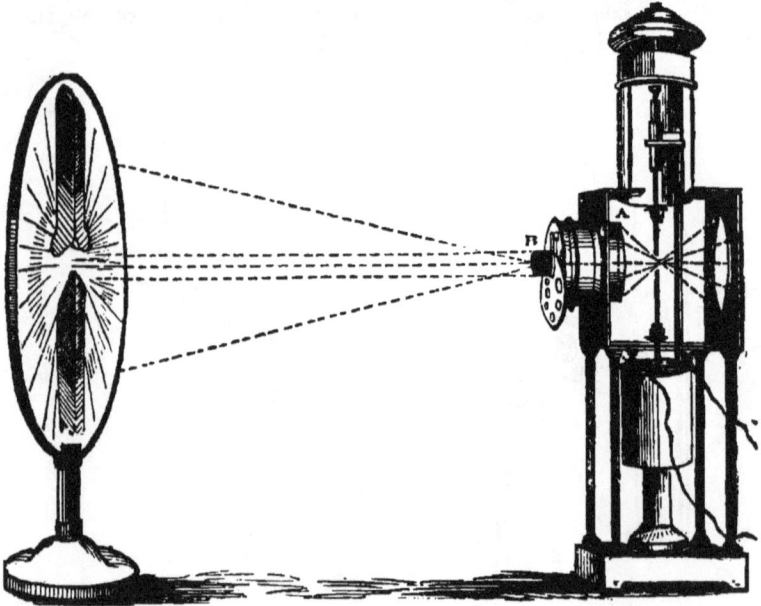

Fig. 334. The electric lamp and lantern of Duboscq, showing the projection of the carbon poles on the disc. This experiment is performed with the help of the plano-convex lens, ᴀ, and the rays pass through a very narrow aperture at ʙ.

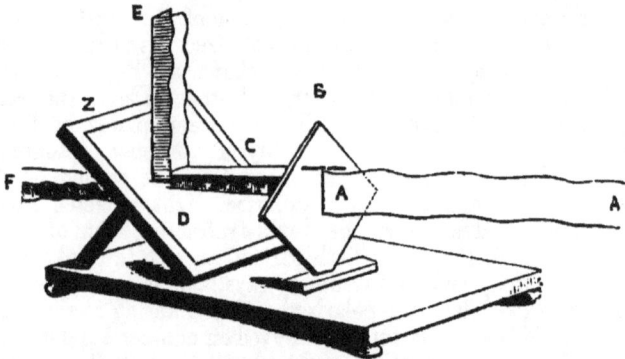

Fig. 335. ᴀ ᴀ. Card model of a beam of polarized light coming from the first bundle of plates of glass, shown at Fig. 326, p. 339. ʙ. Model of the film of selenite, which divides or dipolarizes the ray ᴀ ᴀ into ᴄ and ᴅ, which, interfering by means of the second bundle of plates of glass called the analyser ᴢ, produce reflected chromatic effects by interference at ᴇ, and refracted effects at ʏ.

effected by means of the second bundle of glass plates called the analyser; and when this is brought into a position at right angles to the first set of reflecting glass plates, half the ordinary wave interferes with half the extraordinary wave; and being transmitted through the analyser, produces, say red and orange, whilst the remaining halves also interfere, and being reflected, afford the complementary colours green and blue. (Fig. 334.) The term *complementary* is intended to define any two colours containing red, yellow, and blue, because the three combined together produce white light; for example, the complementary colour to red would be green, because the latter contains yellow and blue; the complementary colour to orange would be blue, because the former contains red and yellow. Any two colours, therefore, which together contain red, yellow, and blue are said to be *complementary;* and if this principle was better understood, ladies would never commit such egregious blunders as they occasionally do in the choice of colours for bonnets and dresses, and select a blue bonnet to be worn with a green dress, or *vice versâ.* By rotating the analyser, the reflected and refracted rays change colours, and if the former is red and the latter green, by moving the analyser round 90°, the reflected rays change to green and the refracted to red; at 180° the colours again change places; at 270° the reflected ray will be again green, and the refracted red; to be once more brought back at 360° to the original position, viz., reflected rays red, refracted green. The thickness of the films of selenite determines the particular colour produced.

If the selenite is of a uniform thickness, one colour only is obtained, and by ingeniously connecting pieces of various thicknesses (in the same forms as stained glass for cathedral windows), the most beautiful designs were made by the late Mr. J. T. Cooper, jun., which have since been manufactured in great quantity and variety by Mr. Darker, of Paradise-street, Lambeth. The colours of these selenite objects are seen by placing them in front of a piece of black glass, fixed at the polarizing angle, and then examining the design with a slice of tourmaline, or still better with a single-image Nicol prism, when the most brilliant colours are obtained, and varied at every change of the angle of the analyser.

Selenite, or sparry-gypsum, is the native crystallized sulphate of lime, which contains water of crystallization ($CaO, SO_3, 2HO$). It frequently occurs imbedded in London clay, and is called *quarry glass* by the labourers who find it at Shotover Hill, near Oxford, and also in the Isle of Sheppey.

At a very early period, before the discovery of glass, selenite was used for windows; and we are told that in the time of Seneca, it was imported into Rome from Spain, Cyprus, Cappadocia, and even from Africa. It continued to be used for this purpose until the middle ages, for Albinus informs us, that in his time, the windows of the dome of Merseburg were of this mineral. The first greenhouses, those invented by Tiberius, were covered with selenite. According to Pliny, beehives were encased in selenite, in order that the bees might be seen at work.

The late Dr. Pereira has placed the phenomena already described in the form of a most instructive diagram, which we borrow from his elaborate work on "Polarized Light." (Fig. 336.)

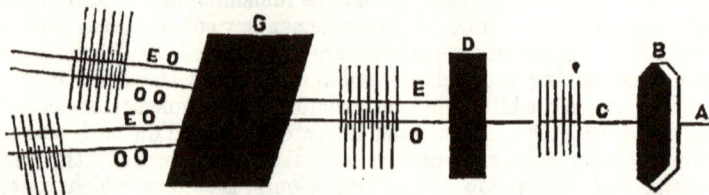

Fig. 336. A. A ray of common or unpolarized light, incident on B. B. The polarizer (a plate of tourmaline). C. A ray of plane polarized light, incident on D. D. The doubly-refracting film of selenite. E. The extraordinary ray. O. The ordinary ray, produced by the double refraction of the ray C. G. The analyser (or doubly-refracting or Nicol's prism). E O. The ordinary ray. E E. The extraordinary ray, produced by the double refraction of the extraordinary ray, E. O O. The ordinary ray. O E. The extraordinary ray, produced by the double refraction of the ordinary ray, O.

The chromatic effects described are not confined to selenite objects only, but are obtained from glass, provided the particles are in a state of unequal tension, as in masses of unannealed glass of various forms. (Fig. 337.) Consequently, polarized light becomes a most valuable

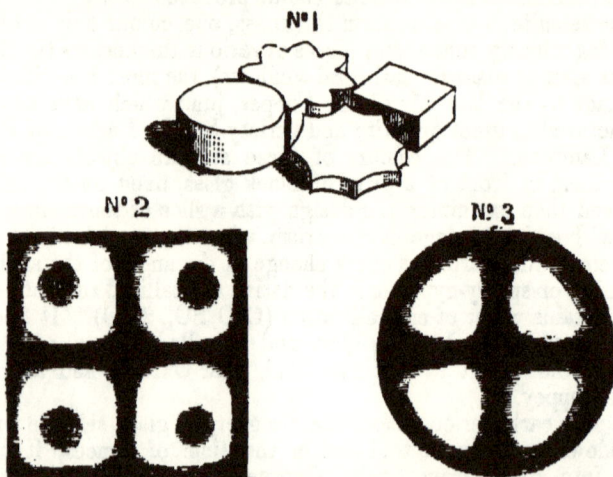

Fig. 337. No. 1. Unannealed glass for the polariscope. Nos. 2 and 3. Appearance of the black cross and coloured circles in a square and circular piece of unannealed glass in the polariscope.

means for ascertaining the condition of particles otherwise invisible and inappreciable. One of the most beautiful experiments can be made with a bar of plate-glass, which refracts light singly until pressure is applied to the centre, in order to bend it into an arch or curve, when the appearance presented in Fig. 338 is apparent.

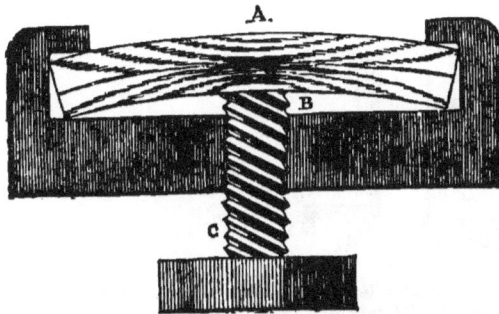

Fig. 338. A B. Bar of glass under the pressure of the screw C, and appearance of bands or fringes of coloured light, which entirely disappear on the removal of the screw. An effect, of course, only visible by polarized light.

A quill placed in the polarizing apparatus is also discovered to be in a state of unequal tension by the appearance of coloured fringes within it, which change colour at every movement of the analyser.

Another series of beautiful appearances present themselves when a ray of white polarized light is made to pass perpendicularly through a slice of any crystallized substance with a single axis; if the analyser consist of a slice of tourmaline, a number of concentric coloured rings are rendered visible with a black cross in the centre, which is replaced with a white one on moving the tourmaline through each quadrant of the circle.

Crystals of Iceland spar present this phenomenon in great beauty; and if the crystal (such as nitre) has two axes of double-refraction, a double-system of coloured rings is apparent, with the most curious changes and combina-

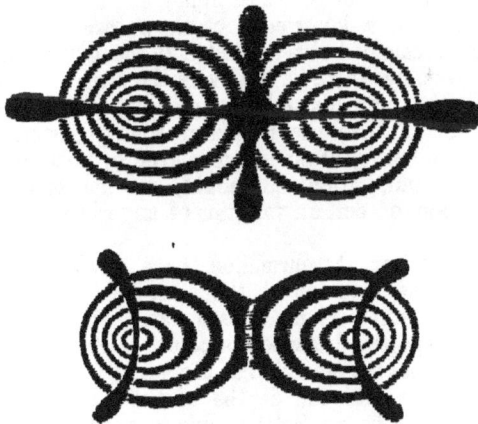

Fig. 339. Crystal of nitre with two axes, as seen in polarized light.

tions of the black and white crosses with them. (Fig. 339.)

Mr. Goddard has recommended the optical arrangement (Fig. 340) for showing the rings with great perfection, as also the number of rings that increase in some crystals (the topaz, for example), with the divergence of the rays of polarized light passing through them.

Mr. Woodward's table and oxy-hydrogen polariscope and microscope, made by Smith and Beck, of Coleman-street, is well adapted, from its simplicity and perfection, to exhibit all the varied and beautiful effects of polarized light; and we only regret that want of space prevents us

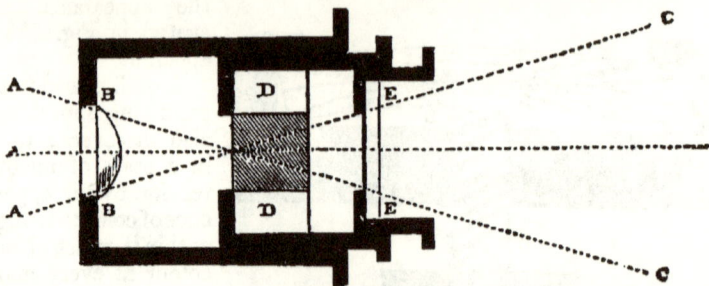

Fig. 340. A A A. Polarized light. B B. A lens of short focus, transmitting a cone of light with an angle of divergence for its rays, c o, of 45°. D D. The crystal of topaz, Iceland spar, or nitre. E E. The slice of blue tourmaline for analysing.

describing it in detail, although the reader may see the body of the apparatus at page 123, where the modifications of the oxy-hydrogen light are described and figured; and the polarizing apparatus would be placed, of course, in front of the light issuing from the lantern.

Finally, the question of utility (the *cui bono*) may be considered in answer to the query, What is the use of polarized light?

The value to scientific men of a knowledge of the nature of this modification of common light cannot be overrated. It has given the philosopher a new kind of test, by which he discovers the structure of things that would otherwise be perfectly unknown; it has given the astronomer increased data for the exercise of his reasoning powers; whilst to the microscopist the beauty of objects displayed by polarized light has long been a theme of admiration and delight, and has served as a guide for the identification of certain varieties of any given substance, such as starch.

A tube provided with a polarizer of tourmaline, or a single-image Nicol prism, is invaluable to the look-out at the mast-head in cases where vessels are navigating either inland or sea water, where the presence of hidden rocks is suspected, because the polarizer rejects all the glare of light arising from unequal reflection at the surface of water, and enables the observer to gaze into the depths of the sea and to examine the rocks, which can only be perfectly visible by the refracted light coming from their surfaces through the water.

Professor Wheatstone has invented an ingenious polarizing clock for showing the hour of the day by the polarizing power of the atmosphere. Birt, Powell, and Leeson have each invented instruments for examining the circular polarization of fluids, by which a more intimate knowledge of the relative values of saccharine solutions may be obtained, besides unfolding other truths important to investigators in this branch of science.

And last, but not least, it was with the assistance of polarized light that Dr. Faraday established the relation that exists between light and magnetism, and through the latter, with the force of electricity; and the next figure indicates the necessary apparatus required to repeat this highly important physical truth—viz., the deviation of the plane of polarization of light by the influence of the magnetic force from a powerful electro-magnet. (Fig. 341.)

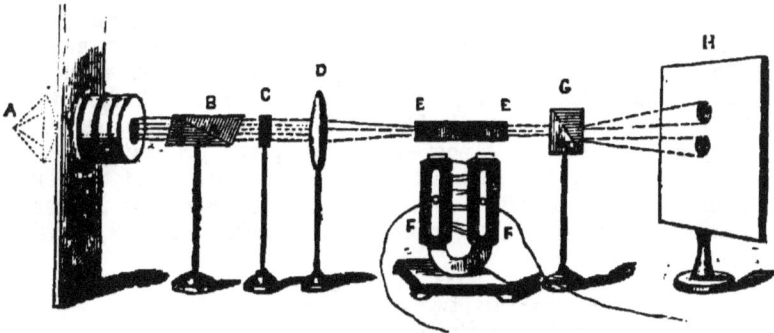

Fig. 341. A. The light and condenser lens. B. Single-image Nicol prism. C. Rock crystal of two rotations. D. A double-convex lens. E E. Faraday's heavy glass. F F. The powerful electro-magnet connected with battery. G. Double-refracting prisms. H. Image, or screen where the deviation of the plane of polarization by the magnetic force is shown.

By another and equally beautiful experiment at the London Institution, Professor Grove demonstrated the production of all the other kinds of force from light, using the following arrangement for the purpose: A prepared daguerréotype plate is enclosed in a box full of water having a glass front with a shutter over it; between this glass and the plate is a gridiron of silver wire; the plate is connected with one extremity of a galvanometer coil, and the gridiron of wire with one extremity of a Breguet's helix; the other extremities of the galvanometer and helix are connected by a wire, and the needles brought to zero. As soon as a beam of either daylight or the oxy-hydrogen light is, by raising the shutter, permitted to impinge upon the plate, the needles are deflected. Thus, light being the initiatory force, we get

Chemical action on the plate,
Electricity circulating through the wires,
Magnetism in the coil,
Heat in the helix,
Motion in the needle.

Such, then, are some of the glorious phenomena that we have endeavoured to explain in this and the preceding chapters on light. Here we have noticed specially how completely we owe their appreciation to the sense of sight operating through the eye, the organ of vision. Well may those who have lost this divine gift speak of their darkness as of a lost world of beauty to be irradiated only by better and more enduring light; and most feelingly does Sir J. Coleridge speak on this point when he says:—

"Conceive to yourselves, for a moment, what is the ordinary entertainment and conversation that passes around any one of your family tables; how many things we talk of as matters of course, as to the understanding and as to the bare conception of which sight is absolutely necessary. Consider, again, what an affliction the loss of sight must be, and that when we talk of the golden sun, the bright stars, the beautiful flowers, the blush of spring, the glow of summer, and the ripening fruit of autumn, we are talking of things of which we do not convey to the minds of these poor creatures who are born blind, anything like an adequate conception. There was once a great man, as we all know, in this country, a poet — and nearly the greatest poet that England has ever had to boast of—who was blind; and there is a passage in his works which is so true and touching that it exactly describes that which I have endeavoured, in feeble language, to paint. Milton says:—

> 'Thus with the year
> Seasons return; but not to me returns
> Day, or the sweet approach of even, or morn,
> Or sight of vernal bloom, or summer's rose,
> Or flocks, or herds, or human face divine;
> But cloud instead, and ever-during dark
> Surrounds me; from the cheerful ways of men
> Cut off, and for the book of knowledge fair
> Presented with a universal blank
> Of Nature's works, to me expunged and rased,
> And wisdom at one entrance quite shut out.
> So much the rather, thou, celestial light,
> Shine inward, and the mind through all her powers
> Irradiate; there plant eyes; all mist from thence
> Purge and disperse, that I may see and tell
> Of things invisible to mortal sight.'

The great poet, when intent upon his work, sought for celestial light to accomplish it. And this brings me to that part of the labours of our Blind Institutions upon which I dwell the most and which, after all, is the greatest compensation we can afford to the inmates for the affliction they suffer; and that is, the means we provide for them to read the blessed Word of God, which they can read by day as well as by night, for light in their case is not an essential."

Fig. 342.   James Watt.

CHAPTER XXVII.

HEAT.

THROUGHOUT the greater number of the preceding chapters it will be evident that the active properties of matter may be summed up under one general head, and may be considered as varieties of attraction—such as the attraction of gravitation, cohesive attraction, adhesive attraction, attraction of composition (or chemical attraction), electrical attraction, magnetical attraction.

The absolute or autocratic system does not, however, prevail in the works of nature; and she seems ever anxious, whilst imparting great and peculiar powers to certain agents, to create other forces which may control and balance them. Thus, for instance, the great force of cohesive attraction is an ever-present power discernible, as has been shown, in solids and liquids; but if this agent were allowed to run riot in its full strength and intensity, it would tyrannically hold in subjection all liquid matter, and every drop of water which is at present kept in the liquid state, would succumb to its iron rule, and retain the solid state of ice. Hence, therefore, the wise creation of an antagonistic force—viz., heat; which is not provided in any niggardly manner, but is liberally bestowed upon the globe from that all-sufficient and enormous source, the sun. And it is by the softening and liquifying influence of his rays that the greater proportion of the water on the surface of the globe is maintained in the fluid condition, and is enabled to resist the power of cohesion, that would otherwise turn it all, as it were, to stone.

Cohesion, electricity, and magnetism fully embody the notion of powers of attraction, or *a drawing together;* whilst heat stands almost alone in nature as the type of repulsion, or *a driving back.*

Mechanically, repulsion is demonstrated by the rebound of a ball from the ground; the parts which touch the earth are for the moment compressed, and it is the subsequent repulsion between the particles in those parts which causes them to expand again and throw off the ball.

The development of heat is produced from various causes, which may be regarded as at least four in number. Thus, it was shown by Sir Humphrey Davy, that even when two lumps of ice are rubbed together, sufficient heat is obtained to melt the two surfaces which are in contact with each other. Friction is therefore an important source of heat, and one of the most interesting machines at the Paris Exposition consisted of an apparatus by which many gallons of water were kept in the boiling state by means of the heat obtained from the friction of two copper discs against each other. The machine attracted a good deal of attention on its own merits, and especially because it supplied boiling water for the preparation of chocolate, which the public was duly informed was boiled by the heat *rubbed out* of the otherwise cold discs of copper. When cannon made on the old system are bored with a drill, it is necessary that the latter should be kept quite cool with a constant supply of water, or else the hard steel might become red-hot, and would then lose its *temper*, and be no longer capable of performing its duty.

Count Rumford endeavoured to ascertain how much heat was actually generated by friction. When a blunt steel bore, three inches and a half in diameter, was driven against the bottom of a brass cannon seven inches and a half in diameter, with a pressure which was equal to the weight of ten thousand pounds, and made to revolve thirty-two times in a minute, in forty-one minutes 837 grains of dust were produced, and the heat generated was sufficient to raise 113 pounds of the metal 70° Fahrenheit—a quantity of heat which is capable of melting six pounds and a half of ice, or of raising five pounds of water from the freezing to the boiling point. When the experiment was repeated under water, two gallons and a half of water, at 60° Fah., were made to boil in two hours and a half.

Chemical affinity has been so often alluded to in these pages, that it

A A

may be sufficient to mention only one good instance of its almost magical power in evoking heat. When a bit of the metal sodium is placed on the tip of a knife, and thrust into some warm quicksilver, or if a pellet of sodium and a few globules of mercury are placed on a hot plate just taken from the oven, and then gently squeezed together, a vivid production of heat and light is apparent; and when the mixture of the two metals is cold, it will be found that the quicksilver has lost its fluidity, and a solid amalgam of sodium and mercury is obtained, which gradually, by exposure to the air, returns to the liquid state, the mercury being set free, whilst the sodium is oxidized, and forms soda. Just as an ordinary alloy of copper and gold used by jewellers would lose its colour and brilliancy by the oxidation of the copper; and when the rusty, dirty film is removed by rubbing and polishing, the surface is again brilliant, and remains so until another film of the exposed copper is attacked: in like manner the sodium is attacked and changed by the oxygen of the air, whilst the mercury being unaffected retains its brilliancy, and at the same time regains its fluidity. The evolution of heat in the above case indicates that a chemical union has taken place between the two metals.

Examples of the production of heat by electricity and magnetism have been abundantly shown in the chapters on these subjects; and one of the best illustrations of this fact has been shown on the occasion of the opening of the telegraphic communication between France and England by means of the submarine cable, when cannon were fired alternately at both ends of the conducting cable by means of electricity, and the event thus inaugurated in both countries.

That heat is a product of living animal organization is shown, as it were, visibly by the marvellous phenomena that proceed in our own bodies. People do not very often trouble themselves to ask where the heat comes from, or even to think that this invisible power must be maintained in the body, and that slow combustion, or, as Liebig terms it, *eremacausis*, must continually go on inside our frail mortal tenements; and more than this, that we cannot afford to waste our heat. If the body is deprived of heat faster than it can be generated, death must inevitably occur; and a very melancholy instance of this remarkable mode of death has lately occurred in Switzerland to a Russian gentleman.

Such another instance of a man being slowly frozen to death within sight and sound of other beings, through whose veins the blood was flowing at its accustomed temperature (about 90° Fahr.), it would be difficult to find, and it stands forth, therefore, as a marked example and illustration of the statement already made, that living animal organisms are truly a source of heat, which is as essential to the well-being of the body as meat, drink, and air.

Heat is of two kinds, and may be either apparent to our senses, and therefore called *sensible* heat; or it may be entirely concealed, although present in solids, liquids, and gases, and is then termed *insensible* or *latent* heat.

Sensible Heat.

The first effect of this force is a demonstration of its repulsive agency, and the dilatation or expansion of the three forms of matter whilst under the influence of heat, admits of very simple illustrations. The expansion of a solid substance, as, for instance, a metal, on the application of heat, is apparent by fitting a solid brass cylinder into a proper metal gauge, which is accurately filed so as to admit the former when perfectly cold. If the brass rod is then heated, either by plunging it into boiling water or by the application of the flame of a spirit lamp, its particles are separated from each other; they now occupy a larger space, and expansion is the result, and this is clearly proved by the application of the gauge,

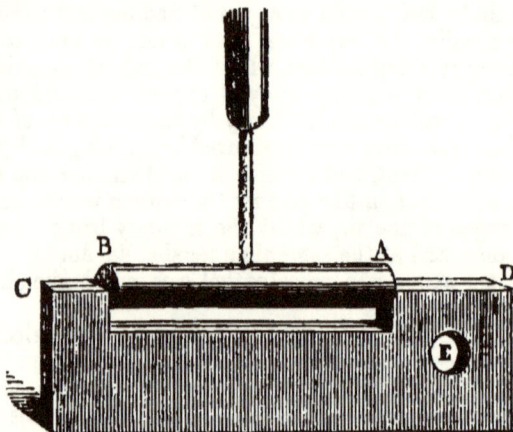

Fig. 343. A B. Cylinder of brass. C D. Iron gauge, admitting A B longitudinally, and also in the hole E when cold, but excluding A B when the latter is heated and expanded.

which is no longer capable of receiving it. (Fig. 343.) When, however, the latter is cooled, the opposite result occurs, the particles of brass return to their old position, and *contraction* takes place; hence it is stated that "Bodies expand by heat and contract by cold;" and it is proper to state here that the term "*cold*" is of a negative character, and simply means the absence of heat.

Solid bodies do not expand equally on the application of the same amount of heat; thus, a bar of glass one inch square and one thousand inches long would only expand one inch whilst heated from the freezing to the boiling point of water. A bar of iron one inch square and eight hundred inches long would expand one inch in length, through the same degrees of heat; and a bar of lead one inch square and three hundred and fifty inches long would also dilate one inch in length. Hence,

Lead expands in volume . . . . $\frac{1}{350}$th.
Iron . . . . $\frac{1}{800}$th.
Glass . . . . $\frac{1}{1000}$th.

The unequal expansion of the metals is well illustrated by an experiment devised by Dr. Tyndal, the respected Professor of Natural Philosophy in the Royal Institution of Great Britain, and is arranged as follows:—A long bar of brass and another of iron are supported on the

A A 2

edges of two pieces of wood placed at an angle, and resting against the sides of a mahogany framework. The metallic bars only touch one end of the frame, and are in metallic communication with a piece of brass inserted there, and forming part of a conducting chain connected with a voltaic battery; when heat is applied to both bars they expand unequally; the brass bar dilates first, and filling up the minute space left between the two ends of the frame, touches another brass plate and instantly completes the voltaic circuit, when a coil of platinum wire becomes ignited, showing the fact of expansion; and secondly, the difference in the power of dilatation possessed by each is clearly shown by removing the two angular supports of wood, when the iron falls away, whilst the brass remains and still completes the voltaic circuit. (Fig. 344.)

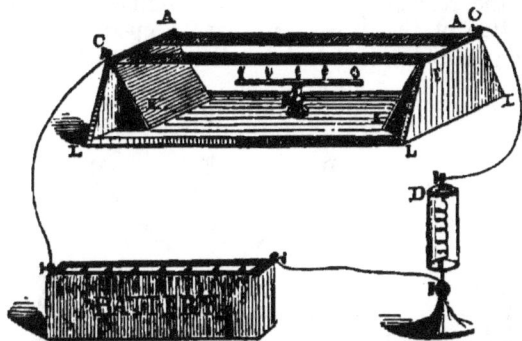

Fig. 344. A A. The brass bar which has expanded by the heat from the gas jet B, and making the contact between the brass plates in connexion with the binding screws c c, the voltaic circuit is completed, and a coil of platinum wire in the glass tube D, is immediately ignited. The iron bar at E B has not expanded sufficiently, which is shown afterwards by removing the angular wooden supports x x, when the iron falls off, and the brass remains on the two ledges of the mahogany framework L L L.

The force exerted by the expansion of solids is enormous, and reminds us again of the amazing power of all the imponderable agents; and it is truly wonderful to notice how the entry of a certain amount of heat into and between the particles of metals, or other solids, endues them with a mechanical force which is almost irresistible, and is capable of working much harm. Kussné made an experiment with an iron sphere, which he heated from a temperature of 32° Fahr. to 212° Fahr., and he found that the expansion of the ball exerted a force equal to 4000 atmospheres —i.e. 4000×15—on every square inch of surface, or a pressure equal to thirty millions of pounds; the entry of only 180° of heat into the iron sphere produced this remarkable result, just as Faraday has calculated that a single drop of water contains a sufficient quantity of electricity to produce a result equal to the most powerful flash of lightning, provided the electricity of quantity in the drop of water is converted into electricity of high tension or intensity.

The practical applications of this well-known property of solids with respect to heat are very numerous; thus, the iron bullet-moulds are always made a little larger than the requisite size, in order to allow for the expansion of the hot liquid lead, and the contraction of the cold metal. The tires of wheels and the hoops of casks are usually placed on whilst hot, in order that the subsequent contraction may bind the spokes and fellies, or the staves, closely together. If an allowance was not made for the expansion and contraction of the iron rails on the permanent ways of railroads, the regularity of the level would be constantly destroyed, and the position of the rails, chairs, and sleepers would be most seriously deranged; indeed it is calculated that the railway bars between London and Manchester are five hundred feet longer in the summer than in the winter.

The walls of the Cathedral of Armagh, as also those of the Conservatoire des Art et Mètiers, were brought back to a nearly perpendicular position, by the insertion(through the opposite walls) of great bars of iron, which being alternately heated, expanded, and screwed up tight, then cooled and contracted, gradually corrected the bulging out of the walls or main supports of these buildings.

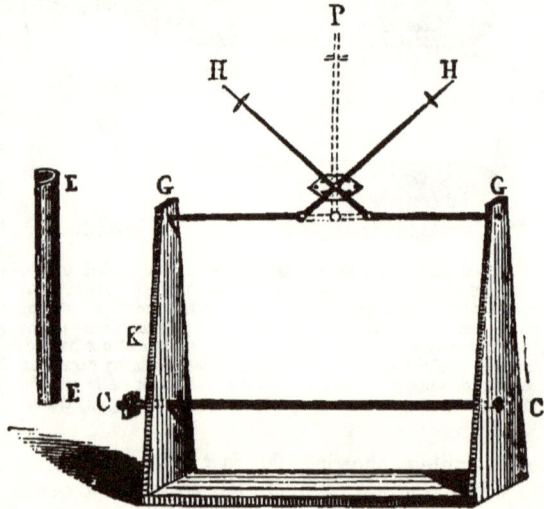

Fig. 345. The iron frame, with C C, wrought-iron bar heated by putting on the semicircular piece of iron E E, which is first made red-hot, and as the heat is communicated to the wrought iron rod C C, it is screwed up tight by the nut K. G G. The index rod attached to the iron frame screwed up when hot; the arms come together at P, and separate further to H H as the contraction takes place by cooling the bar C D.

The principle of these famous practical experiments is neatly illustrated by means of an iron framework with a bar of iron placed through both its uprights, and screwed tight when hot; on cooling, contraction occurs, which is shown by a simple index. (Fig. 345.)

It has often been remarked that there is no rule without an exception, and this applies in a particular instance to the law that "bodies expand by heat and contract by cold"—viz., in the case of Rose's fusible metal, which consists of

Two parts by weight of bismuth,
One part   ,,   lead,
One part   ,,   tin.

To make the alloy properly, the lead is first melted in an iron ladle, and to this are added first the tin, and secondly the bismuth; the whole is then well stirred with a wooden rod, and cast into the shape of a bar.

When placed in the pyrometer and heated, the bar expands progressively till it reaches a temperature of 111° Fahr.; it then begins to
contract, and is rapidly shortened, until it arrives at 156° Fahr., when it
attains a maximum density, and occupies no more space than it would
do at the freezing-point of water. The bar, after passing 156°, again
expands, and finally melts at about 201°, which is 11° below the
boiling-point of water. Fusible metal is sometimes made into teaspoons, which soften and melt down when stirred in a cup of hot tea or
basin of soup, to the great surprise and bewilderment of the victim of
the practical joke.

Unequal expansion is familiarly demonstrated with a bit of toasted
bread, which curls up in consequence of the surface exposed to the fire
contracting more rapidly than the other; and the same fact is illustrated with compound flat and thin bars of iron and brass, which are
fixed and rivetted together; when heated, the compound bar curves,
because the iron does not expand so rapidly as the brass, and of course
forms the interior of the curve, whilst the brass is on the exterior.

The experiment with the compound bar is made more conclusive and
interesting by arranging it with a voltaic battery and platinum lamp. One
of the wires from the battery is connected with the extremity of the
compound bar, and as long as it remains cold, no curve or arch is produced, but when heat is applied, the bar curves upwards, and touching
the other wire of the battery, the circuit is completed, and the platinum
lamp is immediately ignited. (Fig. 346.)

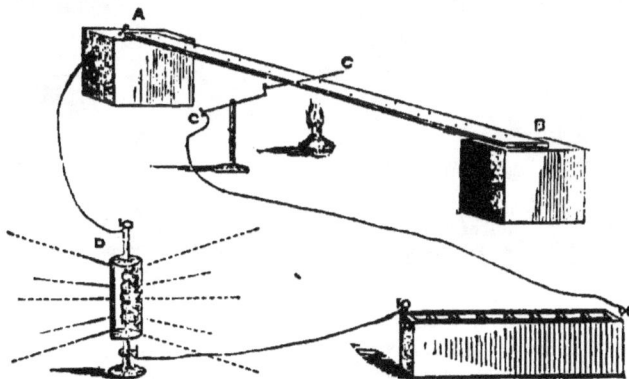

Fig. 346. A B. Compound bar resting on two blocks of wood. The end A is connected
with one of the wires from the battery. The circuit is completed and the platinum lamp
D ignited directly the bar curves upwards by the heat of the spirit lamp, and touches the
wire c c connected with the opposite pole of the battery.

The expansion and contraction of liquids by heat and cold is also
another elementary truth which admits of ample illustration, and
indeed introduces us to that most useful instrument called the thermometer.

If a flask is fitted with a cork through which a long glass tube, open at both ends, is passed, and then carefully filled with water coloured with a little solution of indigo, so that when the cork and tube are placed in the neck, all the air is excluded, a rough thermometer is thus constructed, which, if placed in boiling water, quickly indicates the increased temperature by the rising or expansion of the coloured water inside the flask. (Fig. 347.)

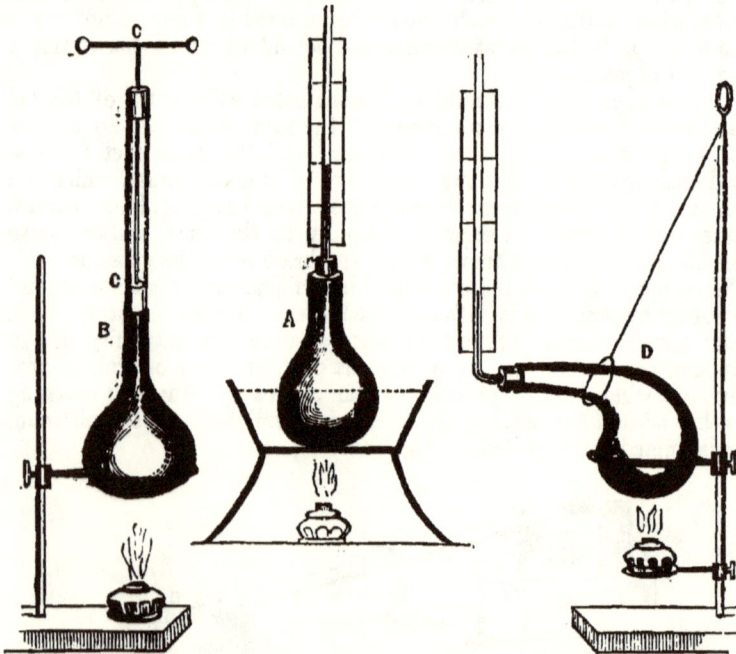

Fig. 347. Expansion of liquids shown at A by the coloured water rising in the tube from the flask, which is quite full of liquid, and heated by boiling water. B. The expansion of the water heated by the spirit-lamp is shown by the rising of the piston and rod c c. D represents a retort filled up like A to show the expansion of a liquid by heat.

The thermometer embraces precisely the same principle as that already described in Fig. 347, with this difference only, that the tube is of a much finer bore, and the liquid employed, whether alcohol or mercury, is boiled and hermetically sealed in the tube, so that the air is entirely excluded. To make a thermometer, a tube with a capillary bore is selected of the proper length; it is then dipped into a glass containing mercury, so that the tube is filled to the length of half an inch with that metal. The half-inch is carefully measured on a scale, and the place the mercury fills in the tube marked with a scratching diamond; the mercury is then shaken half an inch higher, and again marked, and this proceeding is continued until the whole tube is divided into half inches. The object of doing this is to correct any inequalities in the diameter of the bore of the glass tube, because if wider at one part than another, the spaces filled with the mercury are not equal; as the bore is usually conical, the careful measurement of the tube with the half inch of mercury in the first place gives the operator at once a view of the interior of his tube, and enables him to graduate it correctly afterwards. (Fig. 348.)

Fig. 348. A B. Magnified view of the bore of one of the thermometer tubes which are made by rapidly drawing out a hollow mass of hot glass whilst soft and ductile, consequently the bore must be conical, and larger at one end than the other.

The next step is to heat one extremity by the lamp and blowpipe, and whilst hot, to blow out a ball upon it; if this operation were performed with the mouth, moisture from the breath would deposit inside the fine bore of the glass tube, and injure the perfection of the thermometer afterwards. In order to prevent any deposit of water, the bulb is blown out, whilst red-hot, with the air from a small caoutchouc

Fig. 349 a.—No. 1. First bulb. The intended length of the thermometer is shown at the little cross.—No. 2 is the second bulb placed above the cross.

Fig. 349 b. Heating and expanding the air in the top bulb, so that when cool the mercury in the glass A, may rise into the tube and fill the bulb B.

bag fitted on to the other extremity of the tube. The operator now marks off the intended length of his thermometer, and above that point the tube is again softened with the flame and blowpipe, and a second bulb blown out. (Fig. 349 a.)

The open end of the tube is now placed under the surface of some pure, clean, dry quicksilver, and heat being applied to the upper bulb, the air expands and escapes through the mercury, and as the tube cools a vacuum is produced, into which the mercury passes. By this simple method, the mercury is easily forced into the tube, as otherwise it would be impossible to *pour* the quicksilver into the capillary bore of the intended thermometer. (Fig. 349 b.)

The tube is now taken from the glass containing the mercury, and simply inverted; but in consequence of the very narrow diameter of the bore the air will not pass out of the first bulb until heat is applied, when the air expands, and the mercury, first stationary in the second bulb, will now displace the air, and fall into the first bulb when the tube is again cool.

The ball, No. 1 (Fig. 349 *a*), is now full of mercury, and there is also some left in No. 2; in the next place, the tube is supported by a wire, and held over a charcoal fire, when it is heated throughout its entire length, and the mercury being boiled expels the *whole of the air,* so that there is nothing inside the bulbs and capillary bore but mercury and its vapour. (No. 1, Fig. 350.) The open end of the intended thermometer is now temporarily closed with sealing-wax, and the whole allowed again to cool with the sealed end uppermost, so that the ball No. 2, Fig. 350, and the tube above it, are quite filled with quicksilver.

After cooling, the tube is placed at an angle with the sealed end uppermost, and, guided by experience, the operator heats the lower bulb so as to expand enough mercury into the upper one to leave space for the future expansion and contraction of the mercury in the tube, which has now to be hermetically sealed. This is done by dexterously heating the tube at the cross whilst the mercury in the first bulb is still expanded; and by drawing it out rapidly with the help of the heat obtained from the lamp and blowpipe, the second bulb is separated from the first at the little cross (B, No. 3, Fig. 350), and the thermometer tube at last properly filled with quicksilver, and hermetically closed. (No. 4, Fig. 350.)

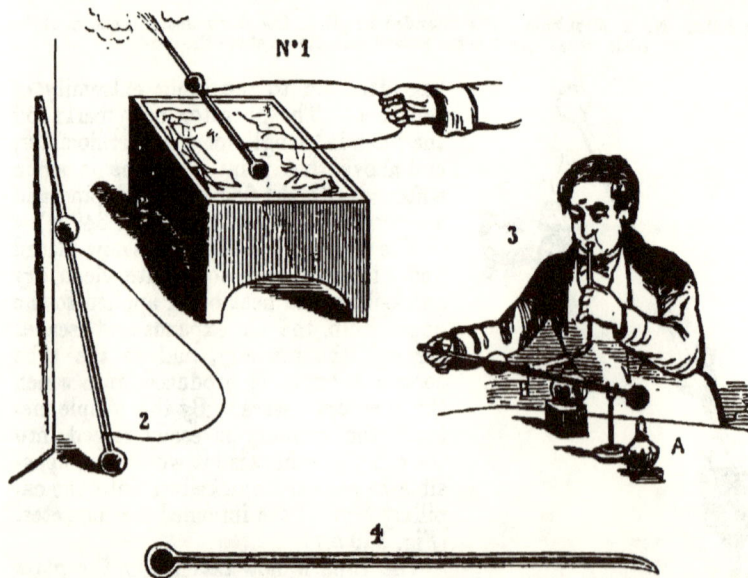

Fig. 350.—No. 1. Boiling quicksilver in the tube with two bulbs.—No. 2. Tube cooled, with the sealed end uppermost.—No. 3. Mercury in first bulb expanded by lamp A, and at the proper moment hermetically sealed by the flame urged by the blowpipe at B. The upper bulb and tube to the cross being drawn away and separated.—No. 4. Thermometer tube containing the requisite quantity of mercury, hermetically sealed, and now ready for graduation.

In order to procure a fixed starting-point, the thermometer tube is placed in ice, with a scale attached; the temperature of ice never varies, it is always at 32 degrees. When, therefore, the mercury has sunk to the lowest point it can do by exposure to this degree of cold, the place is marked off in the scale, and represents that position in the graduated scale where the freezing point of water is indicated.

The tube is placed in the next place in a vessel of boiling water, care being taken that the whole tube is subject to the heat of the water and the steam issuing from it, and when the mercury has risen to the highest position attainable by the heat of boiling water, another graduation is made which indicates 212 degrees—viz., the boiling point of water. This graduation should be made when the barometer stands at 30 inches, because the boiling point of water varies according to the weight of the superincumbent air pressing upon it.

Between the graduation of the freezing and the boiling point of water the space is divided into 180 parts, which added to 32 make up the boiling point of water to 212 degrees, being the graduation of Fahrenheit, who was an instrument-maker of Hamburg. Why he divided the space between the freezing and boiling point of water nobody appears to know, unless he took a half circle of 180 degrees as the best division of space. If the thermometer contains air the mercury divides itself frequently into two or three slender threads, each separated from the other in the capillary bore, and thus the instrument is rendered useless until the threads again coalesce. If the thermometer has been well made, and is quite free from air, it may be tied to a string and swung violently round, when the centrifugal force drives the slender threads of mercury to their common source—viz., the bulb containing the quicksilver, and the whole is again united. The string must be attached, of course, to the top of the thermometer scale.

When travelling on the Continent it is sometimes desirable to be able to read the thermometers which are graduated in a different manner to that of Fahrenheit. In France the Centigrade scale is preferred, and in many parts of Germany Reaumur's graduation The difference of the graduation is seen at a glance.

In the Centigrade the freezing point is 0, the boiling point 100°.
,, Reaumur ,, 0, ,, 80°.
,, Fahrenheit ,, 32°, ,, 212°.

The number of degrees, therefore, between boiling and freezing is 100 in the Centigrade, 80 in Reaumur, and (212 —32, that is) 180 in Fahrenheit.

If, then, the letters C, R, F, be taken to denote the *number* of degrees from the freezing point at which the mercury stands in the Centigrade, Reaumur, and Fahrenheit thermometers, we have the following proportions:—

(1.) $100 : 80 :: C : R$, whence $C = \frac{5}{4}$ of R, or $R = \frac{4}{5}$ of C.
(2.) $180 : 100 :: F : C$, whence $F = \frac{9}{5}$ of C, or $C = \frac{5}{9}$ of F.
(3.) $180 : 80 :: F . R$, whence $F = \frac{9}{4}$ of R, or $R = \frac{4}{9}$ of F.

The following examples will show how to apply these formulæ:—

(1).—Suppose the Reaumur stands at 28°, at what height does the Centigrade stand? We have $C = \frac{5}{4}$ of R (in this case), $\frac{5}{4}$ of $28 = 35$: that is, the Centigrade stands at 35°.

(2).—Suppose Fahrenheit to stand at 41°, what will Reaumur stand at? $R = \frac{4}{9}$ of $(41 - 32)$ (that is, the number above freezing in Fahr.) $= \frac{4}{9}$ of $9 = 4$. Reaumur stands at 4.

(3).—Suppose Fahrenheit stands at 23°, what will the Centigrade stand at? $C = \frac{5}{9}$ of $F = \frac{5}{9}$ of $(32 - 23) = \frac{5}{9}$ of $9 = 5$ below freezing (or —5).

(4).—If Fahrenheit stands at 4 below 0, what will Reaumur indicate? $R = \frac{4}{9}$ of $F = \frac{4}{9}$ of $(32 + 4) = \frac{4}{9}$ of $36 = 16$ below 0 (or —16).

The only liquid which has the exceptional property of expanding by cold is water, and it will be seen presently that this curious anomaly is of the greatest importance in the economy of nature.

If a box containing a mixture of ice and salt is placed round the top of a long cylindrical glass containing water at a temperature of 60° Fahr., the intense cold of the freezing mixture, which is zero—that is to say, 32° below the freezing point of water—very soon reduces the temperature of the water contained in the glass, and as it becomes colder it contracts, is rendered heavier, and sinks to the bottom of the vessel, and its place is taken by other and warmer water. This circulation commencing downwards, proceeds till the water has attained a temperature of about 40° Fahr., when the maximum density is obtained and the circulation stops, because after sinking below 40° the cold water becomes *lighter*, and continues to be so until it freezes, and of course, being of a less specific gravity than the warmer water, it floats (like oil on water) upon its surface; so that a small thermometer placed at the bottom of the jar indicates only 40° Fahr., whilst the solid ice enveloping the other or second thermometer placed at the top may be as low as 29°, or even lower, according to the quantity of ice and salt used in the box surrounding the top of the glass. (Fig. 351.)

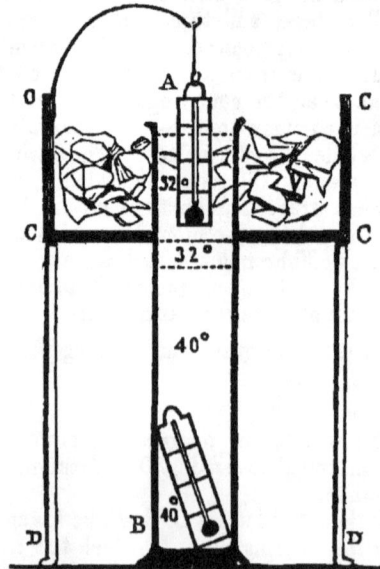

Fig. 351. A B. Long cylindrical glass containing water and two thermometers; the one at the bottom shows a temperature of 40°; the other at the top 32°, or even lower. C C C C. Section of box containing the ice and salt, and standing on four legs, two of which are shown at D D.

The importance of this curious anomaly cannot be overrated. If water did not possess this rare property, all the seas, rivers, canals, lakes, &c., would gradually become impassable from the presence of enormous blocks of ice formed during the winter. The whole bulk of water contained in them would have to sink below 32° before it could solidify, provided water increased in density or continued to contract by cold. Having once solidified, the warmth of the rays from a summer's sun would certainly melt a great deal of the ice, but not the whole, and winter would come again before the solid masses had disappeared. The ocean could not be navigated in safety even near our own shores, in consequence of the vast icebergs that would be formed, and float about and jostle each other even in the British Channel.

The earth has been wonderfully prepared for God's highest work— Man, and in nothing is this supreme wisdom more apparent than in the fact that water offers the only known exception to the law " that bodies expand by heat and contract by cold."

The expansion of gases by heat and contraction by cold take place in obedience to a law to which there is no exception, except in degree. It was discovered in 1801 by M. Gay Lussac, of Paris, and also about the same period by the famous English philosopher who established the atomic theory—viz., by Dr. Dalton. Since these experiments and calculations Rudberg, Magnus, and Regnault have made other researches, and their successive experiments give the following results:—

	Vols. of air.				Volumes.
Dalton, Gay Lussac	1000 heated from 32° to 212° became				1375
Rudberg	1000	"	"	"	1366
Magnus, Regnault	1000	"	"	"	1365·5

As a natural result, air at 32° Fahr. expands $\frac{1}{491}$ part of its volume for every degree of heat on the scale of Fahrenheit; and a volume of air which measures 491 cubic inches at 32° will measure 492 at 33°, 493 at 34°, and so on. The exception is only in degree, and Magnus and Regnault discovered by their searching experiments that the gases easily liquified are more expansible by heat than air and those gases (such as oxygen, hydrogen, and nitrogen) which have never been liquified.

The expansion of air is easily shown by placing the open end of a tube with a large bulb blown at the other extremity, under the surface of a little coloured water; on the application of heat the air expands and escapes, and its place is taken, when cool, by the coloured liquid. Such an arrangement represents the first thermometer constructed by Sanctorio about A.D. 1600, which might certainly answer for rough purposes, but as the ascent and descent of the fluid depend on the bulk of air contained in the bulb, and as this is affected by every change of the height of the barometer, no satisfactory indication of an increase or decrease of temperature could be obtained with it, although the instrument itself is interesting in an historical point of view, and is a modified form as an air thermometer has been employed by Sir John Leslie, under the name of the "Differential Thermometer," in his refined and delicate experiments with heat. (Fig. 352.)

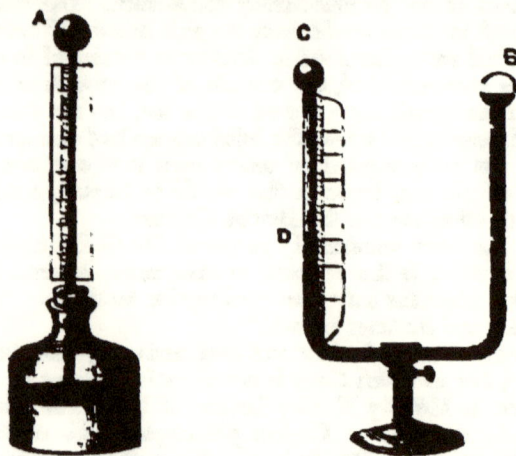

Fig. 352. A. Sanctorio's original air thermometer; the expansion and contraction of the air in the bulb indicate the rise or fall of the temperature. The cork is merely a support, and is not fitted into the bottle air-tight. D C. The differential thermometer. When both bulbs are subjected to a uniform temperature, no movement of the fluid shown at D occurs; but if the bulb D is put into any place warmer than the position of the bulb C, then the air expands in D, and drives the coloured liquid, which consists of carmine dissolved in oil of vitriol, up the scale attached to the stem of the bulb C.

Fire balloons are a good example of the expansion of gases, and the levity of the air thus increased in bulk was taken advantage of by Montgolfier in the construction of his famous balloon, which, with a cage containing various animals, ascended, in the presence of the King and royal family of France, at Versailles; and in spite of huge rents in two places, it rose to a height of 1440 feet, and after remaining in the air for eight minutes, fell to the ground at the distance of 10,200 feet from the place whence it started, without injury to the animals. When it is considered that a volume of air heated from 32° to 491° is doubled, and tripled when heated to 982°, it will at once be understood how great must be the ascending power of such balloons, provided the air within them is kept sufficiently hot.

That gallant aëronaut, Pilâtre de Rozier, offered himself to be the first aërial navigator; and having joined Montgolfier, they made three successful ascents and descents with a large oval-shaped balloon, forty-eight feet in diameter, and seventy-four feet high. On the fourth occasion he ascended to a height of 262 feet, but in the descent a gust of wind having blown the machine over some large trees of an adjoining garden, the situation of the brave aëronaut was extremely dangerous, and if he had not possessed the strongest presence of mind, and at once given the balloon a greater ascending power, by rapidly supplying his stove with some straw and chipped wood, he might on this occasion have met with that untimely end which subsequently, in another rash aëronautic adventure, befel this brave but foolhardy Frenchman.

On descending again, he once more, and without the slightest fear, raised himself to a considerable height by feeding his fire with chopped straw. Some time after he ascended, in company with M. Giroud de Vilette, to the height of 330 feet, hovering over Paris at least nine minutes, in sight of all the inhabitants, and the machine keeping all the while perfectly steady.

The danger in using this method of inflating the balloon arises from the possibility of generating gas, which escaping unburnt into the body of the balloon, may accumulate and blow up, or burn afterwards.

Fire balloons, as usually made, are very dangerous toys, and may sometimes prove rather costly to the person who may send them off, in consequence of their being blown by the wind on a hay or corn rick, or other combustible substances. The safest mode of using fire balloons is to fill them with hot air from a lighted gas stove (Wessel's, for instance); the balloons may then be used in large rooms, or out in the air, without fear of doing any harm to neighbouring property, as of course the stove and the fire remain behind, and will fill any number of air balloons. (Fig. 353.)

Fig. 353.   A B. Wessel's gas stove, with ring of gas jets lighted inside; the air rushes in the direction of the arrows, C C, and escaping at the top of the chimney, D D, soon fills the air or fire balloon, which is usually made of paper.

After all the fuss made about the novelty of the American hot-air engine, it is somewhat amusing to look back to the records of civil engineering, and in the "Transactions of the Institution of Civil Engineers," to read Mr. James Stirling's account of his improved air engine, in which the great expansion of air mentioned at p. 365 has been successfully applied. The engine was constructed about the year

1843, and the principle, discovered thirty years before by Mr. R. Stirling, will be comprehended by reference to the cut. (Fig. 354.)

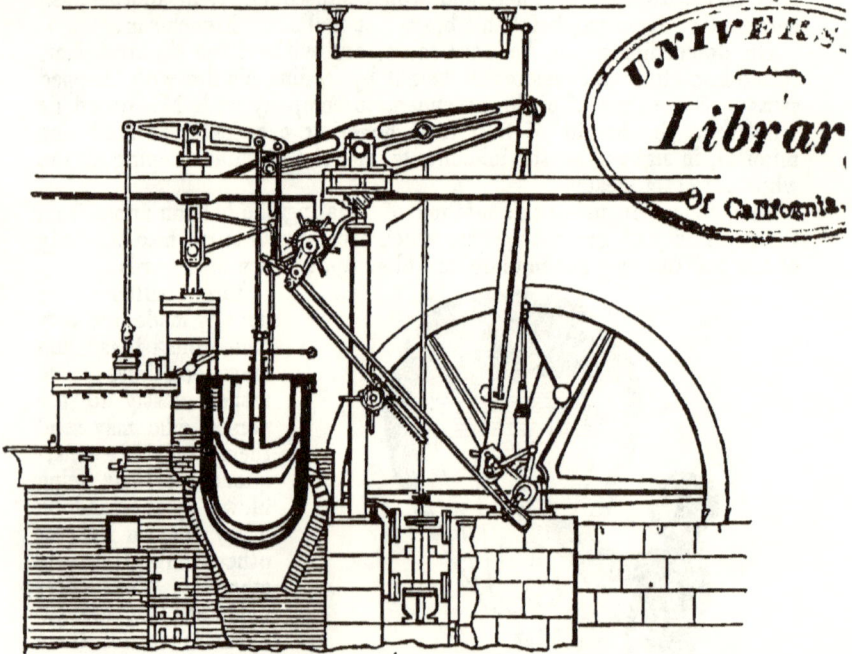

Fig. 354. Stirling's air engine.

Two strong air-tight vessels are connected with the opposite ends of a cylinder, in which a piston works in the usual manner. About four-fifths of the interior space in these vessels is occupied by two similar air-tight vessels or plungers, which are suspended to the opposite extremities of a beam, and capable of being alternately moved up and down to the extent of the remaining fifth. By the motion of these interior vessels, which are filled with non-conducting substances, the air to be operated upon is moved from one end of the exterior vessel to the cther, and as one end is kept at a high temperature, and the other as cold as possible, when the air is brought to the hot end it becomes heated, and has its pressure increased; and when it is brought to the cold end, its heat and pressure are diminished. Now, as the interior vessels necessarily move in opposite directions, it follows that the pressure of the enclosed air in the one vessel is increased, while that of the other is diminished. A difference of pressure is thus produced upon the opposite sides of the piston, which is thereby made to move from the one end of the cylinder to the other, and by continually reversing the motion of the suspended bodies or plungers, the greater pressure is successively thrown upon a different side, and a reciprocating motion of the piston is kept up. The piston is connected with a fly-wheel in any of the usual modes; and the plungers, by whose motion the air is heated and cooled, are moved in the same manner, and nearly at the same relative time, with the valves of a steam engine.

The pressure is greatly increased and made more economical by using somewhat highly-compressed air, which is at first introduced, and is afterwards maintained, by the continued action of an air-pump. The pump is also employed in filling a separate magazine with compressed air, from which the engine can be at once charged to the working pressure. Mr. Stirling's chief improvement consists *in saving all or nearly all the heat of the expanded air after it has done its work*, by passing it from the hot to the cold end of the air vessel through a multitude of narrow passages, whose temperature is at the beginning of the tubes nearly as great as that of the hot air, but gradually declines till it becomes nearly as low as the coldest part of the air vessel. The heat is therefore retained by these passages, so that when the mechanism is reversed, the cold air returns again through these hot pipes, and is thus made nearly hot enough by the time it reaches the heating vessel to do its work. Thus, instead of being obliged to supply at every stroke of the engine as much heat as would be sufficient to raise the air from its lowest to its highest temperature, it is necessary to furnish only as much as will heat it the same number of degrees by which the hottest part of the air vessel exceeds the hottest part of the intermediate passages. This portion of the engine may be called the *economical process*, and represents the foundation of all the success to which it has attained in producing power with a small expenditure of fuel. No boiler being required, of course the danger of explosions is much lessened. The higher the pressure under which the engine was worked the greater was the effect produced. A small engine on this principle was worked to a pressure of 360 pounds on the square inch; and perhaps the best popular notion of the novelty in the arrangement is that suggested by Mr. George Lowe, who compared the economical part of the machine to a "Jeffrey's Respirator" used by consumptive patients. The heat from the air *expired* being retained by the laminæ, and again used when cold air is inspired or drawn into the lungs. Mr. Stirling states that the consumption of fuel as compared to the steam engine which the air engine had replaced was as 6 to 26; the same amount of work being now performed by about six cwt. of coals which had formerly required about twenty-six cwt., though he ought to have stated that the steam engine removed was not of the best construction, nor had the boiler any close covering. (Fig. 354.)

Conduction of Heat.

This property of heat with reference to matter, and the consideration of the curious manner in which it creeps, as it were, through solid substances, brings the thoughtful mind at once to the bold question of What is heat? Is it to be regarded as something real or material? or must it be considered only as a property or state of matter? These questions are not to be solved easily, and they demand a considerable amount of experiment and reasoning even to appreciate their meaning.

If a red-hot ball is placed in the focus of a concave metallic speculum, it gives out certain emanations that are quite invisible, but which are reflected from the surface of the mirror in the same manner as visible rays of light, and may be collected in the focus of another and second concave speculum, when they can be concentrated on to a bit of phosphorus, and will cause the combustion of that substance. If the air from a pair of bellows is blown forcibly across the rays of heat as they are being concentrated upon the phosphorus, the rays are not moved from their course, they are no more blown away than a sunbeam darting through an aperture in a cloud on a stormy, windy day. The heat has, therefore nothing to do with the air, and is wholly independent of that medium in its passage from one mirror to the other. Such an experiment as that described would at once suggest the idea that heat is a matter *sui generis*, a component part of all bodies, and given off from incandescent matter, the sun, &c., and that it may be propagated through space much in the same manner as light. (Fig. 355.) The mechanism may be very much like the corpuscular movement

Fig. 355. Heat reflected by mirror, but not blown away by air from bellows.

of light as defined by Sir Isaac Newton, and already explained in another portion of this book. Hence it has been supposed that heat is propagated through the air, water, and solid substances by a direct emission of material particles from the heat-giving agent, and that these molecules of heat force their way into, or along, or through them, according to circumstances.

Certain bodies are almost transparent to heat rays, such as air, whilst others take an intermedial position, and only stop a certain quantity of the heat molecules, such as rock crystals, mirror glass, and alum. A third class of bodies absorbs the heat plentifully, such as charcoal, black cloth, &c.; and a fourth, when polished and placed at the proper angle, reflects or throws off the heat, as in the case of polished mirrors. The transparency or opacity of substances (so far as light is concerned) does not affect the transmission of heat. Light of every colour and from all sources is equally transmitted by all transparent bodies in the liquid or solid form, but this is not the case with heat.

The rays of heat emitted by the sun and other luminous bodies have properties quite different to the rays of light with which they are accompanied. From these statements it will be evident that the *material theory* of heat is surrounded with difficulties and anomalies that cannot be reconciled the one with the other, or neatly adapted, fitted in, and dovetailed with all the puzzling phenomena that arise. Our knowledge of the theory of heat has been greatly assisted by the researches of Melloni, who has demonstrated that different *species* of rays of heat are given off by the same body at different temperatures, which may be distinctly sifted and separated from each other. Long before the experiments of Melloni philosophers had endeavoured to weigh heat; trains of the most delicate levers were exposed, without effect, to the action of heat rays; and all attempts, experimental as well as theoretical, to define heat by the *material* theory, are imperfect, crude, and unsatisfactory. We are perforce obliged to adopt another theory, and the one that obtains the greatest favour, as offering the best definition of heat, is the *dynamical* theory, which is more or less analogous to the undulatory theory of light. At pages 262, 328, 335, this theory has been partly explained, and in speaking of it again, great care must be taken not to confuse the undulations of heat with those of light. The sun and the stars swim in a molecular medium, and 39,180 vibrations or waves must occur in one inch to produce the sensation of red light, and 57,490 undulations in the space of one inch to produce a violet light. As vibrations of the ethereal molecules affect the eye, so there may be other nerves in our bodies which are peculiarly sensitive to the waves of heat. It requires eight vibrations of the air to occur in a second to produce an audible sound; whilst if the vibrations of the air amount to 25,000 per second they cannot be appreciated by the human ear, although it is possible to conceive that the ears of certain animals may be so susceptible of rapid vibrations that they may be able, for certain wise purposes of the Creator, to appreciate sounds which are inaudible to human ears.

Melloni exhibited a spectrum to a number of persons, and found that there was more light apparent to some eyes than to others. Lubeck put a scarlet cloth on a donkey, and found that the two were frequently confounded together by the eyes of many spectators. These facts indicate that there may be vibrations of molecules that produce the sensation of heat, but which do not affect the nerves that are sensitive to the action of light waves, and *vice versá;* and it is also probable that all these different undulations, some affording heat and some light, may be generated and propagated through space, as from the sun; or through shorter distances, as from burning lamps and fires, without in any way interfering with or impeding each other's progress.

The dynamical theory seems to offer the best idea of the transmission of heat which is carried, conducted, or propagated through solids with variable rapidity, either by the vibration of the constituent molecules of the body itself, or by the undulation of a rare subtle fluid which pervades them. If a copper and iron wire of the same length and diameter are bound together and heated at the point of union, the waves of heat travel faster through the copper than the iron, and the former is said to be the best conductor of heat; and the fact itself is demonstrated by placing a bit of phosphorus at the end of each metallic wire, and it will be found by experiment that the combustible substance melts first and takes fire on the copper, and that a considerable interval of time elapses before the phosphorus ignites on the iron.

Fig. 356. c. Copper wire bound at A to I, an iron wire. After the heat of the lamp has been applied for about five minutes the heat travels to c first, and ignites the bit of phosphorus placed there. After some time has elapsed the phosphorus at I also ignites.

The same fact is exhibited in a most striking manner by inserting a series of rods of equal lengths and thicknesses in the side of a rectangular box, allowing them to pass across the interior to the opposite side. The rods are composed of wood, porcelain, glass, lead, iron, zinc, copper, and silver, and have attached to each of their extremities, by wax or tallow, a clay marble. When the water placed in the box is made to boil, the heat passes along the different rods, and melting the wax or tallow, allows the marble to drop off. Consequently the first marble would drop from the silver rod, the next from the copper, the third from the iron, the fourth from the zinc, the fifth from the lead, whilst the porcelain, glass, and wooden rods would hardly conduct (in several hours) sufficient heat to melt the wax or tallow, and discharge the marbles.

Conduction of Metals.

Gold	1000
Silver	973
Copper	898·2
Iron	374·3
Zinc	363
Lead	179·6

Fig. 357. A B. Trough containing boiling water, heated by gas jets below. C. The eight rods and marbles attached, one of which has fallen. D. The tray to receive the marbles.

The experiment is made more striking if the marbles are allowed to fall on a lever connected with the detent of a clock alarum, which rings every time a marble falls from one of the rods. (Fig. 357.)

During a cold frosty day, if the hand is placed in contact with various substances, some appear to be colder than others, although all may be precisely the same temperature; this circumstance is due to their conducting power: and a piece of slate seems colder than a bit of chalk, because the former is a much better conductor than the latter, and carries away the heat from the body with greater rapidity, and diffuses it through its own substance.

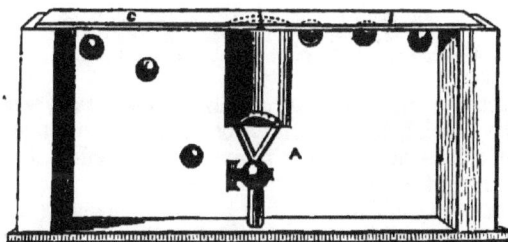

Fig. 358. A. Section of an argand gas lamp, with a copper chimney supporting the ends of the bars of copper and iron marked C and I. The balls have fallen from C, the copper bar.

The gradual passage of heat along a bar of iron as compared with one of copper, is well illustrated by supporting the ends of the two bars on the top of the chimney of an argand lamp, whilst the other extremities are held in a horizontal position by little blocks of wood. If marbles are attached by wax to the under side, they fall off as the heat travels along the metallic bars, and more rapidly from the copper than the iron, because the former is a better conductor of heat than the latter. (Fig. 358.)

From the experiments of Mayer, of Erlangen (" Ann. de Ch.," xxx.), it would appear that the conducting powers of different woods are to a certain extent to be regarded as in the inverse proportion to their specific gravities—i.e., the greater the density of the wood the less conducting power, and the contrary.

If a cylindrical bar or thick tube of brass, six inches long, and about two inches in diameter, is attached to a wooden cylinder of the same size, the conducting powers of the two substances are well displayed by first straining a sheet of white paper over the brass, and then holding it in the flame of a spirit lamp. The heat being conducted rapidly away by the metal will not scorch the paper, until the whole arrives at a uniform high temperature; whereas the paper is rapidly burnt when strained over the wooden cylinder, because the heat of the flame of the lamp is concentrated upon one point, and is not diffused through the mass of the wood. (Fig. 359.)

In the course of the highly philosophical experiments of Sir H. Davy, which led him gradually to the discovery of the construction of the safety lamp, he connected together, by a copper tube of a small bore, two vessels, each containing an explosive mixture composed of fire damp and air. When the mixture was fired in one vessel he found that the flame did not appear to be able to travel, as it were, across the bridge—viz., the

Fig. 359. Cylinder, half brass and half wood. The paper strained over the wood is taking fire. The other extremity, shaded, is the brass portion.

copper tube—and communicate with the other magazine, because it was deprived of its heat whilst passing through the tube, and was no longer flame, but simply gaseous matter at too low a temperature to effect the inflammation of the mixture in the second box.

A mass of cold metal may be suddenly applied to a small flame, such as that of a night light, and depriving it rapidly of heat (like the case of the unfortunate Russian described at page 354), it is almost immediately extinguished (fig. 360), not by the mere exclusion of the oxygen

Fig. 360. A. Small flame from night light. B C. Large mass of cold copper wire open at both ends to place over flame, and by conduction of the heat to extinguish it.

of the air, but on account of the withdrawal of the heat necessary for the maintenance of the combustion.

Sir H. Davy first thought of making his safety lamp with small tubes, which would supply fresh air, and carry off the burnt or foul air, at the same time they were to be so narrow that no flame could pass out of his lamp to communicate with an outer explosive atmosphere; and in speaking of his lamp with tubes he says :—"I soon discovered that a *few apertures*, even of very small diameter, were not safe unless their *sides* were very *deep;* that a single tube of one-twenty-eighth of an inch in diameter, and two inches long, suffered the explosion to pass through it; and that a *great number* of small tubes, or of apertures, stopped explosion, even when the depths of their sides was only equal to their diameters. And at last I arrived at the conclusion that a *metallic tissue*, however thin and fine, of which the apertures filled more space than the cooling surface, so as to be permeable to air and light, offered a *perfect barrier* to *explosion*, from the force being divided *between*, and the heat communicated to an *immense number* of *surfaces*. I made several attempts to construct safety lamps which should give light in all explosive mixtures of fire damp, and after complicated combinations, I at length arrived at one evidently the most simple, that of *surrounding the light entirely by wire gauze, and making the same tissue feed the flame with air and emit light.*"

If a number of square metallic tubes of a fine bore are placed upright side by side, and a section cut off horizontally, it would represent the wire gauze which possesses such marvellous powers of sifting away the heat from a flame, so that it is destroyed in its attempted passage through the metallic meshes; and of this fact a number of proofs may be adduced.

A gas jet delivering coal gas may be placed under a sheet of wire gauze, the gas permeates the gauze, and may be set on fire at the upper side, but the flame is cut off from the mouth of the jet by the cooling action of the wire gauze. The same experiment reversed, by holding the gauze over the gas burning from the jet, shows still more decidedly that flame will not pass through the metallic tissue. (Fig. 361.)

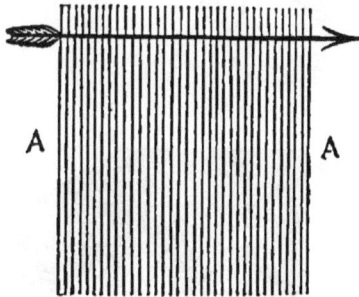

Fig. 361. ▲ ▲. A number of square tubes placed upright. The arrow shows the direction of the section to obtain a figure like wire gauze.

Sir H. Davy again says : "Though all the specimens of fire damp which I had examined consisted of carburetted hydrogen mixed with different small proportions of carbonic acid and common air, yet some phenomena I observed in the combustion of a *blower* induced me to believe that small quantities of olefiant gas may be sometimes evolved in coal mines with the carburetted hydrogen. I therefore resolved to make all lamps safe to the test of the *gas produced by the distillation of coal*, which, when it has not been exposed to water, always contains olefiant gas. I placed my lighted lamps in a large glass receiver through which there was a current of atmospherical air, and by means of a gasometer filled with coal gas, I made the current of air which passed into the lamp more or less explosive, and caused it to change rapidly or slowly at pleasure, so as to produce all possible varieties of inflammable and explosive mixtures, and I found that iron gauze wire composed of wires from one-fortieth to one-sixtieth of an inch in diameter, and containing twenty-eight wires or seven hundred and eighty-four apertures to the inch, *was safe under all circumstances in atmospheres of this kind;* and I consequently adopted this material in guarding lamps for the coal-mines, when in January, 1816, they were immediately adopted, and have long been in general use."

The remarkable conducting power of wire gauze is further shown by placing some lumps of camphor on a piece of this material, and when the heat of a spirit-lamp is applied on the under side of the gauze, the camphor volatilizes, and as the vapour is remarkably heavy, it falls through the meshes of the gauze, and takes fire; but the most curious and further illustration of the conducting power of the wire meshes is shown in the fact that the fire does not communicate through the thin film of gauze to the lumps of camphor placed upon it.

The camphor may be ignited by applying flame to the upper side of the gauze, showing that, although this substance is so exceedingly combustible, it will not take fire even if placed at no greater distance from flame than the thickness of the wire gauze, provided the latter material is interposed between it and the flame.

A square box made of wire gauze, with a hole at the bottom to admit a candle or spirit-lamp, may have a considerable jet of coal gas forced upon it from the outside, or a large jug of ether vapour poured upon it; and although the box may be full of flame, arising from the combustion of the gas or ether, the fire does not come out of the wire box or communicate with the jet or the ether vapour as it is poured from the jug. (Fig. 362.)

Sir Humphrey Davy's safety lamp consists of a common oil-lamp, *f*, with a wire through the cistern for the purpose of raising or depressing the cotton wick without unscrewing the wire gauze; *b* is the male screw fitting the screw attached to the cylinder of wire gauze, which is made double at the top. The entire lamp is shown at A, whilst the platinum coil which Sir H. Davy recommends should be wound round the wick is shown at *h*. The small

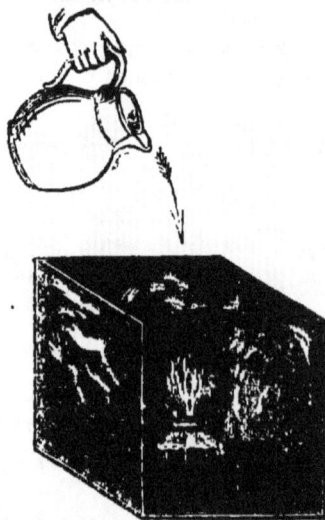

Fig. 362. A box made of wire gauze, with a hole in the bottom to admit a spirit lamp lighted. A hot jug full of the vapour of ether may be poured on to the flame, but it only burns inside the box, and does not communicate with that in the jug.

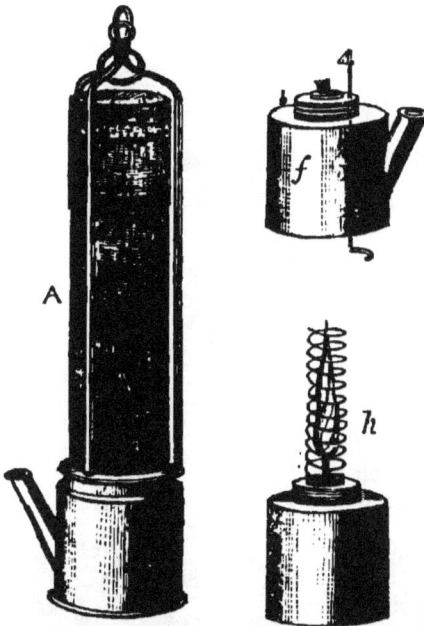

cage of platinum consists of wire of one-seventieth to one-eightieth of an inch in thickness, fastened to the wire for raising or depressing the cotton wick, and should the lamp be extinguished in an explosive mixture, the little coil of platinum begins to glow, and will afford sufficient light to guide the miner to a safe part of the mine. With respect to this platinum coil, Sir H. Davy gives a careful charge, and says: — "The greatest care must be taken that no filament or wire of platinum protrudes on the exterior of the lamp, *for this would fire externally an explosive mixture.*"

Since the invention of the Davy lamp, a great number of modifications have been brought forward, some of which for a short time have occupied the public attention, but whether from increased cost or a sort of inertia that arrests improvement, it is certain that the lamp originally devised by Sir Humphrey Davy is still the favourite. It was perhaps unfortunate that the lamp was called the *safety* lamp, because it is not so under every circumstance that may arise, unless it happens to be in the hands of persons who have taken the trouble to study it and understand how to correct the faults. The lamp might have escaped the incessant attacks that have been made upon its just merits, if the name had simply been that of its illustrious inventor—"a Davy lamp." No one could carp at that, whilst "safety" was held to mean perfect immunity from every possible and probable danger that might arise in the coal-pits. The lamps are now usually placed under the charge of one man, who trims them and ascertains that the wire gauze is in perfect order; this latter is usually locked upon the lamp, and as it is a penal offence, and punishable by a heavy fine and imprisonment, to remove the wire gauze from safety lamps in dangerous parts of the mine, of course the miners are being gradually brought to a sense of the obligations they owe themselves and their brother-miners, and the rash, ignorant, and foolhardy offences of breaking open safety lamps for more illumination, or to light pipes, are becoming much less frequent than formerly. One of the most ingenious "detector lamps" is that of Mr. Symons, of Birmingham. (Fig. 364.) It consisted of the old-fashioned Davy, but inside the rim of the wire gauze is placed a small extinguisher and spring, which does not move so long as the gauze is screwed *on* to the lamp, but directly the gauze is unscrewed, the reversed movement releases the detent, and the extinguisher falls upon the light. In spite of the manifest ingenuity of this lamp, it is not adopted, because it costs a trifle more than the ordinary "Davy." To show the remarkable perfection of the wire gauze principle, some turpentine may be poured upon a lighted safety lamp, when a great smoke is produced by the evaporation of the spirit, but no flame passes through to the outside, although the turpentine burns inside the lamp. If some coarse gunpowder is laid upon two thicknesses of fine wire gauze, it may be heated from below with the flame of the spirit lamp, and the sulphur will gradually volatilize without setting fire to the mass of powder. To show tne security of the Davy lamp, it may be lighted and hung in a large box with glass sides, open at the top, and a jet of coal gas supplied at the bottom; as this rises and diffuses in the air, the mixture becomes explosive, and the fact is at once evident by the alteration in the appearance of the flame

Fig. 364. Symons' self-extinguishing Davy lamp.

of the lamp, which enlarges, flickers, and frequently goes out, in consequence of the suddenness with which the explosion of the mixture takes place inside the lamp, producing a concussion that extinguishes the flame. In this case the utility of the platinum coil is very apparent, and it continues to glow with a red heat until the explosive character of the air in the box is changed.

If a large washhand-basin is first warmed by some boiling water, which is then poured away, and a drachm of ether thrown in, a highly-combustible atmosphere is obtained, and when a lighted Davy lamp is placed into the basin so prepared, the flame inside the lamp immediately enlarges and flickers, but is not extinguished, and does not communicate to the combustible vapour outside. The contrast between the safety lamp and an unprotected flame is very striking; if a lighted taper is thrust into the basin, the ether catches fire, and burns with a very large flame. The solid conductors of heat, which are said to enjoy this property in the highest degree, are the metals, marble, stone, slate, and other dense and compact solid substances; whilst the opposite quality of being non-conductors, or nearly so, is possessed by fur, wood, silk, cotton, wool, eider and swansdown, paper, sand, charcoal, and every substance which is of a light or porous nature. The practical application of this knowledge is very apparent in the affairs of every-day life. Thus we rise in the morning, and immediately after the necessary ablutions, if it is winter time, proceed to encase the body in non-conductors, such as flannel and wool. When we sit down to the breakfast table to make tea, we may notice the contrivances for preventing the handle of the top of the urn, or that of the teapot, from becoming too hot for the fingers, by the interposition of ivory or wood. If asked to place water in the teapot from the kettle, we instinctively seek for the well-worn kettle-holder made of Berlin wool, and therefore a bad conductor. As we cut our meat or fish at the same meal, we may shiver with cold, but our fingers are not quite frozen by contact with the steel knives, as we hold them by ivory handles; and we are agreeably reminded that some metals are good conductors of heat, by the pleasant warmth of the silver teaspoons, as we stir our tea or coffee.

Even the polish of the well-rubbed mahogany is protected from the heat of the dishes by non-conducting mats, and plates are handed about, if " nice and hot," with a carefully-wrapped non-conducting linen napkin. Supposing we prefer a bit of fresh-made toast, the fork is provided with a non-conducting handle; and should we peep out of window some wintry morn whilst the baker delivers his early work in the shape of hot rolls, we notice they come out of nicely-wrapped flannel or baize, which being a bad conductor is employed to retain their heat. We read, occasionally, in the military intelligence, statements respecting some newly-constructed shells which are to burst and scatter melted iron (! !); and of course the idea of the interposition of a good non-conductor of heat between the bursting charge and the molten metal must be realized in their construction.

The *central heat* of our globe is a reality that cannot be disputed, and after digging beyond a depth of twenty feet the thermometer gradually rises at the rate of one degree of Fahrenheit's scale for every fifteen yards. The bad conducting power of the crust of the earth must, therefore, be apparent, as it is easy, knowing the diameter of our globe, to calculate that the increase of heat downwards amounts to 116° for each mile, consequently at a depth of thirty and a half miles below the surface, there will be a temperature most likely equal to 3500°, or a heat that might easily melt cast-iron, and would help to account for the earthquakes and eruptions of volcanoes, which still remind us by their terrible warnings, that we live only on the bad conducting upper crust of a globe, the inside of which is still, perhaps, in a liquid and molten state. Monsieur Fourier has demonstrated the non-conducting power of this shell by calculating that, supposing the globe was wholly composed of cast-iron, the central heat would require myriads of years to be transmitted to the surface from a depth of 150 miles; and by inverting the process of reasoning, we may come to the conclusion that the internal heat must be excessive, because it is confined and shut out from those influences that would carry off and weaken the intensity.

There are no two words, says Tyndal, with which we are more familiar than *matter* and *force*. The system of the universe embraces two things, an object *acted upon*, and an agent *by which* it is acted upon; the object we call matter and the agent we call force. Matter, in certain respects, may be regarded as the *vehicle* of force; thus, the luminiferous ether is the vehicle or medium by which the pulsations of the sun are transmitted to our organs of vision. Or, to take a plainer case, if we set a number of billiard balls in a row, and impart a shock to one end of the series in the direction of its length, we know what will take place; the *last ball* will fly away, the *intervening* balls having served for the transmission of the shock from one end of the series to the other. Or we might refer to the conduction of heat. If, for example, it be required to transmit heat from the fire to a point at some distance from the fire, this may oe effected by means of a conducting body—by a poker, for instance; thrusting one end of a poker into the fire, it becomes heated, the heat makes its way through the mass, and finally manifests itself at the other end. Let us endeavour to get a distinct idea of what we here call heat; let us first picture it to ourselves as an agent apart from the mass of the conductor, making its way among the particles of the latter, jumping from atom to atom, and thus converting them into a kind of *stepping stones* to assist its progress. It is a probable conclusion, even had we not a single experiment to support it, that the mode of transmission must, in some measure, depend upon the manner in which those little molecular stepping stones are arranged. But we must not confine ourselves to the molecular theory of heat. Assuming the hypothesis, which is now gaining ground, that heat, instead of being an agent apart from ordinary matter, consists *in a motion of the material particles;* the conclusion is equally probable that the transmission of the motion must be influenced by the manner in which the particles are arranged. Does experimental science furnish us with any corroboration of this inference ? It does. More than twenty years ago MM. De la Rive and De Candolle proved that heat is transmitted through wood with a velocity almost twice as great along the fibre as across it. This result has been recently expanded, and it has been proved that this substance possesses three axes of calorific conduction; the first and greatest axis being parallel to the fibre; the second axis perpendicular to the fibre and to the ligneous layers ; while the third axis, which marks the direction in which the greatest resistance is offered to the passage of the heat, is perpendicular to the fibre and parallel to the layers.

If many solids are bad conductors of heat, they are at all events greatly surpassed by fluids, and especially by water. The conduction of heat by that fluid is almost imperceptible, so much so, that it has even been questioned whether liquids do really conduct heat downwards at all. It has, however, been found that liquid mercury will conduct heat downwards, and therefore by analogy it may be assumed that other liquids must possess a conducting power, although it may be exceedingly limited.

In order to prove that water is an exceeding bad conductor of heat, a tube with a large glass bulb blown at one end is partly filled with tincture of litmus, until it will just sink below the surface of water placed in a tall cylindrical or open jar. If a copper basin, containing burning ether, is now floated on the top of the water, so as to leave about a quarter of an inch between the top of the air thermometer—viz., the bulb containing the coloured liquid—and the bottom of the copper pan, it will be noticed that whilst the water surrounding the latter almost boils, not the slightest effect arising from the conduction of heat can be perceived in a downward direction. After the ether has burnt out of the copper vessel, it may be removed, and the boiling water stirred down and around the air thermometer, when the air within it expands, drives out the colouring liquid, and the bulb becoming specifically lighter, rises to the top of the containing glass. (Fig. 365.)

Fig. 365. A A. Cylindrical glass full of water. B. The glass air thermometer containing the coloured liquid just standing upright, the mouth of the tube at c being open. D D is the copper basin containing the burning ether. E shows how the glass bulb and tube rise after the upper basin is removed, and the hot water comes in contact with and expands the air, making the thermometer light, and causing it to rise.

Again, if the tube of an air thermometer is placed through a cork in the neck of a gas jar, inverted and standing on a ring stand, and the jar is then filled with water, and boiled at the top with a red-hot iron heater, the heat does not pass downwards and affect the thermometer. By introducing a syphon the water surrounding the thermometer at the bottom of the jar may be drawn off, until the hot water is within a fraction of an inch of the air thermometer, and still no heat is conducted, and the liquid in the latter remains stationary. (Fig. 366.)

The diffusion of heat through water does not take place like that of solids, but is effected by the motion of the particles of the water. When heat is applied to the bottom of a vessel containing water, such as an inverted glass shade, the first effect is to expand the layer of water which is first affected by the heat; this expanded layer being specifically lighter than the cold water above, it rises to the upper part of the glass shade, and its place is immediately taken by other,

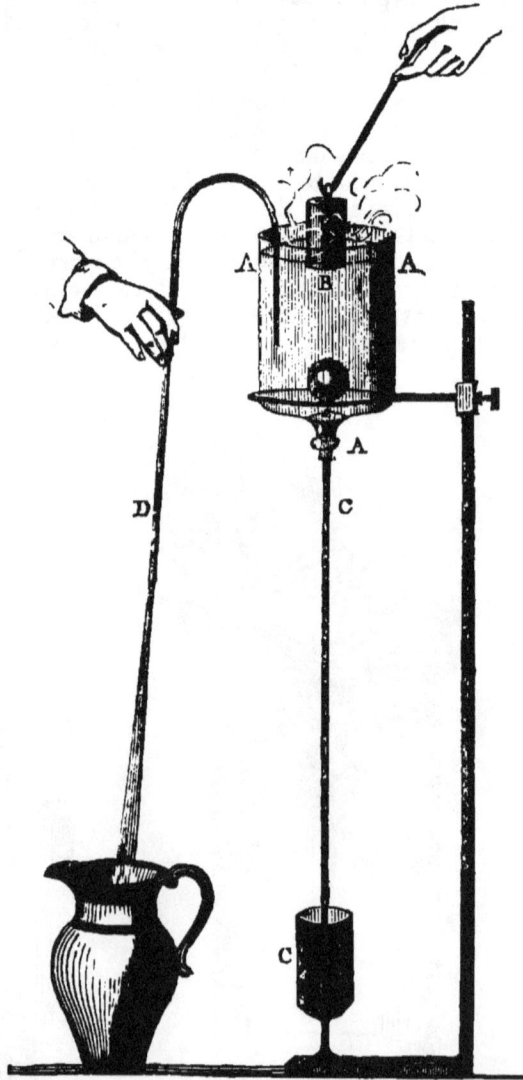

Fig. 366. A A A. Inverted gas jar supported by the ring stand. B. The red-hot urn heater. C C. The air thermometer, with the coloured liquid stationary at c. D. The syphon for drawing off the cold water, and bringing the hot down close to the bulb of c c.

colder and heavier, water, which in like manner moves upwards, and is again succeeded by a fresh portion.   Now, the first and succeeding strata

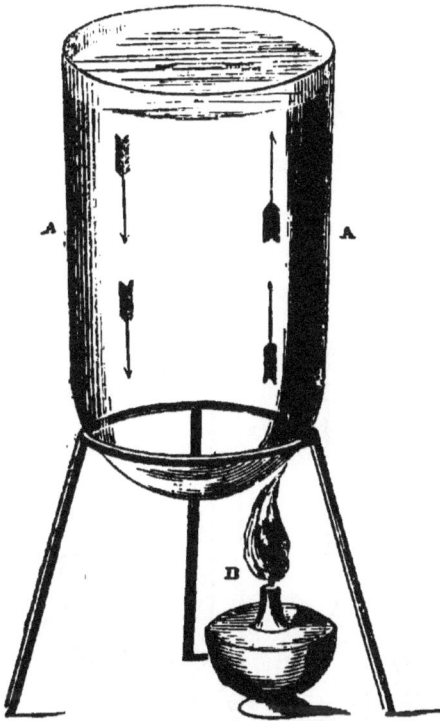

Fig. 367. A A. Inverted glass shade containing water and some paper pulp. B. Burning spirit lamp placed under *one* side of the glass; the pulp shows the rising of the heated water and the sinking of the cold, in the direction indicated by the arrows.

of water all carry off so much heat, and thus by the con-vective or carrying power of the water the heat is diffused finally in the most perfect manner through the whole bulk of fluid; and indeed, the movement itself of the particles of water may easily be watched by putting a little paper pulp at the bottom of the inverted glass shade con-taining the water. (Fig. 367.)

This bad conducting power is not merely confined to water, but is likewise appa-rent with oil and other fluids, and if some water is frozen at the bottom of a long test-tube by means of a freezing mixture, oil may then be poured upon it, and some alcohol above the latter. If the flame of a spirit-lamp is now applied to the alcohol at the top of the tube it may be entirely boiled away, and no heat will travel down the oil and communicate with the ice, and even after the alcohol has been evaporated away the tube can be filled up with water; this may also be boiled, and whilst demonstrating the bad conducting power of the oil, the curious anomaly is observed of a vessel or tube containing ice at the bottom and boiling water at the top, and further showing the wisdom of the Supreme Creator in preventing the freezing of the water of lakes, rivers, and seas, by the exceptional law of the expansion of water by cold. It is evident from what has been stated that liquids acquire and lose their heat by means of those cur-rents and movements of the particles of water which have already been partly explained. Whatever interferes with this movement must pre-vent the passage of heat, and consequently thick viscous liquids are always difficult to boil, and in consequence of their motion being im-peded they rise to too high a temperature and are burnt. This fact is remarkably apparent in the manufacture of nice white lump sugar; as the syrup is evaporated it becomes very thick, and if boiled over a fire might frequently be burnt, but it is boiled by the heat of steam, and under a vacuum produced by an air-pump, and thus the sugar-boiler is enabled to avert all danger from burning.

It is, then, by a continual and perpetual motion, involving circulation of the particles, that heat travels through water; and the fact already described is still further elucidated by one of Professor Griffith's simple but telling experiments. A glass tube, about three feet in length and half an inch in diameter, is bent as at A (Fig. 368), and then being filled with water, is suspended by a string attached to any convenient support inside a copper dish containing water, so that the straight end is at the top of the water, and the curved end at the bottom. Just before it is used some ink or other colouring matter is poured into the copper pan of water; and it should not be added till the moment the experiment is to begin, as any rise of temperature in the room promotes circulation, and interferes with the colourlessness of the water in the tube, which is compared with the inky fluid in the basin. Directly heat is applied the hot water rises to the top of the copper vessel, and thence gradually up the tube; and

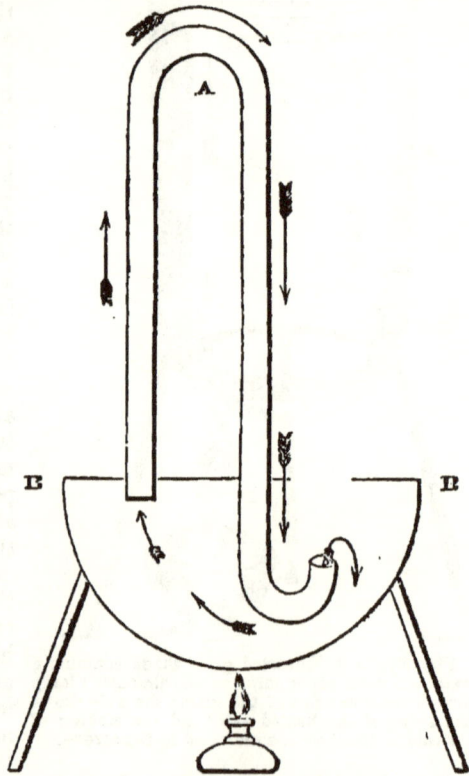

Fig. 368. A. The bent glass tube full of water. B B. The copper pan containing coloured water. The arrows show the circulation of the water.

this movement is rendered visible by the hot coloured liquid matter creeping slowly up the tube, and displacing the colourless water, which falls gradually into the copper pan. (Fig. 368.)

The principle of the circulation of the particles of water being once understood, it is easy to comprehend how it is applied to the heating of buildings by what is called the "Hot Water Apparatus." A coil of pipe is enclosed in a proper furnace, and the bottom end communicates with a pipe coming from a second tube or set of coils, placed above it in another apartment, whilst the top of the latter coil communicates with the top pipe of the first coil. When the fire is lighted, the circulation through the first coil of pipe commences, and is communicated to the second, and from that back again to the first; so that the "hot water system" involves an endless chain of pipes of water, provided with proper safety valves to allow for the escape of any expanded air or steam; and serious accidents have occurred in consequence of persons neglecting to look after the perfection of this safety valve. The fearful accident which occurred to the hot water casing around one of the funnels of the *Great Eastern* offers a painful but memorable example of the heating of water, and of the dangers that must arise if the pipe, casing, or other vessel which contains it, is not provided with an escape or safety valve, which must always be in *good working order*.

Mr. Jacob Perkins, in 1824, made his name remarkable for experiments with the circulation of water through tubes, and his account of the invention and improvement of the "Steam Gun," in which the improvement consists chiefly in the circulation of water through coils of pipe, is so important that we give it verbatim, with a drawing of the steam gun; and the author is enabled to vouch for the accuracy of the statements made in the description of the apparatus, as he purchased one of the improved steam guns, and exhibited it at the Polytechnic Institution, where it discharged three hundred bullets per minute.

"The expansive power of steam has often been proposed as a substitute for gunpowder, for discharging balls and other projectiles; the great danger, however, which was formerly thought to be inseparably connected with the generation and use of steam, at so extraordinary a pressure as appeared necessary to produce an effect approximating to that of gunpowder, prevented scientific men from testing the power of this new agent by experiment. It was also apparent that the apparatus which was ordinarily used for generating steam for steam-engines was wholly inadequate to sustain the necessary pressure, and that one

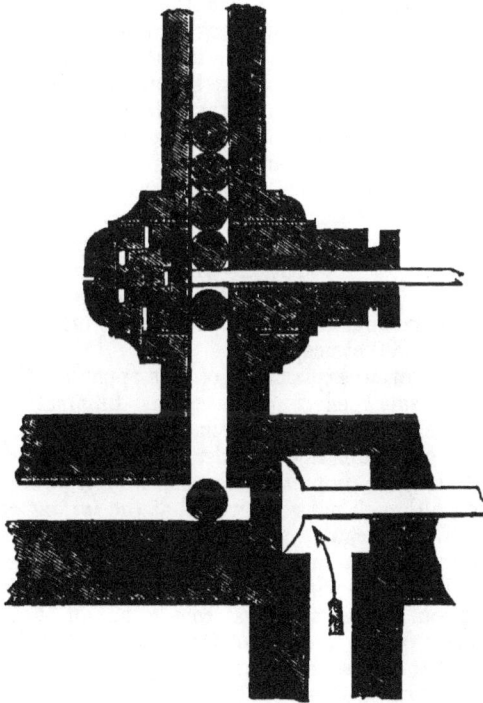

Fig. 369.   The charging tube and gun-barrel of steam gun.

of a totally different character .must be contrived before steam could be sufficiently confined to come into competition with its powerful rival.

"In the year 1824, Mr. Jacob Perkins succeeded in constructing a generator of such form and strength, as allowed him to carry on his experiments with highly elastic steam without danger, although subjected to a pressure of 100 atmospheres. The principle of its safety consisted in subdividing the vessel containing the water and steam into chambers or compartments, so small, that the bursting of one of them was perfectly harmless in its effects, and only served as an outlet, or safety valve, to relieve the rest.

"Although Mr. Perkins' generator was originally intended for working steam engines (it having long been evident to him that highly elastic steam used expansively would be attended with considerable economy), the idea occurred to him, in the course of his experiments, that he had already solved the problem of safely generating steam of sufficient power for the purposes of *steam gunnery;* and that the steam which daily worked his engine possessed an elastic force quite adequate to the projection of musket balls. He therefore caused a gun to be immediately constructed, and connected by a pipe to the generator, the first trial of which fully realized his most sanguine anticipations. Its performance, indeed, was so extraordinary and unexpected, that it gave rise to a paradox, which was difficult of explanation—viz., that *steam, at a pressure of only forty atmospheres, produced an effect equal to gunpowder;* whereas it was known that the combustion of gunpowder was attended with a pressure of from 500 to 1000 atmospheres.

"Mr. Perkins gives the following explanation of this apparent discrepancy, by referring to the small effect produced by fulminating powder, compared to gunpowder, although many times more powerful; he supposes that the action of fulminating powder, however intense, does not continue sufficiently long to impart to the ball its full power. The explosion of gunpowder, although not so powerful at the *instant of ignition,* is nevertheless, in the aggregate, productive of greater effect than that of fulminating powder, because the *subsequent expansion continues* in action upon the ball (but with decreasing effect), until it has left the barrel. The action of steam differs from either of these agents, inasmuch as it *continues in full force until the ball has left the barrel;* and to this is assigned the cause of its superiority.

"In the year 1826, Mr. Perkins had so perfected the mechanism of the gun and generator that, at an exhibition and trial of its power, in the presence of the Duke of Wellington and other distinguished officers of the Ordnance Department, balls of an ounce weight were propelled, at the distance of thirty-five yards, through an iron plate one-fourth of an inch in thickness; also, through eleven hard planks, one inch in thickness, placed at distances of an inch from each other. Continuous showers of balls were also projected with such rapidity, that when the barrel of the gun was slowly swept round in a horizontal direction, a plank, twelve feet in length, was so completely perforated, that the line of holes nearly resembled a groove cut from one of its ends to the other.

c c

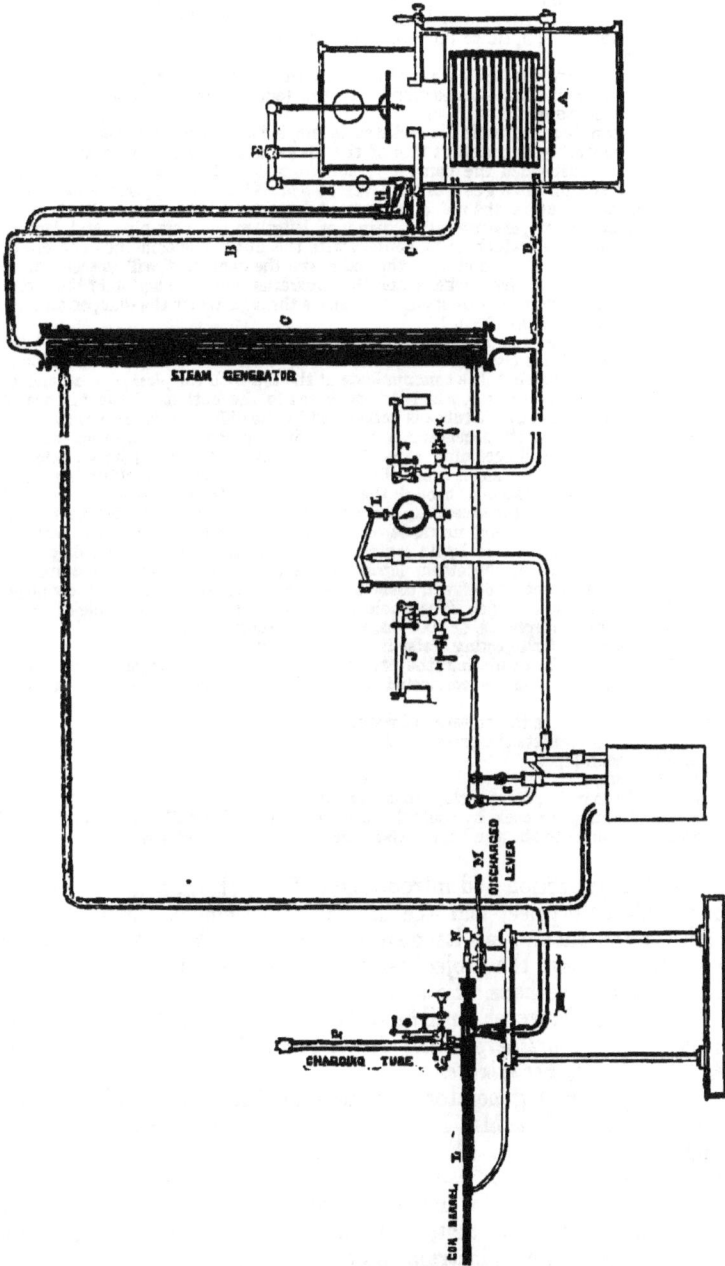

Fig. 370. Perkins's steam gun.

"A is an *iron furnace*, containing a continuous coil of iron tubing, 80 feet in length, 1 inch of external and ⅛th inch of internal diameter, within which the fire is made; the upper end of this tube, B, called the flow-pipe, is extended any required distance to the top of the generator.

"The furnace is provided with a very ingenious *heat governor or regulator*, by which the intensity of the fire is always proportionate to the temperature which it may be requisite to maintain in the tubes.

"H is an iron box, containing a series of levers, *b b b*; *c*, a nut screwed upon the flow-pipe, and in contact with the short arm of the lowest of the levers. E. A lever, from one end of which is suspended the damper *f*, and from the other end the rod *g*, which rests upon the long arm of the highest of the levers, *b b b*. When the apparatus has arrived at the required temperature, the nut *c* is screwed down until it bears upon the lever. Any further increase of temperature will expand or lengthen the flow-pipe, and depress the short arm of the lever, which is in contact with the nut. The combined and multiplied action of the levers will then elevate the rod *g*, and the damper *f* will descend to check the draught. When the fire slackens, and the apparatus cools, the action of the levers will be reversed, and the damper will open. The space through which the damper moves, compared with the nut *c*, is as 200 to 1.

"o is the *generator*, composed of a strong iron tube, 3 inches diameter and 6 feet in length, within which are eight smaller tubes, having their ends welded to the ends of the larger tube. These small tubes communicate at the top with the *flow-pipe* B, and at the bottom with the *return-pipe* D, which is continued to the bottom of the furnace-coil of tubing. The circulation in the tubes is occasioned by the difference in the specific gravities of the water composing the ascending and descending currents; the portion contained in the flow-pipe and fire coil becoming expanded by the heat, ascends by its superior levity; while that contained in the small tubes of the generator, having given off its heat, acquires increased density, and descends through the return-pipe D to the bottom of the furnace-coil, to take the place of the ascending current. When the hot-water current has arrived at a temperature of 212° and upwards, cold water is injected into the generator, and becomes converted into steam by its contact with the small tubes; the rapidity of evaporation and the pressure of the steam depending, of course, upon the temperature of the hot-water current, which at 500° will cause a pressure within the tubes of 50 atmospheres, or 750 lbs. upon the square inch. The whole apparatus is proved to be capable of sustaining a pressure of 200 atmospheres, or 3000 lbs. upon the square inch.

"G. A force pump for injecting water into the generator.

"I. The indicator for exhibiting the pressure of the steam in the generator, and of the water in the boiler; it may be connected with either by means of the valves attached to the levers.

"J. Valve to regulate the pressure of water.

"J 1. Valve to regulate the pressure of steam.

"K. The steam pipe.

"L. The gun.

"M. The discharging lever acting upon the valve N.

"o. The discharging cock, by a simple adjustment in which balls are transferred from the charging tube P to the gun barrel, *singly* or in a *continuous shower*.

"As the perfection and introduction of the steam gun was not a field for private enterprise, and the British Government having declined to institute experiments at its own expense, Mr. Perkins was reluctantly compelled to leave the project, and to engage in others of a more lucrative, although, perhaps, of a less important nature. He did not suspend his operations, however, until he had constructed for the French Government *a piece of artillery which discharged balls weighing five pounds at the rate of sixty per minute.*

"The gun and generator exhibited at the Polytechnic Institution during the time that Mr. Pepper was the Resident Director were the production of Mr. A. M. Perkins, of London, who has invented an entirely *new method of generating steam*, which has been successfully applied to steam engines, and is at once so simple, safe, and economical, as to leave little doubt that, with its aid, the steam gun will ere long rank amongst the first instruments of warfare.

c c 2

"The gun, except in a few·minor mechanical details, does not differ from that originally constructed by Mr. Jacob Perkins.

"The novelty which distinguishes the generator from all others, consists in the manner of conveying the heat from the fire to the water, *without exposing the generator to the action of the fire.* This is accomplished by means of the circulation, in iron tubes, of a current of hot water, which is entirely separate from, and independent of, that to be evaporated in the generator.

"The following are the principal advantages which this generator possesses over all others: *Freedom from all wear or deterioration consequent upon exposure to the fire,* an important quality in a generator that is to be subjected to great pressure, inasmuch as its original strength remains unimpaired; *no accident can arise from want of water in the ·generator,* and the precautions indispensably requisite when a generator is in contact with the fire are quite unnecessary, as the water may be drawn off with impunity without producing the least injurious effect, and the grossest neglect is followed by no worse consequences than an inefficient supply of steam; *an explosion of the generator is impossible,* as the temperature of the furnace-coil always exceeds that of any other part of the apparatus, and consequently, being the weakest part, is invariably the first to yield when the pressure is carried beyond the strength of the pipes; *economy of fuel is also obtained, with a small amount of fire surface.* The circulation of the water has likewise the effect of preserving the fire-coil from the decay to which boilers are liable; many such coils, which have been in constant use for eight years, being apparently as good as when first erected.

"The whole apparatus is exceedingly simple, and will be readily understood by reference to the accompanying diagram. (Fig. 370.)

"The steam has often been raised to a pressure of 700 lbs. on the square inch, but *one-third* of that pressure is sufficient to completely *flatten the balls* when discharged against an iron target one hundred feet distant from the gun; and a pressure of 400 lbs. per square inch, at the same distance, *shivers the ball to atoms,* with the production in a dark room of a visible flash of light. Steam guns are generally mounted upon a ball and socket joint, which allows the barrel to move freely in every direction."

The conduction of heat through gases is also very slow when heat is applied to the upper part of any stratum of air. Heat appears to be diffused through air only by the circulation and rising of the heated and lighter strata, and the sinking of the colder currents which take their places; hence the danger of sitting in a room under an open skylight. A current of cold air may descend upon the head of the individual, whilst the warmer air takes some other opening. to escape from. No doubt the movement of heated volumes of air is subject to definite laws, which apply themselves under every case, but are rather difficult to grasp when the subject of ventilation is concerned. The philosophical ventilator is often dreadfully teased by the inversion of all that he had planned, or the total failure of his apparatus. No specific mode of ventilation can be found to suit all rooms and buildings; they are like the patients of a physician who cannot be cured by one medicine only, but must have a treatment adapted properly to each case. If the fires, candles, gas, or oil-lamps, doors, windows, and chimneys, were always under the control of the scientific ventilator, his task would be very simple, but it is well understood that a ventilating system which answers well if certain doors communicating with lobbies are closed, fails directly they are accidentally opened. The watchful care of the ventilator must begin with the lowest area door, and in his calculations he must study the effect of every other door or window that may be opened, so that if a scientific man undertakes to ventilate a house, he must have a well-drawn plan hung up in the hall, and it must be clearly understood by the inmates that any interference with that plan will prejudice the whole.

There are a few common principles which will guide in ventilation, and these are, first, the rise of hot and the fall of cold air; second, that if an aperture is provided at the top of a room for the escape of hot air, an equally large aperture must be left for the entry of cold air; third, the aperture for the escape of hot air must be adapted in size to the number of persons likely to enter the room, and the number of gas or other lights burning in it. During the daytime, moderate apertures for the exit and entrance of air may suffice, but these must be largely increased at night, when the room is filled with people and lighted up. Expanding and contracting openings are therefore desirable, and they are to be regulated by rules stated on the plan of the ventilating system (already alluded to as being hung up in the hall) of the house which has submitted itself to a perfect system of ventilation, and no hall-keeper, footman, or butler should be allowed to remain in his post unless he undertakes to comprehend the system and work it properly by the written rules.

Dr. Angus Smith, in a very able paper "On the Air of Towns," says— "One of the conditions of health, and a most important, if not the most important of all, is to be found in the state of the atmosphere. As to the effect on the inhabitants, the question becomes exceedingly complicated ; but the Registrar-General's returns are an unanswerable reply as to the results of the lethal influences of the district. Few people seem clearly to picture to themselves the meaning of a decimal plan in the per-centage of death, and few clearly see that there are districts of England where the deaths at least in some years, and when no recognised epidemic occurs, are three times greater than in others. When we hear of the annual deaths in some districts being 3·4 per cent., and in the whole of England 2·2, it is simply that 34 die instead of 22, whilst even that is too slightly stated, as the whole of England would show a lower death-rate if the towns were not used to swell it."

This quotation is given here to remind our readers of the important question of a supply of pure air as well as pure water and pure food; and if the agricultural labourer, with all his exposure to variable weather, can take the first place in the scale of mortality, and outlive the members of all other trades and professions, it is evident that the importance of pure air is not overrated.

Every effort ought, therefore, to be made in large schools, hospitals, and barracks, to enforce a rigid system of supply of fresh air, and a sewage or removal of the impure; and in the use of a certain test employed by Dr. Smith for the detection of organic matter in the air a number of approximations were obtained, which clearly demonstrated that 1 grain of organic matter was detected in 72,000 cubic inches of air in a room, and the same quantity in 8000 cubic inches taken from a *crowded* railway carriage.

To show the rising of heated air, a long glass tube, about three-quarters of an inch in diameter, may be provided and held over the flame of a spirit lamp at an angle of sixty degrees. As the tube warms, the heated air rushes past the flame with great rapidity, and pulls it out or elongates it so much, that the sharp point of the spirit-flame

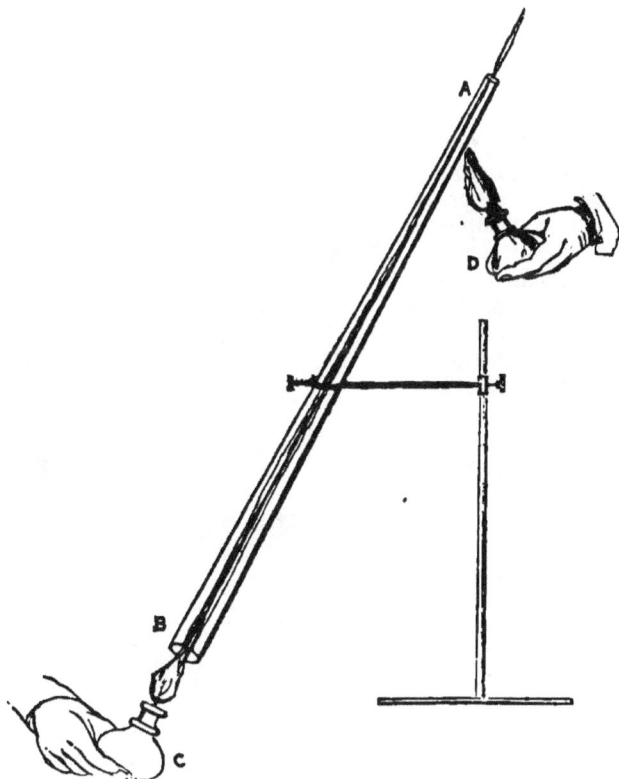

Fig. 371. A B. The glass tube. C. The spirit lamp, with a very large wick; if a little ether is mixed with the spirit in the lamp it increases the length of the flame. D. The effect of the ascension of air, increased by warming the top of the tube with the lamp D.

will frequently be seen at the end of a tube ten feet six inches in length. The flame is, as it were, the sign-post that indicates the path or direction of the air. (Fig. 371.)

Upon the like principle, heated air may be dragged down the short arm of a syphon, provided the other arm is sufficiently long to impart a strong directive tendency to the upward current, and this mode of setting air in motion has been frequently proposed in numerous schemes for ventilation. In order to prove the fact that an inverted syphon will act in this manner, an iron pipe of three inches diameter and six feet long may be bent round during the construction into the form of a syphon, so that the short length is about one foot long, and the long length the remaining four feet, allowing one foot for the bend. If the interior of the long arm is first warmed by burning in it a little spirits of wine from a piece of cotton or tow wetted with the latter (which can be easily done by dropping in such a wetted piece into the bend of tube, so that it is just under the opening of the long part of the tube), the air is soon set in motion up the long pipe, and as it must be supplied with fresh vo-lumes of air to take the place of that which rises, and as the only entrance for the fresh air can be *down* the short arm of the sy-phon, the circulation soon commences, and it pro-ceeds as long as the upper arm is kept sufficiently warm. If a flame is held over the mouth of the short arm, it is immedi-ately dragged downward, whilst, if held at the mouth of the long pipe, the motion of the air is seen by the assistance of the flame to be in the contrary direction. (Fig. 372.)

This plan of ventilation was proposed to be used in rooms in connexion with the chimney and chimney-piece, and in order to give it an ornamental appear-

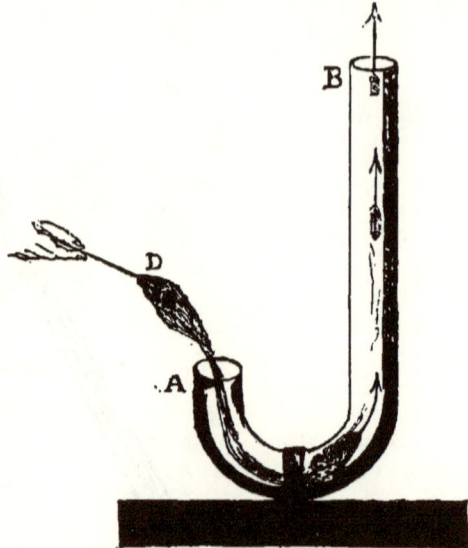

Fig. 372. A B. Inverted sheet iron syphon. At C is seen the piece of tow moistened with alcohol, which, being set on fire, warms the tube B. D. A lighted torch of coloured spirit, the flame of which is dragged down the tube at A by the descending current, and is impelled upwards by the ascending current B.

ance, the chimney-piece was supplied with two ornamental hollow columns, the ends of which were open at the mantelshelf, and the tubes or columns were continued under the hearthstone, proceeding up the back of the grate and entering the chimney, in which there would be a constant current of heated air, and it was expected that

the syphon arrangement would keep a current of air always in
motion, and thus help to ventilate the room. (Fig. 373.) This plan,

Fig. 373. A B. Chimney-piece supported on two hollow ornamental pillars corresponding
with the short arm of a syphon. c c c. The dotted line showing the pipes leading from
each pillar under the hearth, and terminating in a long pipe passing into the chimney.
The arrows show the path of the air descending from the chimney-piece and ascending in
the chimney.

however, does not appear to have been adopted, and wisely so, because
half the time the syphon arrangement might invert itself, and vomit
smoky air out of the chimney into the room; indeed it is surprising what
odd and contradictory freaks are performed by currents of air.   The
author remembers a case where two rooms on the same floor, the one
a dining-room and the other a drawing-room, were always exhibiting the
most absurd phenomena of smoke.   If the fire in one room was lit, then
the other, in a few moments, began to smell exactly like the inside of a
gas manufactory, and was, of course, more or less filled with smoke,
whilst the room in which the fire was actually burning remained quite
free from this annoyance.   The smoke appeared to issue from the
wainscot or moulding which runs round at the bottom of the wall, and
was at first thought to be an escape from the chimney of the kitchen
beneath, the inside of which was duly examined and thoroughly stopped
with cement in every place likely to afford a channel to the smoke, and

the crevice whence the smoke issued was also filled in neatly with cement. But it was all in vain; the smoke then made its way out from another part of the cornice, and at last the rooms exhibited a beautiful reciprocating action. If the drawing-room fire was lighted the dining-room was full of smoke, and if the latter was lighted the former had the agreeable visitation. At last the backs of the two grates were examined, and in each was discovered a hole about one inch in diameter; and it was also found that the spaces at the back of the stoves had not been filled in properly, and, indeed, communicated with the hollow space behind the cornice. When, therefore, the fire was lighted, and coals heaped on just above the hole, the gas and smoke distilled through the orifice and travelled on, where it found the most convenient exit; and the fact is sadly at variance (*apparently*) with theory, because it might be considered that cold air would rush towards a fire, and that the draught ought to have been from the cornice to the chimney instead of *vice versâ*. The fact seems to be that the coal in all grates is, in the act of burning, distilling and giving off inflammable gas; when the coal was, therefore, heaped above the orifice, and was, possibly, caked hard at the top, the gas distilling from it escaped more easily from the little orifice than elsewhere, and chance determined that the channel or delivery pipe should be in the direction of the drawing-room when the fire was burning in the dining-room, and in the contrary direction when the fire was lighted in the latter chamber. The nuisance was stopped by plugging the holes at the back of the grate with clay, and putting a sheet of iron over the orifice.

Before Dr. Faraday was appointed as a scientific counsellor to assist the deliberations of the Trinity Board in connexion with lighthouses, all the lamps were burnt in the lanterns with the smallest and most imperfect arrangement for carrying off the heated air and products of combustion; as a natural consequence, and particularly on cold nights, the windows of the lantern of the lighthouse were covered with ice derived from the condensation of the water produced by the combustion of the hydrogen of the oil, whilst the carbon generated such quantities of carbonic acid that the light-keepers were unable to stay in the lantern, and if obliged to visit the latter (whilst looking to improving the light of any single lamp that might be burning dimly), they were almost overpowered with the excess of carbonic acid, and stated, in their evidence, that it produced headache and sickness, and a tendency to insensibility. Faraday immediately established a system of ventilation; and by attaching a copper tube to the top of each lamp-chimney, and centering them all in one large funnel passing to the top of the lighthouse, the whole of the water which previously condensed on the glass windows and impeded the light, besides injuring the brass and copper fittings, was carried off, as also the poisonous carbonic acid gas; and thus, as Dr. Faraday expressed himself, a complete system of sewage was applied to the lamps of the lighthouses.

If any one of the numerous stories of ships saved by the Eddystone Lighthouse could demonstrate more than another the value of this beacon

The British fleet rounding the Eddystone Lighthouse during the great storm
of October, 1859. *p. 394.*

numerous light-keepers, one of which in plain but striking language related that "*the enemy* (alluding to the water and carbonic acid) *was now driven out.*"

The ingenious invention alluded to was succeeded by another and equally simple but philosophical arrangement, which Dr. Faraday presented to his brother, and it was duly patented. It consisted of an arrangement for ventilating gas burners, and it must be obvious that a necessity exists for such ventilation, because every cubic foot of coal gas when burnt produces a little more than a cubic foot of carbonic acid. A pound weight of ordinary coal gas contains about $\frac{3}{10}$ths of its weight of hydrogen, which when burnt produces two pounds and $\frac{7}{10}$ths of a pound of water. A pound of ordinary coal gas also contains about $\frac{7}{10}$ths of its weight of charcoal, which produces when burnt rather more than two and a half pounds of carbonic acid gas—viz., 2·56. In order to burn this quantity of gas nineteen cubic feet and $\frac{3}{10}$ths of a foot of atmospheric air, containing 4·26 cubic feet of oxygen, are required.

It is not therefore surprising that as common coal gas is sometimes purified carelessly, and contains a minute trace of sulphuretted hydrogen, with some bisulphide of carbon vapour, that it should produce the most prejudicial effects in badly ventilated rooms, and especially in some of those perched up glass boxes in large places of business, where clerks are obliged to sit for many consecutive hours, lighted by gas, and breathing their own breath and the products of combustion from the gas light, thereby rendering themselves liable to diseases of the lungs, and also to very troublesome throat attacks, when leaving their close glass boxes, and passing into the cold night air. The dangerous product of the combustion of ordinary coal gas is sulphurous acid—viz., the

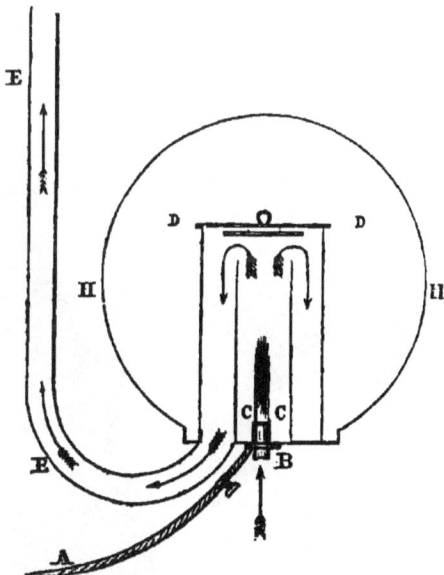

Fig. 374. A B. Gas pipe and argand burner; the air enters, as usual, up the centre of the argand. C C. The first glass chimney open at the top. D D. The second glass chimney closed at the top, with a disc of double talc, and fitting over C C, and leaving a space between the two glasses, down which the air passes, and into the ventilating tube, E E. H H. The ground-glass globe closed at the top, and surrounding the whole.*

* Mr. Faraday, of Wardour-street, supplies this ventilating lamp.

same gas as that generated when a sulphur match is burnt; and if it will attack the bindings of books, and damage furniture, goods in shops, curtains, &c., in consequence of the large quantity of water with which it is accompanied, how much more is it not likely to injure the delicate organism of the breathing apparatus of the lungs? Dr. Faraday's lamp is therefore a great boon, but, like a great many other clever things, it must be adapted to the currents of air and draught from the room; and means must be taken to prevent the draught becoming too powerful in Faraday's lamp, or else the illuminating power is destroyed by the thorough combustion of the carbon of the coal gas, and the heat generated is so intense that the glasses soon crack, and of course become useless. The lamp will answer very well if (as has been already stated) the draught in the ventilating pipe is not too great.

The system already explained and illustrated is likewise carried out on a much larger scale in the ventilation of coal pits, where a shaft is usually sunk into the ground for the admission of air, which, after circulating through the intricate windings and mazes of the coal pit workings, escapes at last from another shaft, at the bottom of which is placed a powerful furnace, and this is kept burning night and day, so

Fig. 375. Section showing the two air-shafts. A. The downcast. B. The upcast. c c. One of the working galleries in connexion with the upcast and downcast. D. The furnace at the bottom of the *upcast*. In this sketch *one* gallery only has been shown, to prevent confusion and to show the principle.

that the movement of the air is maintained in one direction—viz., from the outer air down the shaft called *the downcast*, thence to the galleries, where the coal hewers are working, to the second shaft, near which the furnace is placed, and up this latter the air travels; the shaft, pit, or funnel being very appropriately termed the *upcast*.

Should the furnace at the bottom of the upcast be neglected, the ventilation may be just balanced, or set slightly towards the downcast; under these circumstances the carbonic acid from the fire will begin to circulate in the galleries, and poison those who are not aware of its presence and take the proper means to escape. Such accidents, amongst the host of others that occur in a coal pit, have actually been recorded; and the firemen, whose duty it might be to attend to the proper burning of the furnace, have had to pay the penalty of death for their own carelessness in falling asleep and neglecting to maintain the ventilation of the mine in one direction. (Fig. 375.)

These details are amply sufficient to demonstrate the manner in which heat is diffused through air, whilst the rarefication of the air by heat suggests the cause of those frightful storms of wind that rush from other and colder parts of the surface of the globe, to supply the void produced by the cooling and contraction of the enormous volumes of gaseous matter.

The Radiation of Heat.

When rays of heat are emitted from incandescent matter, they are not necessarily visible, nay, they are generally invisible, and not accompanied with a manifestation of light, and pass with great velocity through a void or vacuum, also through air and certain other bodies. From what has been stated respecting the manner in which air, by continually moving, and by convection, carries off heat, it might be thought that no proof existed that invisible rays of heat are really thrown off from a ball filled with boiling water. But this question is set at rest by the fact, that such a ball will cool rapidly when suspended by a string inside the receiver of an air pump from which the atmospheric air has been removed, so that no conduction of the particles of air could possibly remove the heat.

In the year 1786, Colonel Sir B. Thompson examined the relative conducting powers of air and a Torricellian vacuum—the latter being used because, as the experimenter stated, it was impossible to obtain a perfect vacuum, on account of the moist vapour which exhaled from the wet leather and the oil used in the machine, for at that time carefully *ground* brass plates were not used in air-pumps, but plates only, with a circular piece of wet leather upon them. In a paper which Colonel Sir B. Thompson read before the Royal Society, he stated that "It appears that the Torricellian vacuum, which affords so ready a passage to the electric fluid, so far from being a good conductor of heat, is a much worse one than common air, which of itself is reckoned among the worst; for when the bulb of the thermometer was surrounded with air, and the instrument was plunged into boiling water, the mercury rose from 18° to 27°

in forty-five seconds; but in the former experiment, when it was surrounded by a Torricellian vacuum, it required to remain in the boiling water one minute thirty seconds to acquire that degree of heat. In the vacuum it required five minutes to rise to $48°\frac{2}{16}$ths; but in air it rose to that height in two minutes forty seconds; and the proportion of the times in the other observation was nearly the same.

"It appears, from other experiments, that the conducting power of air to that of the Torricellian vacuum, under the circumstances described, is as 1000 to 702 nearly, for the quantities of heat communicated being equal, the intensity of the communication is as the times inversely. By others it appears that the conducting power of air is to that of the Torricellian vacuum as 1000 to 603."

It is therefore very interesting to discover that the attention of experimentalists was early directed to the fact that heat was independent of the air, and passed either as waves of heat or molecules of heat through space. The velocity with which heat moves through a vacuum is very great, and in an experiment performed by M. Pictet, no perceptible interval took place between the time at which caloric quitted a heated body and its reception by a thermometer at a distance of sixty-nine feet. It appears also, from the experiments of the same philosopher, to be thrown off or radiated in every direction, and not to be diverted (as shown at p. 369) by any strong current of air passing it transversely. Sir Humphrey Davy ignited the charcoal points connected with a battery in a vacuum, taking care to place the charcoal points at the top of the jar, and a concave mirror, with a delicate thermometer in its focus, at the bottom of the vessel placed upon the air-pump plate. The effect of radiation was

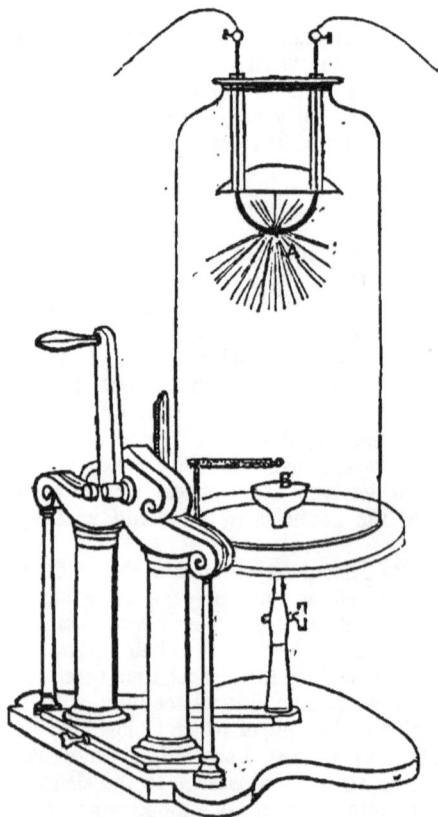

Fig. 376. The air-pump and receiver, containing at A the electric light in the focus of a concave mirror, and at B a delicate thermometer, also in the focus of a concave mirror.

ascertained first when the receiver was full of air, and next when it was exhausted to $\frac{1}{120}$th (*i.e.*, 199 parts pumped out, leaving only one part of air in the receiver). In the latter case, the effect of radiation was found to be three times greater than in an atmosphere of the common density. The greater rise of the thermometer *in vacuo* than in air is to be ascribed to the conducting power of the latter; for this conducting power, by reducing the temperature of the heated body, has a constant tendency to diminish the activity of radiation, which is always proportional to the excess of the temperature of the heated body above that of the surrounding medium. (Fig. 376.)

Count Rumford's experiments with a Torricellian vacuum gives the proportion of five *in vacuo* to three in air for the quantities of heat lost by radiation, and by conduction or diffusion. It is not, perhaps, departing very far from the truth, if it be stated that one half of the heat lost by a heated body escapes by radiation, and that the rest is carried off by the convective power of currents of air.

If the process of radiation was not constantly proceeding, it can easily be imagined that the temperature of our globe would become so elevated by the regular accession of heat from the sun's rays, that the vegetation would be parched up and destroyed, and consequently all animals and the human race must become extinct. The best time to notice the radiation of heat from the earth is at night and after a hot summer's day. If the sky is clear, it will be noticed (with the help of a thermometer,) that the ground is several degrees colder than the air a few feet above it. (Fig. 377.) It is this reduced temperature that causes

Fig. 377. Negretti and Zambra's terrestrial radiation thermometer. The bulb of this instrument is transparent, and the divisions engraved on its glass stem. In use it is placed with its bulb fully exposed to the sky, resting on grass, with its stem supported by little forks of wood, and protected from the wind.

the deposition of dew, and produces the earth-cloud which so nearly resembles a sheet of water as to have been occasionally mistaken for an inundation, the occurrence of the previous night. Mr. Luke Howard has called this cloud, which is the lowest form of these draperies of the sky, "The Stratus," or evening mist; but when permanent, and increased to a depth so as to rise above our heads, it is then called the morning fog, so peculiarly agreeable in London when incorporated with the black smoke, making a fine reddish-yellow ochreous mist. By placing a thermometer, standing at the ordinary temperature of the air, cased with a good radiating material, such as filaments of cotton, in the focus of a concave mirror, and by turning this arrangement towards a clear sky in the evening, it will be noticed that the temperature falls several degrees. Good radiators of heat are black and scratched surfaces, filaments of cotton, grass, twigs, boughs, and certain leaves, especially those with a rough surface.

Bad radiators of heat are bright and polished metallic surfaces, white woollen cloth or flannel, hard and dense substances, such as a gravel path and stone, or those leaves which have a polished surface, such as the common laurel. It is the frozen dew and mist which produce the beautiful effect of hoar-frost and icicles on the trees and bushes, the primary cause being the radiation of heat from the various objects on the surface of the earth, as well as from the latter itself. When the wind is high, dew does not deposit, as it is necessary that the air should be calm, in order to receive the cooling impression of the cold earth, and to deposit the moisture, which it holds in solution as invisible steam. When the wind blows, it mixes all parts of the air together, and prevents that difference of temperature which causes the deposit of dew. Hence the evening mist will be more generally observed in the bosom of a valley surrounded by hills and screened from the winds that may blow from either quarter. The continual presence of moisture in the air is well shown by the condensation of water on the outside of a glass of cold spring water, or especially on the outside of a jug containing iced water. The invisible steam is always ready to bathe the tender plants with dew, which would otherwise perish and be burnt up during a hot summer, if they did not radiate heat at night, and thus condense water upon themselves. The presence of watery vapour in the air becomes therefore a matter of great importance, and hence the construction of hygrometers or measurers of the moisture in the air.

Regnault's condenser hygrometer consists of a tube made of silver, very thin, and perfectly polished; the tube is larger at one end than the other, the large part being 1·8 in depth by 8·10 in diameter. This is fitted tightly to a brass stand, with a telescopic arrangement for adjusting when making an observation. The tube has a small lateral tubulure, to which is attached an India-rubber tube with ivory mouth-piece; this tubulure enters at right angles near the top, and traverses it to the bottom of largest part. A delicate thermometer is inserted in through a cork, or India-rubber washer, at the open end of the tube, the bulb of which descends to the centre of its largest part. A ther-mometer is attached for taking the temperature of the air; also a bottle for containing ether.

To use the condenser hygrometer, a sufficient quantity of sulphuric ether is poured into the silver tube to cover the thermometer bulb. On allowing air to pass bubble by bubble through the ether, by breathing in the tube, an uniform temperature will be obtained; if the ether continues to be agitated by breathing briskly through the tube, a rapid reduction of temperature will be the result. At the moment the ether is cooled down to the dew-point temperature, the external surface of that portion of the silver tube containing the ether will become covered with a coating of moisture, and the degree shown by the thermometer at that instant will be the temperature of the dew-point.

The most simple form of the hygrometer was formerly a very favourite indicator of the state of the weather, and usually consisted of the figure of a monk with his hood, which is attached to a bit of catgut; this covering of paper, painted to represent the hood, falls over the head on the approach of damp weather, and inclines well back during the period that the air is dry or contains less moisture; and simple as it is, this hygrometer, in conjunction with the reading of the barometer, may assist *Paterfamilias* in deciding the fate of a pet bonnet or velvet mantle, which is or is not to be worn on a doubtful day. (Fig. 378.)

A decision on the possible changes of the weather requires considerable experience, and it has been said that one of the most celebrated marshals of France owed his invariable success in military combinations and

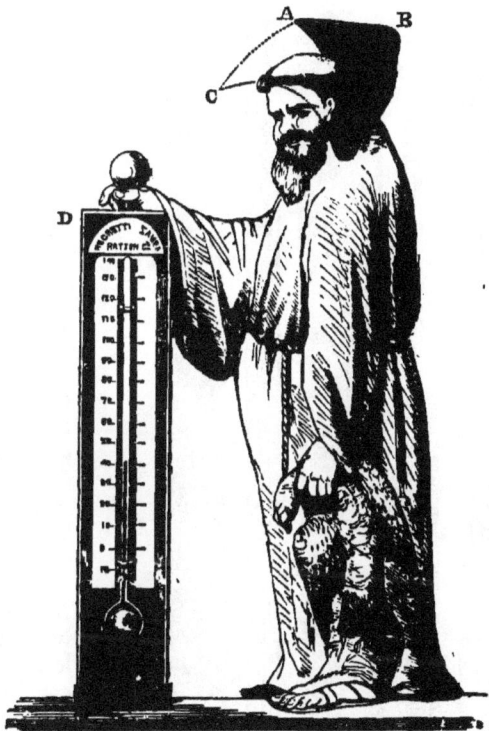

Fig. 378. The monk hygroscope, in which the hood, A B, covers the head to dotted line o in wet weather, and takes various intermediate positions, being quite back and on the shoulders in dry states of the air. A thermometer, D, is usually attached.

attacks to his attention to the signs of the weather, as indicated by the state of the air during the phases of the moon. Inexperienced persons (and by that we mean young persons) may, however, take a certain position in the rank of "weather prophets" by consulting the weathercock, the barometer, and the hygrometer, before committing themselves to an opinion, if asked to say what the weather will be.

The dry and wet bulb hygrometer (as represented in the next engraving) consists of two parallel thermometers, as nearly identical as possible, mounted on a wooden bracket, one marked *dry*, the other *wet*. The bulb of the wet thermometer is covered with thin muslin, round the

D D

neck of which is twisted a conducting thread of lamp-wick, or common darning-cotton; this passes into a vessel of water, placed at such a distance as to allow a length of conducting thread of about three inches; the cup or glass is placed on one side, and a little beneath, so that the water within may not affect the reading of the *dry bulb thermometer*. In observing, the eye should be placed on a level with the top of the mercury in the tube, and the observer should refrain from breathing whilst taking an observation. The temperature of the air and of evaporation is given by the readings of the *two thermometers*, from which can be calculated the dew-point, tables being furnished for that purpose with the instrument. (Fig. 379.)

The colour of the sky at particular times affords the most excellent guidance to doubting members of pic-nic or other out-of-door parties. Not only does a rosy sunset presage fine weather, and a ruddy sunrise bad weather, but there are other tints which speak with equal clearness and accuracy. A bright yellow sky in the evening indicates wind; a pale yellow, wet; a neutral grey colour constitutes a favourable sign in the evening, an unfavourable one in the morning. The clouds, again, are full of meaning in themselves. If their forms are soft, undefined, and feathery, the weather will be fine; if their edges are hard, sharp, and defined, it will be foul. Generally speaking, any deep, unusual hues betoken wind or rain, while the more quiet and delicate tints bespeak fine weather.

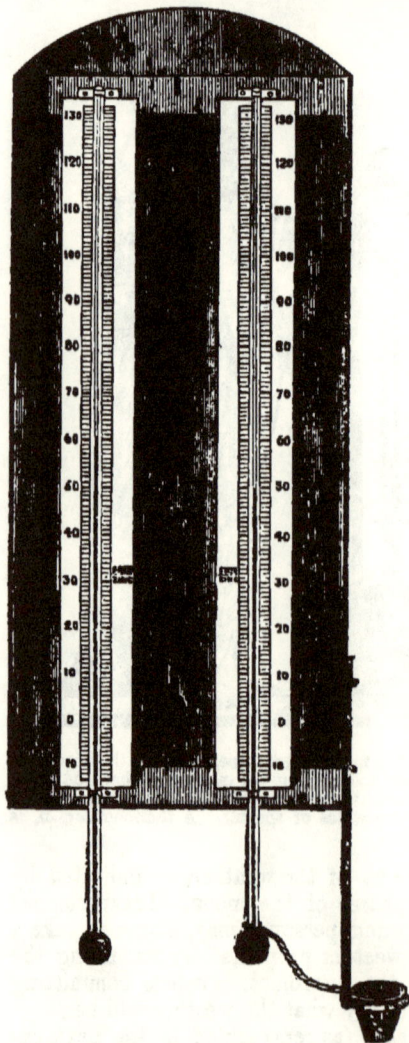

Fig. 379. The dry and wet bulb hygrometer.

The principle of radiation of heat is employed by the Indian natives in the neighbourhood of Calcutta for the purpose of obtaining small quantities of ice. In that climate, the thermometer during the coldest nights does not indicate a lower temperature than about 40° Fahr. The sky, however, is perfectly cloudless, and as heat radiates with great rapidity from the surface of the ground, the Indian natives ingeniously place very shallow earthenware pans on straw, which is a bad conductor of heat, and hence insulates the pans from communication with the parched earth. In a few hours, the water in the pans is covered with a thin sheet of ice, and there can be no doubt of its production by an absolute loss of heat by radiation, because the plan does not succeed on a windy night, and succeeds best even when the pans are sunk in trenches dug in the earth. A windy night prevents that difference of temperature between one portion of the surface of the earth and another, which is so essential to a steady and uniform loss of heat, as it must be evident that the continual mixture of warmer portions of air with that which is colder would tend to prevent the desired lowness of temperature being attained.

The manner in which heat is observed to be radiated has suggested another theory to the fertile brain of philosophical observers, and it has been supposed that the conduction of heat may be nothing more than a radiation from one particle of matter to another, as through a bar of copper, in which the particles, though packed closely together, are not supposed to be in actual contact, so that it is possible to conceive each separate atom of copper receiving and radiating its heat to the neighbouring particle, and so on throughout the length and breadth of the metal. By this theory the radiation of heat through a vacuum is brought into close connexion with that of the radiation of heat through the air and other solid and liquid bodies.

Some of the most interesting phenomena of heat are those discovered by Leslie, who has proved in a very satisfactory manner that the rapidity with which a body cools, depends (like the reflection of light) more on the condition of the surface than on the nature of the material of which the surface is composed. With a globular and bright tin vessel it was observed that water of a certain heat contained in it, required 156 minutes to cool; but when the latter vessel was covered with a thin coating of lamp-black and size, the water fell to the same degree as that noticed in the first experiment in the space of eighty-one minutes.

By very careful observations made with a differential air thermometer, Leslie determined that the power of radiating heat in various substances was as follows:—

Lamp-black	100
Writing paper	98
Sealing wax	95
Crown glass	90
Plumbago	75
Tarnished lead	45
Clean lead	19
Iron, polished	15
Tin plate, gold, silver, copper	12

As in the reflection of light, it was noticed that a piece of charcoal covered with gold leaf, partook of the nature of the precious metal so far as its power of throwing off or scattering the rays of light was concerned, so a piece of glass covered with gold-leaf appears to possess the same power of radiating heat as that of any brilliant metal.

Radiant heat, like light, can be propagated through a great variety of substances, but is stopped by the larger number; and it can be reflected, refracted, polarized, absorbed, or it may undergo a secondary radiation.

The intensity of radiant heat follows the same law as that of light, and decreases as the square of the distance from its source. The same law that governs the reflection of light, also prevails with that of heat; and it may be found by experiment that the angle of incidence is equal to the angle of reflection, so that the heat is disposed of in the same manner as light when it falls upon bright polished planes, convex and concave surfaces; hence the use of bright tin meat screens and Dutch ovens, and of all those simple pieces of culinary furniture which are employed in the kitchen for the purpose of arresting the cold currents of air that set towards burning matter, as also to reflect the heat upon whatever viands may be cooking before the fire. A bright silver teapot retains its heat better than a dirty one, and the fact is determined very readily by pouring boiling water into two teapots, the one being made of bright tin and the other of black japanned tin. A thermometer inserted into each vessel will soon show that the latter radiates, and therefore loses its heat quicker than the former; the relative radiating powers of bright and blackened tin being as 15 to 100. Pipes for the conveyance of hot water or steam should be kept bright, if possible, although this trouble is avoided usually by packing them in bad conductors of heat, whilst the polish of the cylinder of a steam-engine is of great importance as a means of economizing heat.

When the finger is approached within an inch or so of a red-hot ball, the heat radiated from the latter is so intense that it cannot be held there

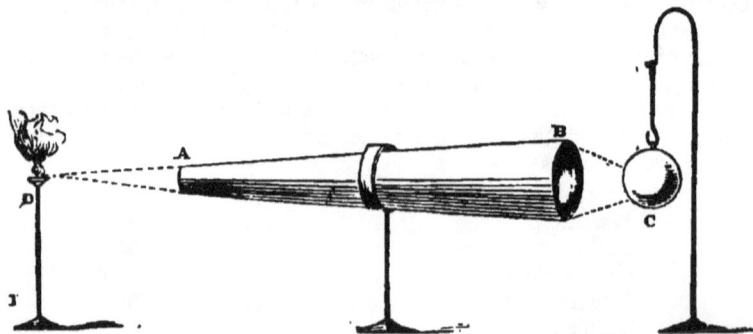

Fig. 390. A B. The cone of paper, gilt inside. c. The red-hot ball. D. Stand with wood supporting a slice of phosphorus, which is brought into the focus of the rays of heat reflected through the cone.

for more than a few seconds. If, however, the finger is coated with gold leaf it may be kept near the iron ball for some considerable time, because the radiant heat is reflected from the surface of the gold. If the word heat is written upon a sheet of paper and the letters afterwards gilt, the whole of the white surface is rapidly toasted and scorched when held before a fire, whilst the surface of the paper under the gold leaf remains perfectly white, which can be ascertained by turning the paper round and observing the other side. A sheet of paper gilt inside and turned round as a cone, being left open at both ends, may be employed as a reflecting surface; and if a bit of phosphorus, placed on paper, is held, say at two feet from a red-hot ball of about two inches diameter, the radial heat from the latter has not sufficient intensity at that distance to set it on fire quickly; if, however, the cone of gilt paper is used between the two, and the phosphorus brought into the focus of the rays of radial heat, it very quickly takes fire. (Fig. 380.)

Dr. Bache has determined by experiments that the radiation of heat from a body is not affected by colour, so that in winter all coloured clothes are alike in that respect, and radiate heat without any appreciable difference. The power of *absorbing* heat, however, is greatly dependent on colour; and as a general rule, good radiators of heat (such as a black cloth, or indeed any surface covered with lamp-black), are also excellent absorbents of heat. Dr. Hooke and Dr. Franklin placed pieces of cloth of similar texture and size on snow, allowing the sun's rays to fall equally upon them. The dark specimen always absorbed more heat than the light ones, and the snow beneath them melted to a greater extent than under the others; and they both remarked that the effect was nearly in proportion to the depth of the shade, as in the following order:—After black, the maximum absorbent quality was possessed by, first, blue; second, green; third, purple; fourth, red; fifth, yellow. The minimum absorbent power was observed to belong to white.

When radiant heat is allowed to pass through glass, the latter substance is not found to be transparent to heat rays as it is to those of light, but a considerable proportion of heat is arrested and stopped; consequently glass fire-screens are to be found in the mansions of the wealthy, because they obstruct the heat but do not exclude the cheerful light and blaze of the fireside.

Melloni's researches on the nature of the rays of heat, and also on the media which affect them, would demand and merit a chapter to themselves; want of space, however, obliges us to omit the consideration of thermo-electricity, and the refined and beautiful experiments of Melloni, whose labours are a model for the imitation of all original seekers after truth.

Fig. 391.   Hancock's steam omnibus, which ran on the common roads.

CHAPTER XXVIII.

THE STEAM-ENGINE.

IT must be apparent to those who read popular works on science, that they possess, at all events, one point of utility—viz., that they are *indicative* of the various subjects that may be selected in science for special, searching and exhaustive study.   The subject of steam and the steam engine is not one that could be thoroughly treated of in the narrow space allowed in this volume, but enough may be said to give some instruction and to impart common principles, whilst the minute details are better examined and learnt in the works of Bourne, Rankine, and other authors who devote themselves specially to the important commercial question of steam.

The first truth to be comprehended is, that all matter contains within its substance the power of creating heat—or as it may be expressed more plainly, solids, fluids, and gases contain what is termed *latent* or insensible heat, in contradistinction to the heat which is apparent when we touch a vessel containing warm water or approach a cheerful fire; this latter is termed *sensible* heat, and has formed the subject of the preceding chapters.

If a cold horse-shoe nail is applied to a thin dry slice of phosphorus laid on a sheet of paper, no combustion of the phosphorus ensues, because the temperature of the iron is not sufficiently high to affect that combustible substance; but if the horse-shoe nail is vigorously hammered on an anvil, the particles of the metal are brought closer together, and if it is applied to the phosphorus, so much heat has been generated, thrust or squeezed out by the hammering or *condensation* of the iron, that it is now sufficiently warm to set fire to it.

The reverse or antithesis to this experiment—viz., the production of cold—would be shown if it were possible to expand a mass of metal suddenly, and this can be effected by first melting together

207 parts by weight of lead.
118   „   „   tin.
284   „   „   bismuth.

When these metals are in the liquid state and perfectly mixed, they are poured from a sufficient height into a pail of cold water, for the purpose of *granulating* or dividing them into small fragments.

If the granulated compound metal is now mixed with 1617 parts by weight of quicksilver, it becomes suddenly liquefied and expanded: liquefaction is the reverse of solidification, and hence cold is produced from the natural heat of the compound metals being rendered latent by the change from the solid to the liquid state; so that a small quantity of water placed in a glass tube, and surrounded with the metals whilst liquefying in the mercury, becomes rapidly converted into ice, the fall of the temperature, as shown by a thermometer, being from 60° Fahr. to 14°, which is 18° degrees below the freezing point of water. In the former case, by hammering the iron the *latent heat* is made *sensible;* whilst in the latter case, by the liquefaction of the compound metal in mercury, the *sensible* heat is rendered *latent.* The heat rendered latent by melting different substances is not a constant quantity, but varies with every special body employed, and the Drs. Irvine have proved this fact by the following experiments :—

	Heat of fluidity.	Ditto, reduced to the specific heat of water.
Sulphur	143·68° Fahr. . . .	27·14.
Spermaceti . . .	145 „	—
Lead	163 „	5·6.
Bees'-wax	175 „	—
Zinc	493 „	48·3.
Tin	500 „	33·
Bismuth	550 „	23·25.

Every one of these substances requires more heat to bring them into the liquid condition than ice, for which 140° of heat are sufficient, or are rendered latent during its conversion into water.

In coining at the Mint, the cold blank pieces of gold, silver, or copper become hot directly they have sustained the violent and sudden pressure of the coining press, and they must be heated again, or annealed, to restore the equilibrium of the heat disturbed by the violent blow, or else they remain hard and unfit to sustain the finishing process of milling.

The condensation of water when it assumes a smaller bulk by union with sulphuric acid, is easily proved by measuring a pint of water and a pint of acid, and mixing them together, when a very great increase of temperature may be perceived; and by placing into the mixture a cold copper wire that previously could not ignite phosphorus, it becomes very hot, and when removed and wiped it will cause phosphorus to fire directly it touches that substance. When the mixture of sulphuric acid and water is measured after it has cooled, it has no longer a bulk of two pints, but is found to have lost bulk equal to one or more ounces by measure. The heat evolved by a mixture of four parts of strong sulphuric acid and one part water is shown by the thermometer to be 300° Fahr., and this mode of obtaining heat has been used by aeronauts for the purpose of obtaining artificial warmth without the danger of setting fire to the gas in the balloon.

Fig. 382. Aeronauts in the car warming their hands by a bottle containing sulphuric acid and water

When alcohol and water are mixed a change of density occurs, and heat is produced; and if equal measures of alcohol of a specific gravity of ·825, and water, each at 50° Fahr., are mixed, a temperature of 70° Fahr. is obtained; if the mixture is made in a glass vessel, as shown in the annexed cut, the combination is very apparent. To perform the experiment properly, water is poured into the lower tube and bulb, and alcohol into the top one; when this is done, the stopper is inserted, and the whole thoroughly shaken and mixed together; the warmth which is thus obtained is apparent to the hand, whilst the contraction is shown after the mixture is cold, as it no longer fills the two bulbs of the instrument. (Fig. 383.)

The latent heat of gases is easily shown by suddenly condensing air in a small syringe or pump, of which the piston contains a minute fragment of amadou (a species of fungus, *Polyporus igniarius ;* this, according to Simmonds, after having been beaten with a mallet, and dipped in a solution of saltpetre, forms the spunk or German tinder of commerce; it is also used as a styptic, and made into razor strops), which takes fire, and before the invention of vesta and other matches, tobacco-smokers were in the habit of obtaining a light for their pipes and cigars in this manner—viz., by the latent heat obtained from the contraction or compression of air. Then, again, an instructive though opposite parallel is afforded by suddenly expanding or rarefying air in a glass receiver provided with a delicate thermometer. By pumping out some of the air, a considerable diminution of the temperature occurs, and equal to several degrees of the thermometer. Every child knows that steam direct from the kettle will scald, but if it issues from a high-pressure boiler, say at fifteen pounds on the square inch, the hand may be held with impunity in the escaping steam, as it merely feels gently warm, and not scalding. This is due partly to the loss of heat rendered latent by the expansion of the high-pressure steam directly it passes into the air, and partly to the currents of air that are dragged into an escaping jet of steam. This

Fig. 383. Glass bulbs and tube to show the contraction in bulk of a mixture of alcohol and water.

tendency of the air to rush into a jet of steam was discovered by Faraday, and explains those curious experiments with a jet of steam by which balls, empty flasks, and globular vessels are sustained and supported either perpendicularly or horizontally.

If steam at a pressure of about sixty pounds per inch is allowed to escape from a proper jet, and a large lighted circular torch composed of tow dipped in turpentine held over it, the course of the external air is shown by the direction of

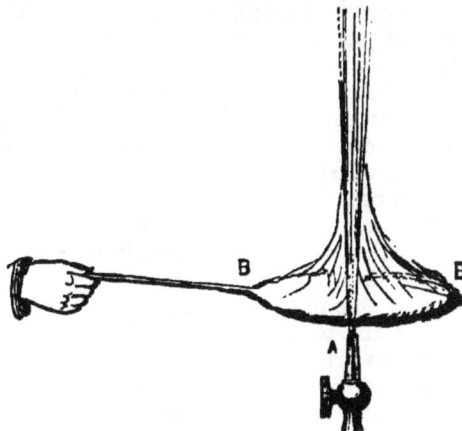

Fig. 384. A. Jet discharging high-pressure steam B B. Lighted torch held round the escaping steam the flames from the former all rush into the latter.

the flames, which are forcibly pulled and blown into the jet of steam with a roaring noise, indicating the rapidity of the blast of air moving to the steam jet. (Fig. 384.)

Egg-shells, empty flasks, india-rubber or light copper and brass balls, are suspended in the most singular manner inside an escaping jet of high-pressure steam ; and before the explanation of Faraday, reams of paper were used in the discussion of the possible theory to account for this effect ; and what made the explanation still more difficult, was the fact that the jet of steam might be inclined at any angle between the horizontal and perpendicular, and still held the ball, egg-shell, or other spherical figure firmly in its vapory grasp. (Fig. 385.)

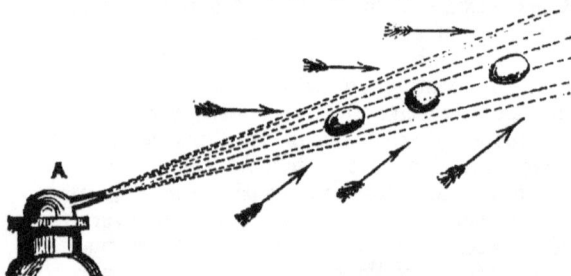

Fig. 385. A. Ball and socket jet at an angle, and discharging steam. The egg-shells are supported by the enormous current of air moving into the jet in the direction of the arrows.

In consequence of the great rush of air towards a jet of escaping high-pressure steam, Mr. Goldsmith Gurney has patented the application of this principle in his ventilating steam jet, which he has already successfully applied ; in one case especially, where a coal-mine had been on fire for several years, and the whole working of the coal-measures in the pit was jeopardized by the spreading of the combustion to new workings ; the fire was first extinguished by a jet of steam blowing into the *downcast*, but placed in connexion with a furnace of burning coke ; and the circulation of the carbonic acid, called *choke-damp*, through the pit workings was further assisted by a jet of high-pressure steam blowing upwards, and placed over the mouth of the *upcast* shaft.

The experiment succeeded perfectly at the South Sauchie Colliery, near Alloa, about seven miles from Stirling, where a fire had raged for about thirty years over an area of twenty-six acres in the waste seam of coal nine feet thick. (Fig. 386.)

For the general purpose of .ventilating the coalmine, Mr. Gurney's plan was tried at the Ebbw Vale Colliery, and very economically, the waste steam alone being used. Experiments have also been satisfactorily made with it for blowing a cupola for smelting iron, and with dry steam—*i. e.*, steam of a very high pressure—escaping through a warm tube, the results were perfectly successful.

With this digression from the subject of latent heat derived from

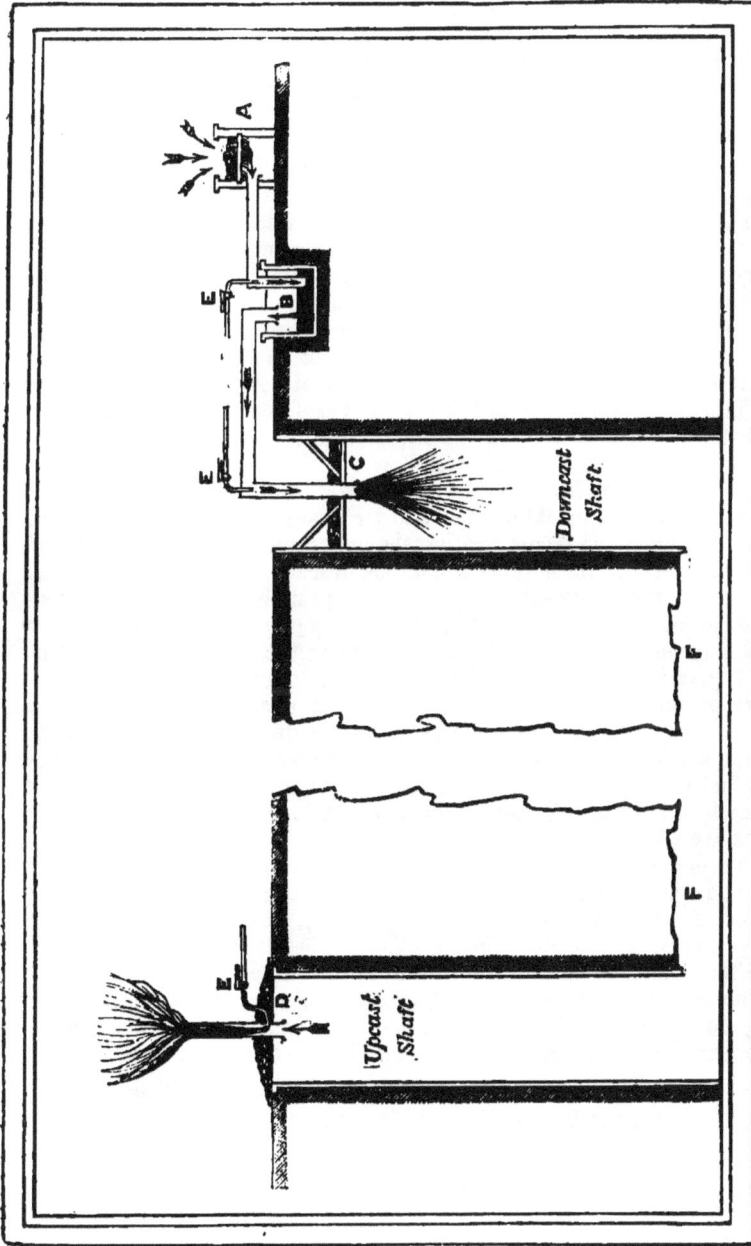

Fig. 386. Gurney's steam jet. A, Furnace. B, Water tank. C, Downcast stopping. D, Upcast stopping. E E E, Steam jets. F F, Galleries from shaft to shaft.

the compression of air, we return again to the subject with another case in point, furnished by the Fountain of Hiero, as it is called, at Schemnitz, in Hungary, described by Professor Brande; and it may be observed that all the phenomena related would apply to the great pressure of the water from the water-towers at the Crystal Palace, if fitted with a similar air-vessel.

"A part of the machinery for working these mines is a perpendicular column of water 260 feet high (the Crystal Palace water-towers are each 284 feet high), which presses upon a quantity of air enclosed in a tight reservoir; the air is consequently condensed to an enormous degree by this height of water, which is equal to between eight and nine atmospheres; and when a pipe communicating with this reservoir of condensed air is suddenly opened, it rushes out with extreme velocity, instantly expands, and in so doing it absorbs so much heat as to precipitate the moisture it contains in a shower of snow, which may readily be gathered on a hat held in the blast. The force of this is so great, that the workman who holds the hat is obliged to lean his back against the wall to retain it in its position."

The best examples of latent heat are furnished by ice, water, and steam, and we are indebted chiefly to Dr. Black for the elegant and conclusive experiments demonstrating the important truths connected with the latent heat of these three conditions of matter. When various solids are heated, they frequently pass through certain intermediate conditions of softness, terminating in perfect liquidity; but ice and many other bodies change at once to the liquid state on the application of a sufficient quantity of heat. The process of melting ice is very slow, because every portion must absorb or render latent a certain quantity of heat before it can take the liquid state—hence the difficulty of melting blocks of ice when they are surrounded with non-conducting materials; and this fact the author has proposed to take advantage of in keeping water cool which is to be supplied to the ova of salmon whilst taking them to stock the rivers of Australia.

In order to prove that heat is rendered latent by the liquefaction of ice, it is only necessary to weigh a pound of finely-powdered ice and a pound of water at 212° Fahr. (*boiling water*), and mix them together; when the ice is all melted, the resulting temperature is only 52°, therefore the boiling water has lost 160° of temperature, of which 20° can be accounted for, because the resulting temperature of the melted ice is 52°; but in the liquefaction of the pound of ice, 140° have disappeared or become latent, or, as Dr. Black termed it, have become *combined*.

1 lb. of ice at 32° + 20° = 52°, the resulting temperature.
1 lb. of water at 212° − 52° = 160° − 20 = 140°, rendered latent.

140° represents the result obtained from innumerable experiments made by mixing equal parts of ice and boiling water, and it is this large quantity of latent heat required by ice and snow that prevents their sudden liquefaction, and the disastrous circumstances that would arise from the floods that must otherwise always be produced.

To put the fact beyond all doubt, it is advisable to mix together equal weights of water at 32° and boiling water at 212°, and the result is found by the thermometer to be the mean between the two, because half the extremes are always equal to the mean; and if the two temperatures are added together and divided by two, the result is a temperature of 122°, as shown below:—

1 lb. of ice water at °32+1 lb. of water at 212°=244°÷2=122°.

From similar experiments Dr. Black deduced the important truth, "that in all cases of liquefaction a quantity of heat *not indicated by*, *or sensible to*, the thermometer, is *absorbed* or disappears, and that this heat is *withdrawn* from the *surrounding bodies*, leaving them *comparatively cold.*" At p. 79 it is shown how the sudden solution or liquefaction of certain salts produces cold, and hence numerous freezing mixtures have been devised. In olden times, when officials in authority did what they pleased, without being troubled with disagreeable returns, and colonels clothed their men, and were merchant tailors on the grand scale, gun cartridges were not confined to practice on the enemy, but they did duty frequently in the absence of ice as refrigerators of the officers' wine, in consequence of the gunpowder containing nitre or saltpetre; as a mere solution of this salt finely powdered will lower the temperature of water from 50° Fah. to 35°; whilst a mixture of four ounces of carbonate of soda and four ounces of nitrate of ammonia dissolved in four ounces of water at 60°, will in three hours freeze ten ounces of water in a metallic vessel immersed in the mixture during the liquefaction or solution of the salts.

Fahrenheit imagined he had attained the lowest possible temperature by mixing ice and salt together, and it is by this means that confectioners usually freeze their ices, or ice puddings; the materials are first incorporated, and being placed in metallic vessels or moulds, and surrounded with ice and salt placed in alternate layers, and then well stirred with a stick, they soon solidify into the forms which are so agreeable, and so frequently presented at the tables of the opulent. The temperature obtained is Fahrenheit's *zero*—viz., thirty-two degrees *below* the freezing point of water. According to the very wise police regulation observed in London, all householders are required to sweep or remove the snow from the pavement in front of their houses, and this is frequently done with salt; should an unfortunate shoeless beggar, tramp past whilst the sudden liquefaction is in progress, the effect on the soles of his feet is evidently very disagreeable, and the rapidity with which he retires from the *zero* affords a thermometric illustration of the most lively description.

Heat the Cause of Vapour.

Every liquid, when of the same degree of chemical purity, and under equal circumstances of atmospheric pressure, has one peculiar point of temperature at which it invariably boils. Thus, ether boils at 96° Fahr., and if some of this highly inflammable liquid is placed carefully in a flask, by pouring it in with a funnel, and flame applied within one inch of the orifice, no vapour escapes that will take fire; but if the flame of a spirit lamp is applied, the ether soon boils, and if the lighted taper is again brought near the mouth of the flask, the vapour takes fire, and produces a flame of about two feet in length. This fire only continues as long as the flame of the spirit-lamp is retained at the bottom of the flask, and on removing it the vessel rapidly cools. The length of the flame is reduced, and is gradually extinguished for the want of that essence of its vitality, as it were— viz., heat. (Fig. 387.) If a thermometer is introduced into the flask, however rapid may be the ebullition or boiling of the ether, it is found to be invariably at 96°.

Fig. 387. Heat the cause of vapour.

The heat carried off by evaporation is most elegantly displayed by placing a little water in a watch glass, and surrounded by charcoal saturated with sulphuric acid, in the vacuum of an air-pump. The rapid evaporation and condensation of the water by its affinity for the sulphuric quickly produces ice; and the pumps and other apparatus of Knight and Co., Foster-lane, City, are greatly to be recommended for this and other illustrations.

The illustration of the determination of the fixed and invariable boiling point belonging to every liquid is further carried out by introducing some water into a second flask standing above a lighted spirit-lamp, with a small thermometer, graduated, of course, properly to degrees above the boiling point of water; when the water boils, it will be found to remain steadily at a temperature of 212°. And however rapidly the water may be boiled, provided there is ample room for the steam to escape, the heat indicated by the thermometer is like the law of the Medes and Persians, which altereth not, and it remains standing at the number 212°. The only exception (if it may be so termed) to this law is brought about by the shape and nature of the containing vessel; under a mean pressure the boiling point of water in a metallic vessel is generally 212°; in a glass vessel it may rise as high as 214° or 216°, but if some metallic filings are dropped in, the escape of steam is increased, and the temperature may then drop immediately to 212°.

When a thermometer is inserted in a flask containing water in a state of ebullition or boiling, so that the bulb does not touch the fluid, but is wholly surrounded with steam, it will be found that the temperature of the latter is exactly the same as that of the former; and if the liquid boils at 96°, the vapour will be 96°, if at 212°, the steam is 212°. Steam has therefore exactly the same temperature as the boiling water that produces it. (Fig. 388.)

Whilst performing the last experiment, it may be noticed that the steam inside the neck of the flask is invisible, and that it only becomes apparent in that kind of intermediate condition between the vaporous and liquid state called *vesicular vapour*— a state corresponding with the "earth fog," and called by Howard the *stratus*. When a flask containing boiling water is placed under the receiver of an air pump (as soon after the ebullition has ceased as may be possible), and the air pumped out, it will be noticed that the water again begins boiling as the vacuum is obtained, showing that the boiling point of the same fluid varies under different degrees of atmospheric pressure, and according to the height of the barometer.

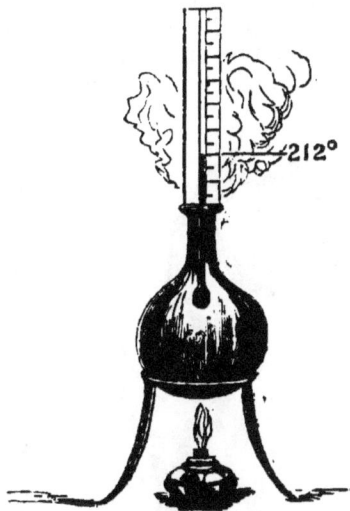

Fig. 388. Thermometer in the steam escaping from boiling water.

Height of barometer.	Boiling point of water.	Height of barometer.	Boiling point of water.
26	204·91°	29	210·19°
26·5	205·79	29·5	211·07
27	206·67	30	212
27·5	207·55	30·5	212·88
28	208·43	31	213·76
28·5	209·31		

Alcohol and ether confined under an exhausted receiver boil violently at the ordinary temperature of the atmosphere, and in general liquids boil with 124° less of heat than are required under a mean pressure of the air; water, therefore, in a vacuum must boil at 88° and alcohol at 49°.

On ascending considerable heights, as to the tops of mountains, the boiling point of water gradually falls in the scale of the thermometer. Thus, on the summit of Mont Blanc water was found by Saussure to boil at 187° Fahr. In Mr. Albert Smith's delightful narrative of his ascent of Mont Blanc, he mentions the violent commotion and escape of the whole of the champagne in froth directly the bottle was opened at the summit of this king of mountains.

Dr. Wollaston's instrument for measuring the heights of mountains by the variations of the boiling point of water has long been known and used for this purpose.

If a Florence flask is first fitted with a nice soft cork, and this latter removed, and the former half filled with water, which is then boiled over a gas or spirit flame, the same fact already mentioned and illustrated in the preceding table may be rendered apparent when the flask is corked and removed from the heat. If it is now inverted, and cold water poured over it, an ebullition immediately commences, because the cold water condenses the steam in the space above the hot water in the flask, and producing a vacuum, the water boils as readily as it would do under an exhausted receiver on an air-pump plate. (Fig. 389.)

Water may be heated considerably higher than 212°, if it is enclosed in a strong boiler, and shut off from communication with the air; by this means steam of great pressure is obtained.

Dr. Marcet has invented a very instructive form of a miniature boiler, supplied with a thermometer and barometric pressure gauge, which can be purchased at any of the instrument makers, and is figured and described in nearly every work on chemistry.

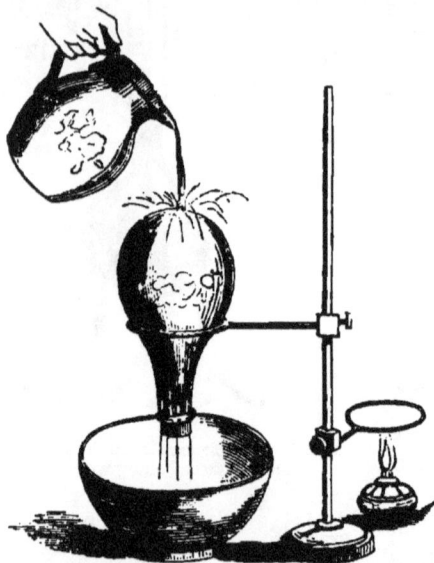

Fig. 389.   The paradoxical experiment of water boiling by the application of *cold* water.

The reason water boiled in an open vessel does not rise to a higher temperature than 212° is because all the excess of heat is carried off by the steam, and is said to be rendered latent in the vapour. The fixation of caloric in water by its conversion into steam may be shown by the following experiment. Let a pound of water at 212° and eight pounds of iron filings at 300° be suddenly mixed together. A large quantity of steam is instantly generated, but the temperature of the water and escaping steam are still only 212°; hence the steam must therefore contain all the degrees of heat between 212° and 300°, or eight times 88. When the water is heated in the hydro-electric machine or other boiler, to 322·7°, it very quickly drops to 212° when the steam is allowed to blow off; yet if the latter is collected, it represents but a very small quantity of water which constituted the steam, and it has carried off and rendered latent the excess of heat in the boiler—viz., the difference between 212° and 322·7°, or 110·7°

If steam can carry off heat, of course it may be compelled, as it were, to surrender it again; and this important elementary truth is shown by adapting a tube, bent at right angles, and a cork, to a flask containing a few ounces of water, and when it boils, the steam issuing from the end of the pipe may now be directed into and below the surface of some water contained in a beaker glass; in a very short time the water in the latter will be raised to the boiling point by the condensation of the steam and the latent heat arising from it. (Fig. 390.) The amount of latent heat is enormous, when it is remembered that water by conversion into steam has its bulk prodigiously enlarged—viz., 1698 times, so that *a cubic inch* of water converted into steam of a temperature of 212°, with the barometer at thirty inches, occupies a space of *one cubic foot*, and its latent heat amounts, according to Hall, to 950°; Southeron, 945°; Dr. Ure, 967°. When we come to the consideration of the steam-engine, it will be noticed that the question of the latent heat of steam is one of the greatest importance.

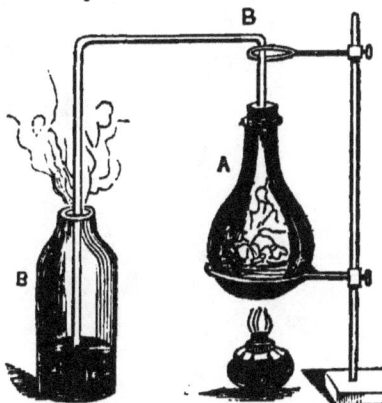

Fig. 390. A. Flask for generating steam. B. Glass pipe bent at right angles to convey the steam into the fluid containing some cold water.

Temperature of Steam.	Elasticity in inches of Mercury.	Latent Heat.
229°	40°	942°
270	80	942
295	120	950

The same weight of steam contains, whatever may be its density, the same quantity of caloric, its latent heat being increased in proportion as its sensible heat is diminished; and the reverse. In consequence of the enormous amount of latent heat contained in steam, it is advantageously employed for the purpose of imparting warmth either for heating rooms or drying goods in certain manufacturing processes. The wet rag-pulp pressed and shaken into form on a wire-gauze frame or *deckle*, passes gradually to cylinders containing steam, and is thoroughly dried before the guillotine knife descends at the end of the paper machine, and cuts it into lengths. In calico stiffening and glazing, also in calico printing, steam-heated cylinders are of great value, because they impart heat without the chance of setting the goods on fire. The elementary principles already described with reference to heat, will prepare the youthful reader for the application of the expansion of water into steam, as the most valuable *motive power* ever employed to assist the labour of man.

Fig. 391.　The first steam-boat, the *Comet*, built by Henry Bell, in 1811, who brought steam navigation into practice in Europe.

CHAPTER XXIX.

THE STEAM-ENGINE—*continued.*

"So shalt thou instant reach the realm assign'd
In wondrous ships, *self-mov'd*, instinct with mind.
　　*　　　　　*　　　　　*　　　　　*
Though clouds and darkness veil the encumbered sky,
Fearless, through darkness and through clouds they fly,
Tho' tempests rage,—tho' rolls the swelling main,
The seas may roll, the tempests swell in vain;
E'en the stern god that o'er the waves presides,
Safe as they pass, and safe repass the tides,
With fury burns; while careless they convey,
Promiscuous, ev'ry guest to ev'ry bay."

THESE lines, from Pope's translation of the "Odyssey," were very aptly quoted twenty-five years ago by Mr. M. A. Alderson, in his treatise on the steam-engine, for which he received from Dr. Birkbeck, the originator of Mechanics' Institutions, the prize of 20*l.*, being the gift of the London Mechanics' Institution, and these lines seem to indicate some sort of rude anticipation by the ancients of that free passage of the ocean by the agency of steam which has rendered ships almost independent of wind and weather.

Homer's description, as above, of the Phœnician fleet of King Alcinous, in the eighth book of the "Odyssey," is certainly an ancient record of an *idea*, but nothing more. In a work written by Hero of Alexandria, about a hundred years B.C., and entitled "Spiritalia seu Pneumatica," a number of contrivances are mentioned for raising liquids and producing motion by means of air and steam, so that the first steam-engine is usually ascribed to Hero; and the annexed cut displays the apparatus. (Fig. 392.)

It is a remarkable circumstance that Sir Isaac Newton applied the same principle in a little ball, mounted on wheels, containing boiling water, and provided with a small orifice; and in his description he says: "And if the ball be opened, the vapours will rush out violently one way, and the wheels and the ball at the same time will be carried the contrary way." From the time of Hero, there does not appear to be any record or mention made of steam apparatus till the year 1002, when, in a work called "Malmesbury's History," mention is made of an organ in which the sounds were produced by the escape of air (query, steam) by means of heated water. It is strange that, in these days of steam application, the Calliope, or steam organ, should be an

Fig. 392. Hero's steam-engine. A. The boiler in which steam is produced, and then passes through the hollow support B, from which there is no outlet but through the two apertures, C C. The reaction of the air on the issuing steam produces a rotatory motion in the jets, C C, attached to a centre but hollow axle.

important feature at the present moment at the Crystal Palace; and it only shows how the same ideas are reproduced as novelties in the everrecurring cycles of years.

On the revival of classical learning throughout Gothic Europe, the work of Hero began to attract attention, and it was translated and printed in black letter, and most likely first from the Arabic character, as in the year 1543 the first fruits appeared in Spain, where Blasco de Garay, a sea captain, propelled a ship of 200 tons burden, at the rate of three miles per hour, before certain commissioners appointed by the Emperor Charles the Fifth. Alas for inquisitorial Spain! had she looked deeper into the matter, and performed her *auto-da-fées* on the boilers of steam-

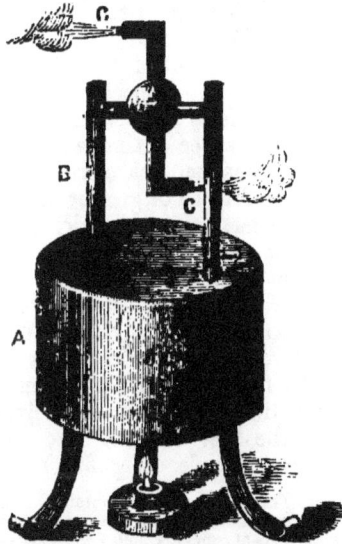

E E 2

engines instead of the bodies of poor human beings, what lasting glories would have been her reward. The invention made its *début* in Spain, the commissioners reported, the worthy inventor was rewarded, but the mighty giant invoked was put to sleep again for at least 150 years. The steam giant was disturbed with dreams; one Mathias, in 1563, gave him a nightmare; Solomon de Caus, in 1624, nearly woke him up; Giovanni Bianca, in 1629, did more; and the Marquis of Worcester, in the middle of the seventeenth century, as the evil genius of Spain, carried off the giant bodily and made him the slave of England; at least, he experimented, and wrote such wondrous tales of his new motive power, that in 1653 we read of steam being fairly tethered to its work, and set to draw water out of the Thames at Vauxhall; and Cosmo de Medici, a foreigner who inspected the apparatus in 1653, says, "It raises water more than forty geometrical feet by the power of one man only, and in a very short space of time will draw up full vessels of water through a tube or channel not more than a span in width, on which account it is considered to be of greater service to the public than the other machine near Somerset House, which last one was driven by *two horses*."

What would the Marquis of Worcester and Cosmo de Medici have thought of Blasco de Garay on the ocean, and ruling 12,000 steam horses? Write the name of the brave and prudent Captain Harrison, in the good ship *Great Eastern*, date 1859, instead of that of the gallant Spaniard, and our brief history is finished.

The first really useful steam-engine was made, not by a plain Mr., but again by a captain—namely, Captain Savery, who appears to have been the first inventor who thoroughly understood and applied the *vacuum* principle. (Fig. 393.)

a a. The furnaces which contain the boiler. B 1 and B 2. The two fireplaces. c. The funnel or chimney, which is common to both furnaces. In these two furnaces are placed two vessels of copper, which I (Savery) call boilers—the one large as at L, the other small as D. D. The small boiler contained in the furnace, which is heated by the fire at B 2. E. The pipe and cock to admit cold water into the small boiler to fill it. F. The screw that covers and confines the cock E to the top of the small boiler. G. A small gauge cock at the top of a pipe, going within eight inches of the bottom of the small boiler. H. A large pipe which goes the same depth into the small boiler. I. A clack or valve at the top of the pipe H (opening upwards). K. A pipe going from the box above the said clack or valve in the great boiler, and passing about one inch into it. L L. The great boiler contained in the other furnace, which is heated by fire at B 1. M. The screw with the regulator, which is moved by the handle Z, and opens or shuts the apertures at which the steam passes out of the great boiler at the steam-pipes o o. N. A small gauge cock at the top of a pipe, which goes half way down into the great boiler. o 1, o 2. Steam pipes, one end of each screwed to the regulator; the other ends to the receivers P P, to convey the steam from the great boiler into those receivers. P 1, P 2. Copper vessels called receivers, which are to receive the water which is to be raised. Q. Screw joints by which the branches of the water-pipes are connected with the lower parts of the receivers. R 1, 2, 3, and 4. Valves or clacks of brass in the water-pipes, two above the branches Q and two below them; they allow the water to pass upwards through the pipes, but prevent its descent; there are screw-plugs to take out on occasions to get at the valves R. s. The forcing-pump which conveys the water upwards to its place of delivery, when it is forced out from the receivers by the impelled steam. T. The sucking-pipe, which conveys the water up from the bottom of the pit to fill the receivers by suction. V. A square frame of wood, or a box, with holes round its bottom in the water, to enclose the lower end of the sucking-pipe to keep away dirt and obstructions. x is a cistern with a bung cock coming from the force-pipe, so as it shall always be kept filled with cold water. Y Y. A cock and pipe coming from the bottom of the said cistern, with a spout to let the cold run down on the outside of either of the receivers, P P. Z. The handle of the regulator to move it by, either open or shut, so as to let the steam out of the great boiler into either of the receivers.

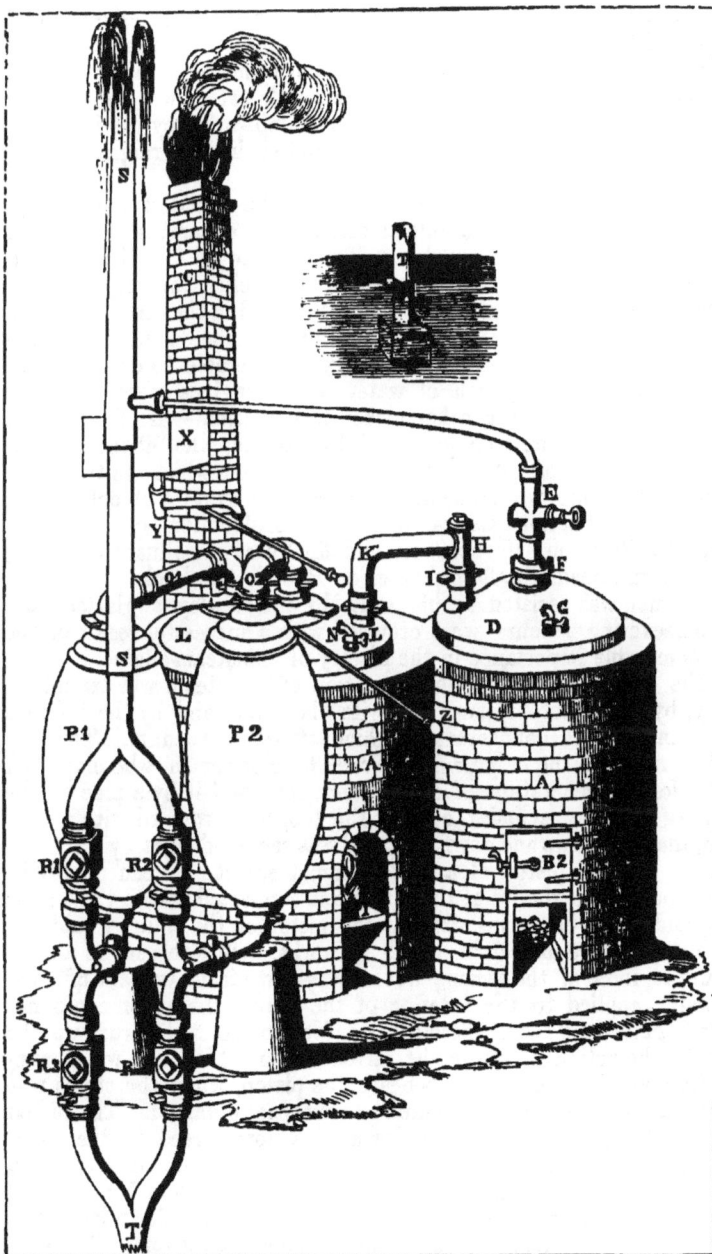

Fig. 393. Savery's engine.

This is Savery's own description (taken from the "Miner's Friend," printed in 1702), of his water-engine, which differs from that suggested by the Marquis of Worcester, in the fact that he made the *pressure of the air* carry the water up the first stage. Savery's patent was "for raising water and occasioning motion to all sorts of mill-work by the impellant force of fire;" and the patent was granted in the reign of King William the Third of glorious memory.

Thus Savery overcame, as he remarks, the "oddest and almost insuperable difficulties," and introduced a steam apparatus or engine, a good many of which were constructed, and employed for raising water. The mechanical skill required to construct the boiler, the very *heart* (as it were) of the iron engine, had not been acquired in the time of Captain Savery, and hence the weakness of the boilers, and the danger of working them. As the pressure required was very considerable to overcome the resistance of a lofty column of water, these engines were gradually relinquished for those of another clever mechanician—viz., for those of Thomas Newcomen, an ironmonger of Dartmouth, who, about the year 1705, constructed and introduced the *cylinder*, from which the transition was gradually made to the mode of condensing by a jet of cold water, the use of self-acting valves, and the construction of self-acting engines by Smeaton, Hornblower, and finally by the illustrious Watt, whose portrait heads the first chapter on Heat in this book.

Newcomen was assisted in his work by one Cawley, a glazier; and their persevering labours were crowned with a successful result of the most memorable importance in the history of the steam-engine.

In the engine by Savery, the operation of the steam was twofold— namely, by the direct pressure from its elasticity, and by the indirect consequence of its condensation, which affords a vacuum. This last may be said to be the only principle used by Newcomen, who employed a boiler for the generation of steam, and conveyed it by a pipe to the bottom of a hollow cylinder, open at the top, but provided with a solid piston, that moved up and down in it, and was rendered tight by a stuffing of hemp, like the piston of a boy's common squirt. It can readily be understood, that if the jet of the latter was connected with a tight little boiler, and steam blown into it, that the piston of the squirt would rise to the top of the barrel in which it works, being thrust up by the pressure or force of the steam; but unless the steam was cut off, and cold water applied to the interior of the barrel, the piston could not descend again. As soon, therefore, as Newcomen had thrust up the piston by the action of steam, he introduced a jet of cold water, sup-plied from an elevated cistern beneath the piston, when the steam was condensed into water, and a vacuum or void space obtained. The piston being free to move either up or down, was now forced in the latter direction by the pressure of the air, which is a constant force equal to fifteen pounds on the square inch; and thus the piston in Newcomen's engine was raised by *heat*—viz., by steam, and thrust down by *cold*— i.e., by the condensation of the steam producing a vacuum. The void obtained in this manner was very considerable, because one cubic *foot* of steam at 212° condenses into one cubic *inch* of water. The production of a vacuum with the aid of steam is quickly effected by boiling some water in a clean camphine can, and when the steam is issuing freely from the mouth of the latter it is then corked, and cold water thrown over the exterior. Directly the temperature is lowered, the steam inside the tin vessel is condensed suddenly into water, and a void space being suddenly obtained, the whole pressure of a column of air of a breadth equal to the area of the vessel, and of a height of forty miles, is brought suddenly down like a sledge-hammer upon the sides of the tin vessel, and as they are not sufficiently strong to offer a proper resistance, they are crushed in like an egg-shell by the giant weight which falls upon them.

The barometer, or measurer of the weight of the air, consists of a glass tube about thirty-three inches in length, hermetically sealed at one end, and containing mercury that has been carefully boiled within it, and being perfectly filled the tube is inserted in a cistern of clean mercury, when it gravitates to a height equal to the pressure of the air, leaving a space at the top called the torricellian vacuum. As the atmospheric air decreases in density by admixture with invisible steam or vapour, any given volume becomes specifically lighter: hence the column of mercury falls to a height of about twenty-eight inches; whilst if the aqueous vapour diminishes, the weight of the air becomes greater, and the barometer may rise to a height of about thirty-one inches.

Having thus secured a "reciprocating motion," Newcomen applied it to the working of a force-pump by the intervention of a great beam or lever suspended on gudgeons (an iron pin on which a wheel or shaft of a machine turns) at the middle, and suspended like the beam of a pair of scales; and, in fact, he invented that method of supporting the beam which is in use to the present day. Supposing we compare Newcomen's beam to a scale beam, he attached to the extremities (instead of scale pans) a water pump and his steam cylinder—the latter being at one end, and the former at the other. The beam played at "see-saw:" by the primary action of the steam on the bottom of the piston in the *cylinder* it was pushed up at this end, and of course suffered an equal fall at the other, to which the pump piston was attached; and when the motion was reversed by the condensation of the steam, down went the piston again by the pressure of the air, whilst that of the water pump was again raised, and being provided with proper valves, the water was pumped slowly out of the mine, although the steam power used was very moderate, and only just sufficient to counterpoise the weight of the atmosphere. Newcomen made the end attached to the water pump purposely heavier than the steam piston of the other end of the beam, and by this means the work of the steam, by its elasticity, was very moderate, whilst the actual lift of the water from the mine was performed by the pressure of the air, equal (as already stated) to fifteen pounds on every square inch of the surface of the steam piston. This engine is called the atmospheric engine, and in the next cut we have a picture taken from a photograph by the "Watt Club" of the actual model of the Newcomen engine in the Hunterian

Museum of the University of Glasgow; the dimensions being—length, 27 in.; breadth, 12 in.; height, 50½ in.; from which, "in 1765, *James Watt,*

Fig. 394. Model of the Newcomen engine, in which the furnace and boiler, the steam cylinder, beam, water-pump, and elevated cistern of water, are apparent.

in seeking to repair this model, belonging to the Natural Philosophy Class in the University of Glasgow, *made the discovery of a separate condenser,* which has identified his name with that of the steam-engine." (Fig. 394.)

In Newcomen's engine, the opening and shutting of the cocks required the vigilant care of a man or boy, and it is stated on good authority that a boy who preferred (like nearly all other boys) *play* to work, contrived, by means of strings, a brick, and one or two catches on the working beam, to make the engine self-acting.

This poor boy's ingenious contrivance paved the way for the improved methods of opening and shutting the valves, which were brought to a great state of perfection by Beighton, of Newcastle, about 1718. Between that time and the year 1763, we find honourable mention made of Smeaton in connexion with the steam-engine, but the name of the great James Watt at this time began to be appreciated, and by a series of wonderfully simple mechanisms, he at last perfected the machine whose origin could be traced back not only to the time of Blasco de Garay, in 1543, but even to the days of the ancient mechanicians, such as Hero, who lived 130 B.C.

In 1763, James Watt was a maker of mathematical instruments in Glasgow, and his attention was drawn to the subject of the steam-engine by his undertaking to repair a working model of Newcomen's steam-engine, which was used by Professor Anderson, who then filled the Chair of Natural Philosophy, and subsequently founded the Andersonian Institution. The repairs required for this model induced Watt to make another, and by watching its operation, he discovered that a vast quantity of heat, and therefore fuel, was wasted in the constant and successive heating and cooling of the steam cylinder. About two years after, when Watt was twenty-nine years of age, he had made so many experiments, that he was enabled to put into a mechanical shape his original ideas, which are embodied in his patent of 1769, as follows:—

" My method of lessening the consumption of steam, and consequently fuel, in fire-engines, consists of the following *principles:*

" First : That vessel in which the powers of steam are to be employed to work the engine, which is called the cylinder in common fire-engines, and which I call the steam-vessel, must, during the whole time the engine is at work, *be kept as hot as the steam that enters it*—first, by enclosing it in a case of wood or any other materials that transmit heat slowly; secondly, by surrounding it with steam or other heated bodies; and thirdly, by suffering neither water nor any other substance colder than steam to enter or touch it during that time.

" Secondly: In engines that are to be worked wholly or partially by condensation of steam, the steam is to be condensed in vessels *distinct* from the steam-vessels or cylinders, although occasionally communicating with them; *these vessels* I call *condensers;* and whilst the engines are working, these condensers ought at least to be kept as cold as the air in the neighbourhood of the engine, by application of water or other cold bodies.

" Thirdly: Whatever air or other elastic vapour is not condensed by the cold of the condenser, and may impede the working of the engine, is to be drawn out of the steam-vessels or condensers by means of pumps wrought by the engines themselves, or otherwise.

" Fourthly : I intend in many cases to employ the expansive force of steam to press on the pistons, or whatever may be used instead of them, in the same manner as the pressure of the atmosphere is now employed in common fire-engines. In cases where cold water cannot be had in plenty, the engines may be wrought by this force of steam only, by discharging the steam into the open air after it has done its office.

" Lastly : Instead of using water to render the piston or other parts of the engines air and steam-tight, I employ oils, wax, resinous bodies, fat of animals, quicksilver, and other metals in their fluid state.

"And the said James Watt, by a memorandum added to the said specification, declared that he did not intend that anything in the fourth article should be understood to extend to any engine when the water to be raised enters the steam-vessel itself, or any vessel having an open communication with it."

"About the time he obtained his patent, Watt commenced the construction of his first real engine, the cylinder of which was eighteen inches in diameter, and after many impediments in the details of the work he succeeded in bringing it to considerable perfection. The bad boring of the cylinder, and the difficulty of obtaining a substance that would keep the piston tight without enormous friction, and at the same time resist the action of steam, gave him the most trouble, and the employment of a piston rod moving through a stuffing-box was a new feature in steam-engines at that time, and required great nicety of workmanship to make it effectual. While Watt was contending with these difficulties, Roebuck's finances became disarranged, and in 1773 he disposed of his interest in the patent to Mr. Boulton, of Soho. As, however, a considerable part of the term of fourteen years, for which the patent was granted, had already passed away, and as several years more would probably elapse before the improved engines could be brought into operation, it was judged expedient to apply to Parliament for a prolongation of the term, and an Act was passed in 1775 granting an extension of twenty-five years from that date, in consideration of the great merit of the invention." (Bourne's "Treatise on the Steam-engine.")

In Fig. 395 (p. 427) we give an illustration of a low-pressure condensing engine and boiler of eight-horse power, constructed on the principle of Boulton and Watt, as the latter had fortunately united his skill, learning, originality, and experience with Mr. Boulton, of Soho, near Birmingham, whose metal manufactory was already the most celebrated in England.

During the explanation of this eight horse-power engine, the opportunity may be taken to discuss occasionally the special improvements effected by Watt. The steam-pipe A conveys the steam generated in the boiler B to the slide-valve C, which is kept close to the surface, against which it works by the pressure of the steam.

Here we notice some of the valuable improvements of Watt in the admission of steam *above* as well as *below* the piston, by which he increased the power of his engine, and no longer confined it to the force of the atmospheric pressure. It is also necessary to remark the beautifully simple mechanism of the slide-valve, by which steam is admitted alternately above and below the piston. Want of space prevents us tracing out the gradual improvements effected by Watt, and therefore we take his invention as it stood in the year 1780, and refer our readers to Bourne's "Treatise on the Steam-engine" for the full and minute particulars of the improvements to that date.

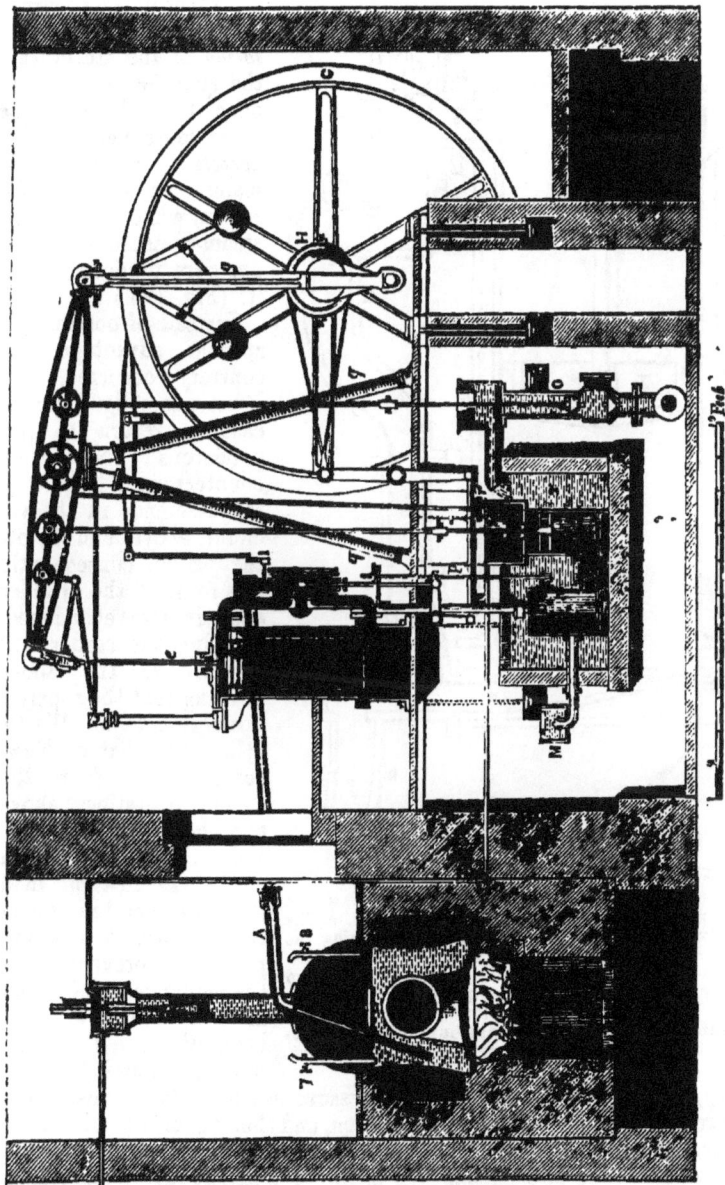

Fig. 395. An eight-horse power condensing steam-engine, after the principle of Boulton and Watt, and explained in pages 426 to 432.

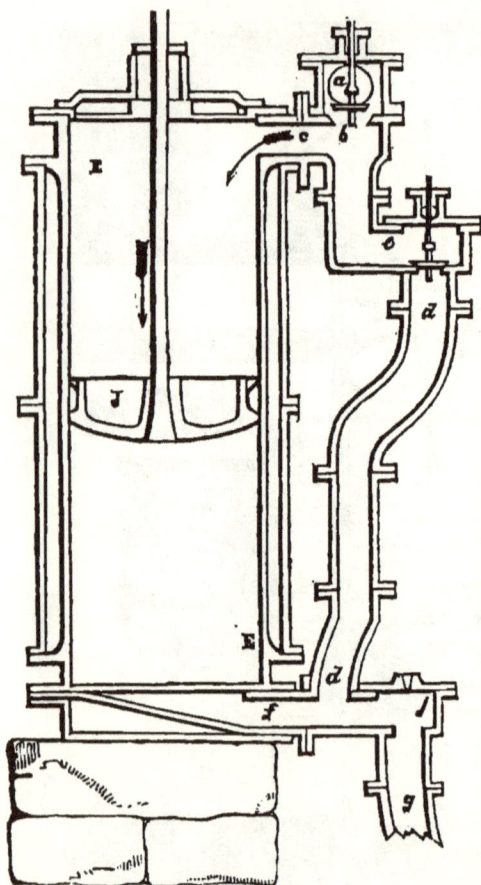

At that time it occurred to Watt that the *condensation* of the steam from the *cylinder* after it had done its work, might be made more perfect if a *perpetual vacuum* was maintained beneath the piston, while an alternate steam-pressure and vacuum were produced above it. (Fig. 396.)

Instead of obtaining a specific advantage the contrary occurred, and Watt was obliged in this case to return to the ponderous Newcomen counterweight to balance the difference in the vacuum above and below the piston, consequently this form of the cylinder and valves was abandoned. The juvenile reader will perceive in the above drawing that the superior arrangement of Watt's cylinder to that of Newcomen arises from the steam operating above and below the piston, and that the piston rod works air-tight in a *stuffing box* at the top of the cylinder. A most important improvement in the employment of steam as a motive power has been discovered in the mode of using it "expan-

Fig. 396. "ɪ ɪ is the cylinder. *ɪ*. The piston. *a*. The steam-pipe. *b*. The regulating or throttle valve. *e*. The eduction and equilibrium single valve, performing the functions of both. *c*. The upper, and *f* the under, portholes, by which passages only the steam can enter and pass away. *d, f, g*. The eduction-pipe by which the steam passes from above the piston during every returning stroke to the condenser, a perpetual exhaustion being maintained beneath it."—From BOURNE *on the Steam-engine*.

sively," by which the steam, at a pressure say of sixty pounds on the square inch, is admitted below the piston, and then cut off and allowed to expand and drive up the latter without the expenditure of any more fuel, and leaving, after lifting the piston to a height say of three feet, an average or mean power of thirty pounds on the square inch.

Returning to the eight-horse condensing engine, D is the steam cylinder surrounded by a case to prevent the steam cooling and to maintain in the cylinder the same, or nearly the same, temperature as that of the steam in the boiler, according to the condition of Art. I. of Watt's Patent, quoted at p. 425 of this book. The same outer case is apparent around the cylinder in Fig. 396; E, the piston, which, by stuffing with hemp or other proper material, fits the interior of the cylinder in the most accurate manner, and prevents the escape of steam by its sides: *e* is the piston rod attached to the parallel motion. This clockwork-like piece of mechanism has often been quoted as one of the masterpieces of Watt, and in its greatest perfection is called the *complete* parallel motion, and may be found in all the best land beam steam-engines. The object of the parallel motion is to cause the piston and pump rods to move always in straight lines, never deviating to either side. (Fig. 397.)

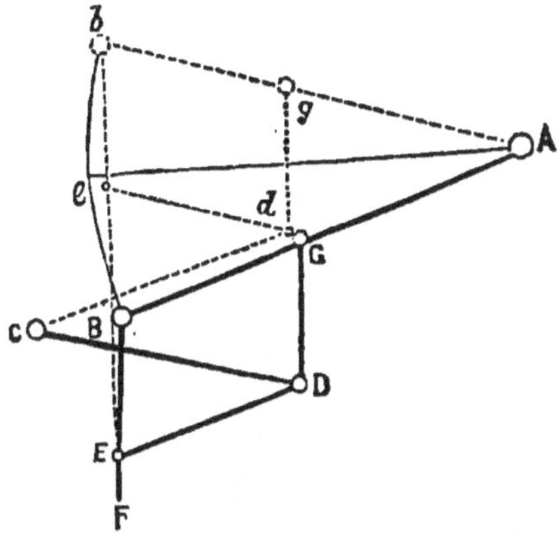

Fig. 397. A B is half the beam, A being the main centre. B E. The main links connecting the piston-rod F with the end of the beam. G D. The air-pump links, from the centre of which the air-pump rod is suspended. C D and E D produce the parallelism, because C D is moveable only round the fixed centre C, whilst E D is not only moveable round the centre D, but the centre itself in the arc described by C D, and by this action E D corrects the distorting influence of its own radius. The dotted lines and letters above enable the observer to see the effect of the movement of the beam on the parallel motion.

In the eight horse-power engine shown in page picture, *e* is also attached to the piston E, which moves the beam F, and the other end of this beam, by the connecting rod *g*, gives motion to the heavy fly wheel G, by means of the crank *h*.

H is an eccentric circle on the axle of the fly wheel G, it gives motion to the slide valve, which admits the steam alternately above and below the piston. The slide valve and its seat are contained within an oblong box or case, large enough to permit the easy motion of the valve within it, and usually forming an enlargement in the course of a pipe.

The valve rod by means of which the valve is opened and shut, passes out through a stuffing box; or, instead of such a rod, a valve of moderate size often has a nut fixed to it, within which works a screw on the end of an axle which passes out through a bush, and has shoulders within and without to prevent it from moving longitudinally, and a square on the outer end on which the key fits that is used in turning it. I is the throttle valve inside the steam pipe and lever connected with a governor for regulating the admission of steam into the cylinder.

Here, again, we pause in the description of our eight horse-power engine to illustrate more·particularly this admirable contrivance of

Watt, which remains to the present day without any material alteration even in the best steam-engines. (Fig. 398.)

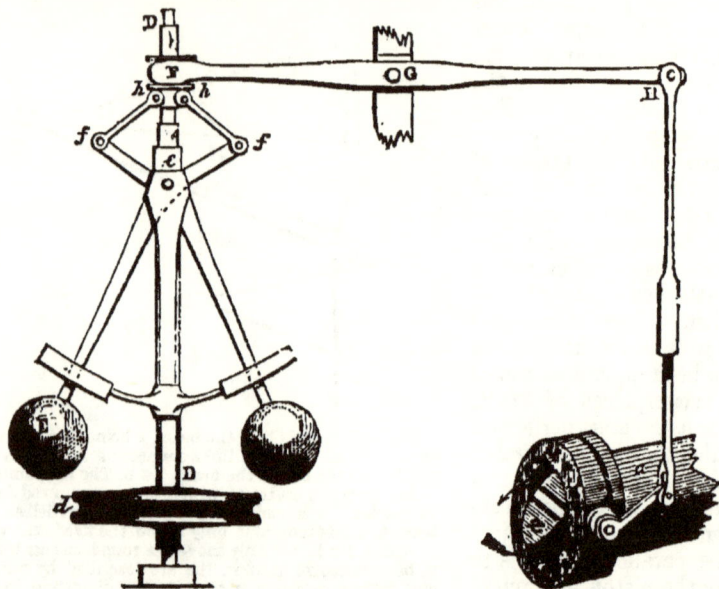

Fig. 399. A. The seat of the throttle valve. z. The valve itself turning on a spindle, which passes through its centre. a is the steam pipe. w. The throttle valve lever on which the rod H, proceeding from the governor, acts. D D. The spindle of the governor revolving by a belt acting on the pulley d. E E. The balls hung on the ends of the arms, which cross each other at e like a pair of scissors. When D D is set in motion, the balls fly out by centrifugal motion, and in doing so draw down the collar into which the lever F works by means of the links f h. When F is depressed, of course H rises, and the valve z is partly closed, and the supply of steam reduced.

In the eight-horse engine already partly explained, k is the cylinder of an air-pump to remove any air, and the water which condenses the steam, from the condenser L. There is also the eduction pipe, which conducts the steam from the cylinder to the condenser L. o is the pump that supplies cold water to the cistern s, in which the condenser and air-pump stand. P is a rod connected with the injection cock for admitting a jet of water into the condenser from the cistern, and which is continually flowing during the working of the engine. Q Q, cast-iron columns, four of which support the principal parts of the engine.

We now come to the boiler of the steam-engine, which is of course of almost equal importance with the engine itself; and the one in our page-picture is a good type of one of the favourite boilers used by Messrs. Boulton and Watt, and is called the "Wagon boiler." The boiler is made of wrought-iron plates rivetted together, and properly strengthened where necessary; and the steam-pipe A conveys the steam to the engine. It may be remarked here that the cylindrical boiler—consisting of two cylinders, one within the other, of which the former contains the fire, whilst the furnace-draught circulates outside the latter, and the space between the two cylinders being filled with water— is the form of boiler which is most highly approved of, and is employed in the famous economical steam-engines of the Cornish mines.

As the water evaporates in the form of steam, the boiler must be continually supplied with fresh water, which comes (as will be noticed by inspecting the page picture) from the *hot well* s, by means of the *hot-water pump* r, attached to the beam F. The water is pumped to the top of a column rising above but connected with the boiler. There is a cylindrical float, inside the column of water, connected with the boiler, suspended over a pulley by a chain passing to the damper of the furnace. The damper and float balance each other, and when the water in the boiler rises to too high a temperature, it causes the float to rise in the column of water, which lowering the damper or shutter that stops the draught of the chimney of the furnace T, diminishes the intensity of the heat, and reduces the formation of steam. On the other hand, as the temperature diminishes, the float descends and the damper rises, and permitting more air to rush to the burning fuel in the fire, a greater quantity of steam is generated.

There is likewise a stone float inside the boiler, for regulating the supply of water by the feed pipe, or column of water, which latter must always be sufficiently lofty to press with greater force than the steam produced in the boiler, or else the power of the steam might, under certain circumstances, eject or blow out the water from the top of the column. The stone is suspended by a brass wire which works through a stuffing box, and is connected with a lever, to which is attached a heavy counterpoise, so adjusted that when the stone is immersed to a certain depth in water (according to the principle of a solid body losing weight in a fluid, explained in the article on specific gravity, page 48), it shall exactly balance the latter, but when the water sinks in the boiler, and the stone is no longer surrounded with water, it becomes heavier, and sinking down opens a conical plug, ground so as to fit water-tight into a hole in the bottom of the column of water or feed pipe, and directly the plug opens, water rushes into the boiler; being cut off again as the stone rises when immersed or surrounded with the proper height of water. Unless our juvenile readers refer to the article on specific gravity, they will not understand the otherwise seeming anomaly of a *stone float*.

A large hole, called the man-hole, covered with an iron plate and securely fastened with screws, is provided for the purpose of allowing the engineer to enter the boiler, when cold, for the purpose of clearing out the incrustation and dirt arising from the water. To prevent the incrustation of lime and other earthy matters, it is sometimes usual, on the principle " *that prevention is better than cure,*" to put a large log of "logwood" inside the boiler, as it is found that the colouring matter curiously prevents the earthy matter, so well known as the "fur" in iron "tea-kettles," sticking to the sides of the boiler. Sal ammoniac and other salts also have the same property, but neither are much used, the mechanical labour of chipping out the boiler and stopping its work for a day or so, being preferred to the *prevention plan* already described.

There is also a valve opening inwards to prevent the consequences of a sudden condensation in the boiler, and also a safety valve and lever with weights opening outwards, and allowing the steam to escape when it reaches a dangerous excess, and in order to look as it were at the state of the pressure inside the iron boiler, a proper steam gauge is provided, also two cocks—viz., a water and steam cock, to enable the engineer to ascertain if the water is up to, and does not exceed, the proper height, because when turned, supposing that all is going on properly, the former, No. 7, should eject water, the latter, No. 8, steam.

It is truly wonderful, considering the number of safeguards and warnings provided, that accidents ever happen to boilers, but the statistics of deaths and annual destruction of property show that science is powerless, nay, absolutely dangerous, when handled by ignorant and careless persons. The great fly-wheel, which is usually such an awe-inspiring and marvellous exhibition of strength in an engine of any great power, is employed for the purpose of storing up force, so that if any parts of the engine work indifferently (they all work with resistance), it shall equalize the wants of the whole, and by its inertia it will continue to move until its motion is stopped by a resistance equal to its momentum.

In starting an engine, the engineer may sometimes be observed labouring to move the "fly-wheel," and when once he succeeds in getting it to move, the resistance of the other parts of the machinery is soon overcome. Mr. Alderson, in his prize essay, remarks that "it is in the property which the steam-engine possesses of regulating itself, and providing for all its wants, that the great beauty of the invention consists. It has been said that nothing made by the hand of man approaches so near to animal life. Heat is the principle of its movement; there is in its tubes circulation, like that of the blood in the veins of animals, having valves which open and shut in proper periods; it feeds itself, evacuates such portions of its food as are useless, and draws from its own labours all that is necessary to its own subsistence. To this may be added, that they are now regulated so as not to exceed the assigned speed, and thus do animals in a state of nature. That the safety valves, like the pores of perspiration, open to permit the escape of superfluous heat in the form of steam. The steam gauge, as a pulse to the boiler, indicates the heat and pressure of the steam within; and the motion of the piston represents the action and the power of which it is capable. The motion of the fluids in the boiler represents the expanding and collapsing of the heart; the fluid that goes to it by one channel is drawn off by another, in part to be returned when condensed by the cold, similar to the operation of veins and arteries. Animals require long and frequent periods of relaxation from fatigue, and any great accumulation of their power is not obtained without great expense and inconvenience. The wind is uncertain; and water, the constancy of which is in few places equal to the wants of the machinist, can seldom be obtained on the spot where other circumstances require machines to be erected. To relieve us from all these difficulties, the last century has given us the steam-engine for a resource, the power of which may be increased to infinitude: it requires but little room; it may be erected in all places, and its mighty services are always at our command, whether in winter or summer, by day or by night, on land or water; it knows no intermission but what our wishes dictate."

The *high-pressure* steam-engine appears to have been first brought into general use by Trevethic and Vivian, although the primary notion of such a modification of the Newcomen or water-engines did not originate with them. As the name implies, the steam is brought to a much higher temperature and pressure than is required in the condensing engines of Boulton and Watt. It consisted, in the first place, of a cylinder open at the top, and provided with a piston. To save heat the cylinder was fixed *inside* the boiler, and was provided with a two-way cock worked by a crank, for the purpose of supplying and cutting off the steam. The downward stroke was produced by the atmosphere, and the steam having done its work, was simply blown away and wasted in the air.

The engine was provided with a fly-wheel, to which the piston-rod was at once attached, producing a continuous rotatory movement without the assistance of the heavier parallel motion, or hot and cold water pumps.

This form of engine was soon adopted for pumping work—such as that of draining fens; and in 1804 Mr. Richard Trevethic used it for propelling the first carriage on the Merthyr Tydvil rail or tram way, and it was then speedily adopted in all the coal districts where the levels were moderate. Stephenson the elder, succeeded by the late lamented Robert Stephenson, followed with inventions and improvements of the locomotive steam-engine; and we are told in "Once a Week" that,

"One of those best qualified to speak to the latter's contributions to the development of the locomotive engine, states that from about five years from his return from America, Robert Stephenson's attention was chiefly directed to its improvement. 'None but those who accompanied him during the period in his incessant experiments can form an idea of the amazing metamorphosis which the machine underwent in it. The most elementary principles of the application of heat, of the mode of calculating the strength of cylindrical and other boilers, of the strength of rivetting and of staying flat portions of the boilers, were then far from being understood, and each step in the improvement of the engine had to be confirmed by the most careful experiments before the brilliant results of the Rocket and Planet engines (the latter being the type of the existing modern locomotive) could be arrived at.'

"Stephenson's time was not, however, so fully taken up during the above interval as to preclude attention to his other civil engineering business, and he executed within it the Leicester and Swannington,

F F

Whitby and Pickering, Canterbury and Whitstable, and Newton and Warrington Railways; while he also erected an extensive manufactory for locomotives at Newton, in Lancashire, in partnership with the Messrs. Tayleur. About the middle of the above period, also, the first surveys and estimates for the London and Birmingham Railway were framed, leading eventually to the obtaining of the Act. Then followed the execution of that line, and here Robert Stephenson had an opportunity of showing his great talent for the management of works on a large scale. This was the first railway of any magnitude executed under the contract system; perfect sets of plans and specifications (which have since served as a type for nearly all the subsequent lines) were prepared —no small matter for a series of works extending over 112 miles, involving tunnels and other works of a then unprecedented magnitude.

" Many other railways in England and abroad were executed by him in rapid succession; the Midland, Blackwall, Northern and Eastern, Norfolk, Chester and Holyhead, together with numerous branch lines, were executed in this country by him; and among railways abroad may be enumerated as works either executed by him or recommended in his capacity of a consulting engineer, the system of lines in Belgium, Italy, Norway, and Egypt, and in France, Holland, Denmark, India, Canada, and New Zealand.

" Robert Stephenson first saw the light in the village of Willington, at a cottage which his father occupied after his marriage with Miss Fanny Henderson—a marriage contracted on the strength of his first appointment as "breaksman" to the engine employed for lifting the ballast brought by the return collier ships to Newcastle. Here Robert was born on the 17th of November, 1803. As the cottage looked out upon a tramway, the eyes of the child were naturally familiarized from infancy with sights and scenes most nearly connected with his future profession."

In locomotive steam-engine boilers, the principal object is to generate steam with the greatest rapidity; hence the boiler consists of two parts —viz., a square box containing the fire, and around which a thin stratum of water circulates, whilst the draught for the fire rushes through a number of copper tubes placed in the second or cylindrical part of the boiler. By the use of these tubes an immense *surface* of water is exposed to the action of the fire, and the steam is not only generated with amazing rapidity, but is also maintained at a very high pressure.

Within the last few years " superheated steam" has been favourably mentioned, and employed economically for driving certain engines. The principle consists in first generating steam, and then passing it through coils of strong wrought-iron pipe, by which it acquires additional heat, and we have therefore combined in steam the ordinary principle of evaporation of water with the heated-air principle of Stirling, described at p. 367. We give a drawing of Scott's patent generator and superheated steam engine. (Fig. 399.)

The apparatus is used as follows :—A fire is made in the furnace, and so soon as a pyrometer connected with that indicates about 800 degrees, a little water is pumped into the coils by hand, which is immediately converted into steam. The donkey engine is then started, which

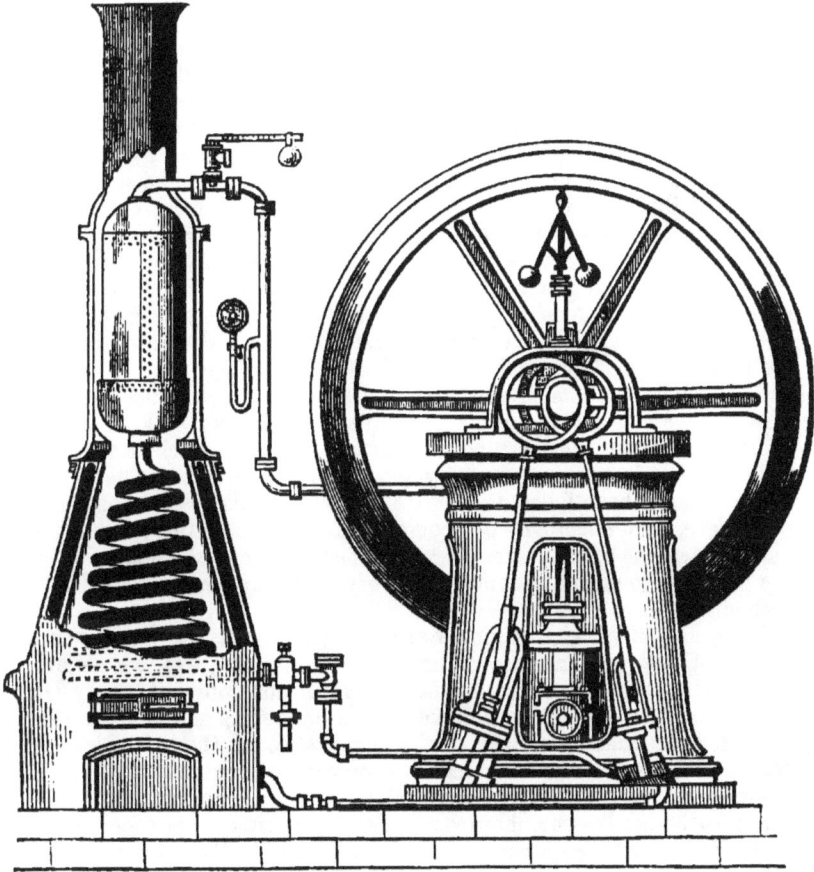

Fig. 399. Scott's patent generator, or new *versus* old steam.

maintains the necessary feed of *air* and water. The generator produces a copious supply of elastic mixed gaseous vapour, at a pressure of 250 pounds on the square inch; and it is stated that this engine works satisfactorily, and is started in the incredibly short time of from three to five minutes, so that for marine engines in war vessels, expecting to to be ordered out suddenly, no fuel need be burnt till the moment required.

Experiments with superheated steam have already been tried most successfully on board the Peninsular and Oriental Company's ship the *Valetta*, whereby it is stated that a saving of thirty per cent. in fuel is obtained. The engine to which the superheated steam was adapted was constructed by Penn and Sons, and the vessel attained a speed of nearly sixteen knots per hour, and under the most adverse circumstances had an abundance of steam to spare.

"A most important experimental improvement in steam machinery was on Thursday last tried for the first time down the river, on board the Peninsular and Oriental Company's ship, the *Valetta*. The actual nature of the improvement may be described in a few words as consisting of a simple apparatus for working marine engines by means of superheated steam; but it is not too much to say that in the success or failure of this experiment are involved results so important as to affect materially all ocean-going steamers, and, indeed, steam machinery of all kinds. To be able to work machinery with superheated steam, means to command increased power with a thirty per cent. reduction in the consumption of fuel. A principle which can effect such important changes in the universal application of steam has not remained undiscovered to the present day. The want of superheated steam has long been felt, and the enormous comparative advantages of working engines on such a plan have long been known. A simple and effective working of the principle, however, has been an engineering difficulty which various expedients—all, however, sufficiently successful to show the value of the improvement—have failed to obviate entirely. This obstacle has now, we believe, been effectually overcome by Mr. Penn, and the value of the improvement so clearly demonstrated, that the general application of the principle to steam machinery of every kind may now be regarded as certain.

"The idea of working engines by superheated steam, and the immense saving of fuel and increase of power it would effect, was, we believe, first started many years ago by Mr. Howard, and subsequently by Dr. Haycraft. The difficulties, however, in the way of its adoption at that time, and the undue estimate of the importance of the principle, prevented those gentlemen from realizing very great practical results. At a later period the matter was again taken up by an American engineer—Mr. Weatherhead—who, however, only superheated a portion of the steam and mixed it with common steam in its way to the cylinders. The success which attended even this partial application of the process again revived the idea, and encouraged other engineers to turn their attention to the subject. The result of these renewed efforts is that several methods of securing the great economy to be effected by superheating the steam are now under trial, and there is no doubt that a most important step in the progress of steam, especially as applied to ocean navigation, is now at last on the point of being successfully accomplished.

"The value of the improvement on the score of economy in working may be best illustrated by a single fact—namely, that the Peninsular and Oriental Company's bill for coal annually amounts to the enormous sum of 700,000*l.*, and that by working their vessels with superheated steam properly applied, it is become almost certain that, without any detriment to the machinery, from 28 to 30 per cent. of this gigantic outlay can be saved. As to the various proposed methods of superheating steam, it may be briefly explained, that the conditions required to be fulfilled are perfect simplicity of arrangement with ready control over the apparatus; that it should be so placed as not to be liable to accidental injury in the engine-room; and that the heat employed for superheating the steam should be waste heat which has already done its duty in the boilers and is passing away.

"All these conditions have been most satisfactorily fulfilled by Mr. Penn in the new engines on board the *Valetta*, which were tried down the Thames for the first time on Thursday. The *Valetta*, as our readers may remember, was for many years the mail-boat between Marseilles, Malta, and Constantinople. While thus employed, she had Penn's engines of 400 horse-power, and to work these up to an average speed of 15 miles an hour required a consumption of fuel of from 70 to 75 tons of coal per day. At no time was it less than from 45 to 55 tons. These engines have now been removed to a vessel nearly double the tonnage of the *Valetta*, and the latter fitted with engines by Mr. Penn on the superheating principle. We may mention that, besides this alteration, the *Valetta* has been considerably improved. A poop and forecastle have been added, increased accommodation given to passengers, and the whole vessel fitted up in the richest style. The saloon is one of the simplest and handsomest things of the kind we have seen, sufficiently lofty and capacious, and above all, admirably ventilated on the system which is now being adopted on all sea-going steamers, and the merit of devising which belongs to Mr. Robinson, of the Peninsular and Oriental Company.

"To return, however, to the engines. Mr. Penn, at the repeated request of Mr. Allen, the Managing Director of the Peninsular and Oriental Company, undertook to apply to them the principle of superheating, to which his attention had many years before been seriously directed by Dr. Haycraft. His method of doing this is to place in the smoke-box of the boiler, through which the hot air from the furnace first passes, as large a number of small pipes as is consistent with allowing a free draught from the furnaces. Through these all the steam from the boilers passes in its way to the cylinders. By this plan an immense heating surface in the pipes is secured, the steam is in a subdivided form, so as to be readily acted on, and the waste heat from the furnace is utilized at the point where its intensity is greatest, and where the greatest conveniences exist for applying the apparatus. By means of three ordinary stop-valves, the whole contrivance can be shut in or off from the engines at pleasure. In ordinary engines steam leaves the boilers at about 250°, but declines from this temperature in its way to the engines to 230°, undergoing from condensation a still greater and more serious diminution of heat in the cylinders. From these causes, and also from the immense quantity of waste heat which escapes through the smoke-box and up the funnels, there has always been a theoretical loss of steam power amounting to forty per cent., as compared with the coal consumed. It is this loss of power and waste of heat which the superheating process is intended to prevent, and which will, of course, allow a reduction of from twenty-eight to thirty per cent. on the fuel now consumed. By the superheating process the steam is raised in passing along the pipes in the smoke-box (where the heat is about 650°) from a temperature of 250° to 350°, and so enters the cylinders at 100° in excess of the temperature due to its pressure. This extra heat is, of course, rapidly communicated to the metals, and prevents the condensation in the cylinders or other parts of the engines, which would otherwise, of course, take place. Singularly enough, a smaller amount of cold water is required to condense the steam at this high temperature of 350° than when at the ordinary heat of common steam.

"The trial trip of the *Valetta* on Thursday was most satisfactory, not only as regards the engines, but still more so as to the application for the superheating process. At the measured mile at the Lower Hope, near the Nore, the result of repeated runs gave an average speed of nearly 14½ knots per hour, thus realizing with engines of 260 horse-power, and a small consumption of fuel, the same rate of speed as had been gained with her previous engines of 400 horse-power, and a consumption of seventy-five tons of coals per day. The superheating apparatus evidently effected a most important saving in fuel, but until an average of many days' working can be obtained, it would be difficult to estimate the exact amount economized. There seems, however, every reason to believe that an average of fourteen knots an hour can be obtained with a consumption of only from twenty-four to twenty-six tons per diem. The thermometer during the trial indicated in the steam pipes an addition to the ordinary temperature of 100°, which Mr. Penn believes to be enough for all practical purposes of superheating. Even when making from thirty-three to thirty-four revolutions per minute, and driving the vessel against a strong head wind and tide, it was impossible to consume all the steam generated, which was blowing off from both boilers all the trip. The engines are remarkable for the extraordinary beauty and simplicity of their proportions, qualities well known in all engines from Penn and Sons, and which, combined with the strength of the materials and perfection of the workmanship, make this firm the foremost in the world for machinery of this description. Both cylinders are oscillating, of sixty-two inches diameter, and with a stroke of four feet six inches. The paddles are on the feathering principle, and the boilers of Lamb and Co.'s patent. During the whole course of the trials, and when going at one time nearly sixteen knots, there was no perceptible vibration, even at the end of the saloon nearest to the engines. When it is remembered that the superheating process which can effect such important results is capable, as we have said, of application to steam machinery of every kind, including even loco-motives, it cannot be doubted that the trial of Thursday and its great success is one of the most important events for the progress of steam which we have had to chronicle for many years." (*The Times*, April 23rd 1859.

Whilst speaking of the application of this somewhat novel condition of steam, it may be observed that many inventors, who have paid little or no attention to *first principles*, have proposed to apply the vapours of alcohol, ether, or turpentine, instead of that of water; and they have founded their notions on the idea that in consequence of the less latent and sensible heat of alcohol, ether, and turpentine vapour, and of the small quantity of fuel required to boil them, that they would compete advantageously with steam. This view of the case, however, is soon proved to be a very shortsighted one, because the *amount* of *expansion* has been quite overlooked; and if it was desirable, by way of comparison, to produce a cubic foot of steam, alcohol, ether, or turpentine, the steam would stand first for cheapness, and would require the least quantity of fuel to produce it, so that if the more expensive of combustible liquids could be obtained for nothing, it would still be cheaper to employ water.

	Latent heat, or equivalent for fuel.
A cubic foot of water yields 1700 cubic feet of steam .	$= 1000°$
A cubic foot of alcohol produces 493 cubic feet=457°. Then, by rule of proportion, 493 cubic inches : 457 :: 1700 :	1575°
A cubic foot of ether yields only 212 cubic feet of vapour=312°, and 212 : 312° :: 1700 :	2500°
A cubic foot of the oil of turpentine affords 192 cubic feet of vapour=183°, and 192 : 183 :: 1700 : . . .	1620°

It will therefore be seen that water, when converted into steam, expands eight times as much as sulphuric ether, and nearly three times and a half as much as alcohol.

The application of steam for the purpose of propelling vessels has already been mentioned in connexion with the Spanish inventor, Blasco de Garay, in the year 1543. The first patent in this kingdom granted for that purpose was that of Mr. Jonathan Hull in 1773. In 1787, Mr. Miller tried a number of important experiments in the propulsion of vessels by steam-engines, and it would appear that Lord Cullen advocated his ideas, and endeavoured to secure the co-operation of the great firm of Boulton and Watt, who, occupied with their land engines, could not pay attention to it; and twenty years elapsed after the reply of Watt to Lord Cullen's application, before the real novelty appeared of a first successful experiment with a steam-boat in "the open sea," by Henry Bell, in 1811. A picture of this boat, called the *Comet*, which was afterwards wrecked, is shown at p. 418. Henry Bell's *novelty* was *success*, and he is fairly entitled to the merit of first introducing steam navigation into Europe.

In 1811, the public stared with mingled astonishment and satisfaction at the realization of that which was called a fable. Only forty-seven years afterwards another generation spontaneously exhibits the liveliest interest in the gigantic private speculation of the *Great Eastern*. Henry

Bell's vessel of 1811 was 40 feet keel, 10 feet 6 inches beam, and 25 tons burthen! The *Great Eastern* of 1859 is 692 feet long, 83 feet wide, 60 feet deep, and 24,000 tons burthen!! The whole nation with one voice wish her God speed in her projected voyage across the Atlantic, as the embodiment of that great goodwill which every generous-hearted Englishman feels towards the enlightened free-born people of the United States.

Should the author's little vessel, with its humble freight of science, meet with the approbation of his good friends, the boys and their advisers, another and another, if health permits, shall be launched .or their benefit. *Vale.*

THE END.

www.ingramcontent.com/pod-product-compliance
Lightning Source LLC
Chambersburg PA
CBHW020908210326
41598CB00018B/1807